Nucleic Acids and Molecular Biology

23

Series Editor

H. J. Gross
Institut für Biochemie
Biozentrum
Am Hubland
97074 Würzburg
Germany

Dieter B. Wildenauer (Ed.)

Molecular Biology of Neuropsychiatric Disorders

With 15 Figures and 8 Tables

Springer

Editor
Dieter B. Wildenauer
Graylands Hospital
Center for Clinical Research in Neuropsychiatry (CCRN)
Claremont WA 6910
Australia

ISBN 978-3-540-85382-4   e-ISBN 978-3-540-85383-1
ISSN 0933-1891

Library of Congress Control Number: 2008933566

© 2009 Springer-Verlag Berlin Heidelberg

This work is subject to copyright. All rights are reserved, whether the whole or part of the material is concerned, specifically the rights of translation, reprinting, reuse of illustrations, recitation, broadcasting, reproduction on microfilm or in any other way, and storage in data banks. Duplication of this publication or parts thereof is permitted only under the provisions of the German Copyright Law of September 9, 1965, in its current version, and permission for use must always be obtained from Springer. Violations are liable to prosecution under the German Copyright Law.

The use of general descriptive names, registered names, trademarks, etc. in this publication does not imply, even in the absence of a specific statement, that such names are exempt from the relevant protective laws and regulations and therefore free for general use.

Cover design: Boekhorst Design BV, The Netherlands

Printed on acid-free paper

9 8 7 6 5 4 3 2 1

springer.com

Preface

The intention of this book is to give an overview about ongoing research into molecular causes for disorders that affect the human brain. These disorders afflict mainly human behavior and are, since borders between "normal" and "abnormal" behaviors are continuous and hard to define, not always easy to diagnose. We have included the major psychoses (schizophrenia and affective disorders), autism, attention deficit hyperactivity, and anxiety disorders, as well as addictive and suicidal behavior. Neuropathological alterations in these disorders are usually not detectable or restricted to sporadic cases and, if present, not easy to define as causative. In contrast, pathological alterations in the brain are present in the two included neuropsychiatric disorders, Alzheimer's and Parkinson's disease. Detection of pathological abnormalities has been helpful as a starting point for studies of the biochemistry and molecular biology of these diseases. Our goal was to describe current research directions using new methods and information provided by molecular biology and molecular genetics. We have included selected, but in our opinion (and therefore biased) most promising, findings rather than attempting to give a complete overview of findings in these disorders. Given the large number of publications appearing each month in the field (up to 50 in one month as estimated from the PubMed entry) it is almost certain that the book would be already out of date while still in print. Specific databases that collect all new information on websites accessible to the public are superior and already available for schizophrenia (http://www.schizophreniaforum.org/res/sczgene/default. asp), Alzheimer's disease (http://www.alzforum.org/res/com/gen/alzgene/default. asp), and Parkinson's disease (http://www.pdgene.org/).

The complex nature of psychiatric disorders with many factors contributing to susceptibility, but not always necessary for expression of the disease, renders the identification of molecular causes extremely difficult. Whilst causes for the two neurodegenerative disorders, Alzheimer's and Parkinson's disease, have been uncovered and their molecular biology is being studied, the search for causes for other neuropsychiatric disorders has not yet led to convincing findings. Complexity and variation in phenotype are major obstacles for these studies. In particular, the difficulty of phenotype definition makes it is hard to establish identical samples for replication or confirmation of findings. As a consequence, most of the published findings are not easy to prove or disprove and may just add to the flood of unconfirmed positive findings in the field. Nevertheless, we are optimistic that a continuing collection of data will be of help in the discovery of the molecular connections and pathways that play a critical role in the development of these disorders.

Finally, I wish to thank all contributors to this book, as well as Professor Dr. Hans J. Gross for encouragement and Frau Gramm from Springer Verlag for help in submission for publication.

Perth

Dieter B. Wildenauer

June 2008

Contents

Decoding the Genetics and Underlying Mechanisms of Mood Disorders 1
Sevilla D. Detera-Wadleigh and Takeo Yoshikawa

1 Introduction 2
2 Bipolar Disorder 5
  2.1 Clinical Presentation and Epidemiology 5
  2.2 Genetic Analysis 6
  2.3 Alternative and Complementary Strategies 15
3 Depression 24
  3.1 Clinical Presentation and Epidemiology 24
  3.2 Genetic Analysis of Depression 25
  3.3 Therapeutics and Associated Mechanisms; Pharmacogenetics 27
  3.4 Neurogenesis 29
  3.5 The Endocrine System (hypothalamic-pituitary-adrenal
    (HPA) axis) 31
  3.6 Animal Studies 33
  3.7 Postmortem Brain Studies 36
4 Perspectives 37
References 38

Dissecting the Molecular Causes of Schizophrenia 51
Dieter B. Wildenauer, Diah Mutiara B. Wildenauer,
and Sibylle G. Schwab

1 Introduction 52
2 Diagnosis 52
3 Therapy 52
4 Incidence, Prevalence 53
5 The Search for Biological Causes 53
6 Epidemiology 54
7 Neuropathology 56
8 Neurotransmitter Hypotheses 57

vii

viii

9 Molecular Genetics 58
    9.1 Localization of Susceptibility Genes
        by Linkage Analysis.. 58
    9.2 Genetic Association.. 59
10 Gene Expression 69
    10.1 mRNA Expression... 70
    10.2 miRNA 70
    10.3 Protein Expression.. 71
    10.4 Epigenetics 71
11 Animal Models 72
12 Conclusion 73
References.. 73

Autism Spectrum Disorders 81
Sabine M Klauck

1 Introduction.. 81
2 Molecular Genetic Screening... 84
    2.1 Genome-Wide Screens and Fine-Mapping
        Approaches 84
    2.2 Chromosomal Aberrations.. 88
    2.3 Candidate Genes 89
3 Future Directions 91
References.. 92

Molecular Genetics of ADHD 99
Virginia L. Misener and Cathy L. Barr

1 Complexities of the ADHD Phenotype 100
    1.1 Developmental Shifts... 101
    1.2 Gender Effects 102
2 Genetic Basis of ADHD 104
3 Environmental Risk Factors.. 105
4 Molecular Approaches to Gene Identification................................. 106
    4.1 Candidate Gene Studies in ADHD...................................... 107
    4.2 Molecular Genetic Studies Using Population-Based
        Samples... 130
    4.3 Genome Scans for ADHD.. 131
5 Relationship of the Molecular Genetic Findings in ADHD
   to Reading Disabilities.. 133
6 Interpretation of the Molecular Genetic Studies 135
7 After the Linkage or Association Finding 135
8 Implications of the Molecular Genetic Findings for Diagnosis
   and Treatment 138
References.. 139

The Genetics of Anxiety Disorders 165
Steven P. Hamilton

1   Anxiety Disorders 165
2   Panic Disorder 166
3   Phobias 169
4   Generalized Anxiety Disorder 170
5   Obsessive Compulsive Disorder 171
6   Genetic Analysis of Anxious Personality Traits 174
7   Conclusion 175
References 176

Molecular Biology of Addiction and Substance Dependence 187
Sibylle G Schwab, Adrian Scott, and Dieter B Wildenauer

1   Introduction 188
2   Heritability of Substance Dependence 188
3   Biology of Drug-Induced Changes in the Brain 191
4   Molecular Genetic Evidence of Substance Dependence 193
    4.1   Heroin 194
    4.2   Nicotine 196
    4.3   Alcohol 198
5   Conclusion 199
References 199

Neurobiology of Suicide 205
Brigitta Bondy and Peter Zill

1   Introduction 206
2   The Clinical Phenotype 206
    2.1   The Link Between Mental Morbidity and Suicide 207
3   Pathophysiological Mechanisms 208
    3.1   The Neurochemical Basis of Suicidality 208
    3.2   Genetic Contribution to Suicidal Behavior 213
    3.3   Suicide as Interplay of Genes and Environment 219
4   Conclusions 220
References 221

Molecular Genetics of Alzheimer's Disease 229
Giuseppe Verdile and Ralph N. Martins

1   Introduction 230
2   Familial Alzheimer's Disease 230
    2.1   The Amyloid Precursor Protein (APP) Gene 231
    2.2   The Presenilin Genes 239
3   Apolipoprotein E (APOE ε4): Strongest Genetic Risk Factor
    for Sporadic AD 246

3.1	The Apolipoprotein ε4 Allele	246
3.2	Apolipoprotein E (ApoE) and Its Role in AD	250
3.3	Mechanisms by Which ApoE May Contribute to AD	252
4	Other Genetic Risk Factors	254
5	Concluding Remarks	257
References		257

Molecular Biology of Parkinson's Disease 277
Abbas Parsian and Biswanath Patra

1	Introduction	278
2	Molecular Mechanisms in the Development of Parkinson's Disease	279
2.1	The Role of Iron and Oxidative Stress	279
2.2	Role of Alpha-Synuclein	280
2.3	Mitochondrial Polymorphism and Deletions in Parkinson's Disease	282
2.4	Heat Shock Protein Chaperone (HSC-70)	283
2.5	Dopamine Neurotransmission and Metabolism	283
2.6	Association of LRRK2 Gene and Inappropriate Phosphorylation in Parkinson's Disease	284
2.7	Role of Glutathione-S-Transferase Gene in the Development of Parkinson's Disease	285
3	Conclusions	285
References		286

Index 291

Contributors

Cathy L. Barr
Room MP14-302, Genetics and Development Division, The Toronto Western Hospital, 399 Bathurst St., Toronto, ON, Canada M5T 2S8, CBarr@uhnres.utoronto.ca.

Brigitta Bondy
Department of Psychiatry, Section of Psychiatric Genetics and Neurochemistry, Ludwig-Maximilians-University Munich, Nussbaumstrasse 7, 80336 Munich, Germany, Brigitta.Bondy@med.uni-muenchen.de

Sevilla D. Detera-Wadleigh
Genetic Basis of Mood and Anxiety Disorders, Mood and Anxiety Disorders Program, National Institute of Mental Health Intramural Research Program, National Institutes of Health, Bethesda, MD 20892, USA, deteras@mail.nih.gov

Steven P. Hamilton
Department of Psychiatry, University of California, San Francisco, 401 Parnassus Avenue, Box NGL, San Francisco, CA 94143 0984, USA, steveh@lppi.ucsf.edu

Sabine M. Klauck
German Cancer Research Center (DKFZ), Division of Molecular Genome Analysis, Im Neuenheimer Feld 580, 69120 Heidelberg, Germany, s.klauck@dkfz.de

Ralph N. Martins
Centre of Excellence for Alzheimer's Disease Research and Care, School of Exercise, Biomedical and Health Sciences, Edith Cowan University, Joondalup, WA 6027, Australia,
Sir James McCusker Alzheimer's Disease Research Unit, Hollywood Private Hospital, Nedlands, WA 6009, Australia,

and School of Psychiatry and Clinical Neurosciences, University of Western Australia, Crawley, WA 6009,
Australia

Virginia L. Misener
Genetics and Development Division, Toronto Western Research Institute,
University Health Network, Toronto, ON, Canada

Abbas Parsian
Department of Pediatrics, College of Medicine, University of Arkansas for
Medical Sciences, ACHRI, 1120 Marshall Street, Little Rock, AR 72202, USA,
parsianabbas@uams.edu

Biswanath Patra
Department of Pediatrics, College of Medicine, University of Arkansas for
Medical Sciences, ACHRI, 1120 Marshall Street, Little Rock, AR 72202, USA

Sibylle G. Schwab
Western Australian Institute for Medical Research and Centre for Medical
Research, University of Western Australia, Nedlands, WA 6009, Australia,
School of Psychiatry and Clinical Neurosciences, University of Western Australia,
Crawley, WA 6009, Australia, and School of Medicine and Pharmacology,
University of Western Australia, Crawley, WA 6009, Australia
sschwab@cyllene.uwa.edu.au

Adrian Scott
Western Australian Institute for Medical Research and Centre for Medical
Research, University of Western Australia, Nedlands, WA 6009, Australia
School of Psychiatry and Clinical Neurosciences, University of Western Australia,
Crawley, WA 6009

Giuseppe Verdile
Centre of Excellence for Alzheimer's Disease Research and Care, School of
Exercise, Biomedical and Health Sciences, Edith Cowan University, Joondalup,
WA 6027, Australia,
Sir James McCusker Alzheimer's Disease Research Unit, Sir James McCusker,
Hollywood Private Hospital, Nedlands, WA 6009, Australia,
and School of Psychiatry and Clinical Neurosciences, University of Western Australia,
Crawley, WA 6009, Australia

Dieter B. Wildenauer
Center for Research in Neuropsychiatry (CRN), University of
Western Australia, Crawley, WA 6009 Australia, Dieter.Wildenauer@uwa.edu.au,
School of Psychiatry and Clinical Neurosciences, University of
Western Australia, Crawley, WA 6009 Australia

Takeo Yoshikawa
Laboratory for Molecular Psychiatry, RIKEN Brain Science Institute, Saitama,
Japan

Diah Mutiara B. Wildenauer

Department of Psychiatry, Faculty of Medicine, University of Indonesia, Jakarta, Indonesia

Peter Zill

Department of Psychiatry, Section of Psychiatric Genetics and Neurochemistry, Ludwig-Maximilians-University Munich, Nussbaumstrasse 7, 80336 Munich, Germany

Decoding the Genetics and Underlying Mechanisms of Mood Disorders

Sevilla D. Detera-Wadleigh and Takeo Yoshikawa

Contents

1 Introduction... 2
2 Bipolar Disorder.. 5
   2.1 Clinical Presentation and Epidemiology... 5
   2.2 Genetic Analysis 6
   2.3 Alternative and Complementary Strategies 15
3 Depression... 24
   3.1 Clinical Presentation and Epidemiology... 24
   3.2 Genetic Analysis of Depression.. 25
   3.3 Therapeutics and Associated Mechanisms; Pharmacogenetics 27
   3.4 Neurogenesis.. 29
   3.5 The Endocrine System (hypothalamic-pituitary-adrenal (HPA) axis)........ 31
   3.6 Animal Studies... 33
   3.7 Postmortem Brain Studies.. 36
4 Perspectives.. 37
References 38

Bipolar disorder and major depression are common debilitating mood disorders. The latest World Health Organization's World Mental Health Survey Initiative estimated the median and inter-quartile lifetime prevalence for mood disorders to be 3.3–21.4% (Kessler et al. 2007). In Global Burden of Disease surveys, depression has been estimated to be a leading cause of disability worldwide and is projected to be a major cause of morbidity in 2030 (first in high income countries, second in medium income countries and third in low income countries) (Mathers and Loncar 2006). A recent study has reported that the rise of high levels of psychological stress in the workplace seems to precipitate depression and generalized anxiety disorders (Melchior et al. 2007). Taken together, these suggest that mood disorders impose an enormous burden on families, society and the health care system. Reducing

---

S.D. Detera-Wadleigh (✉)
Genetic Basis of Mood and Anxiety Disorders, Mood and Anxiety Disorders Program,
National Institute of Mental Health Intramural Research Program, National Institutes of Health,
Bethesda, MD 20892, USA
deteras@mail.nih.gov

D.B. Wildenauer (ed.), *Molecular Biology of Neuropsychiatric Disorders*,
Nucleic Acids and Molecular Biology, © Springer-Verlag Berlin Heidelberg 2009

this burden may require improved diagnostic precision and treatment modalities, preventive measures and greater public awareness. A key element in achieving this goal involves the discovery of genetic and environmental factors that contribute to disease risk. So far, from a combination of approaches evidence for risk-conferring variation in various genes has emerged although further confirmation is needed and functional relevance has to be established. This chapter will focus on strategies designed to clarify the genetics and underlying mechanisms of mood disorders, and discuss progress in this endeavor and the challenge that remains.

1 Introduction

The diagnosis for mood disorders is based on criteria specified in the Diagnostic and Statistical Manual of Mental Disorders (DSM) (American Psychiatric Association 1994). The DSM-IV categories of mood (affective) disorders are shown in Table 1. The third and fourth editions, DSM-III and DSM-IV, have been used in many published genetic studies in mood disorders for the past decade.

Table 1 Mood disorders (DSM-IV)

1. Depressive disorders
   - Major depressive disorder
     - Single episode
     - Recurrent
   - Dysthymic disorder
   - Depressive disorder not otherwise specified (NOS)
   Examples
   - Minor depressive disorder
   - Recurrent brief depressive disorder
   - Postpsychotic depression of schizophrenia
2. Bipolar disorders
   - Bipolar I disorder
     - Single manic episode
     - Most recent episode hypomanic
     - Most recent episode manic
     - Most recent episode mixed
     - Most recent episode depressed
     - Most recent episode unspecified
   - Bipolar II disorder (recurrent major depressive episodes with hypomania)
3. Cyclothymic disorder
4. Bipolar disorder not otherwise specified (NOS)
   Examples
   - Recurrent hypomania without depression
   - Manic episode superimposed on delusional disorder
5. Mood disorder due to a general medical condition
6. Substance-induced mood disorder
7. Mood disorder not otherwise specified (NOS)

Recently, in reviewing the evolution of the DSMs, Hyman (2007) discussed the need to enhance diagnostic precision and the value of incorporating relevant knowledge in neurobiology as they relate to psychiatric disorders. The DSMs define psychiatric diseases as categorical entities but "many mental disorders may be better conceptualized as dimensional traits" (Hyman 2007). The standardized guidelines in the DSMs are based mainly on clinical observation and they lack biological or quantitative measures for specific clinical phenotypes. It follows that there is a need to refine phenotype classification. Furthermore, the DSMs do not address the possibility that either bipolar disorder or major depression is a clinical consequence of various diseases defined by varied or independent sets of etiologic determinants.

Mood disorders display a complex pattern of inheritance and their genetic architecture remains elusive. These diseases are thought to be genetically heterogeneous both at the locus and allelic levels. One well-accepted proposal that attempts to explain etiology is the common disease–common variant (CDCV) hypothesis that predicts the central role of multiple modest-effect variation (Reich and Lander 2001) (Fig. 1). Statistically significant signals from minor-effect loci are inherently more difficult to detect and/or replicate, requiring thousands of samples of identical or similar ethnic background. Furthermore, it is unclear how many modest-effect variants are necessary and sufficient to account for the high heritability of, and impairment of function in individuals with, mood disorders.

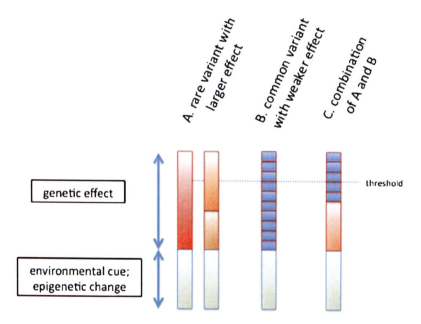

Fig. 1 Proposed genetic architecture for complex diseases, e.g., Mood disorders. Shown are the potential role of common variants, rare variants, a combination of common and rare variants, and the impact of epigenetic factors and environmental cues on these alleles. Boxes and squares are not drawn to scale

Potentially, subtle structural changes in the genome may confer greater sensitivity to environmental cues and thus act as "first responders". Conceivably, what is statistically measured as "small effect" may not be reflected as "minor" inside the cell, as perturbed cascades of reactions and interacting networks magnify the subtle effect and confer a pronounced biological defect.

An alternative hypothesis for the allelic blueprint of complex disease invokes the fundamental role of rare variants (Pritchard 2001). Potentially, infrequently occurring high penetrance alleles may be detectable in large, extended families and/or in population isolates. In these families, one or a few high penetrance variants could account for a large proportion of risk and heritability (Fig. 1). A recent study that evaluated rare single nucleotide polymorphisms (SNPs) in HapMap, ENCODE and SeattleSNPs with minor allele frequency (MAF) of 0.5%, showed that low MAF was predictive of functional alteration and that lower MAF of nonsynonymous SNPs correlated inversely with increased likelihood of disruption of protein function (Gorlov et al. 2008). This led the authors to propose a vital role for "slightly deleterious" rare SNPs, many of which remain to be identified. Simulations showed that thousands of samples are required to achieve reasonable power to detect these SNPs, with MAF ranging from 1 to 5% (Gorlov et al. 2008).

A similar model dubbed "common disease–rare alleles" has been proposed as a possible etiologic scenario for schizophrenia (McClellan et al. 2007). This paradigm draws upon the potential role of multiple rare, new, mutations in individual or few families, variation that might have occurred in the germ line as a consequence of population expansion. As shown in Alzheimer's disease, risk is due to rare major-effect high-penetrance in some families and common alleles in most families (Fig. 1). Also, the collective effect of multiple rare alleles in several candidate genes contribute to low levels of plasma HDL cholesterol, a quantitative trait which is heritable and displays a complex mode of transmission (Cohen et al. 2004). It is further speculated that infrequent, highly penetrant, alleles could be found in genes that have common associated variants or they could be embedded within associated common haplotypes. In this instance, deep resequencing in cases needs to be done to detect the rare-risk-conferring variants.

Several strategies are being employed to begin to untangle the genetic etiology of mood disorders (Fig. 2). In the past decade, the most popular approach has been genetic linkage mapping or positional cloning which has been exceedingly successful in localizing causative genes for monogenic disorders. In mood disorders, linkage analysis has been employed as an initial approach toward vulnerability locus identification but, as will be discussed in later sections, linkage for mood disorders has been only modestly compelling.

Very recently, genome-wide association studies (GWAS) have become feasible and thus the favored approach for genetic analysis of complex disease. The rapid explosion of published articles on GWAS in complex diseases is underpinned by recent technological progress. Advances in high throughput genotyping technology facilitate rapid comprehensive surveys of the entire genome with up to a million SNPs on thousands of samples. Although the cost of high throughput genotyping via chip arrays (e.g., Affymetrix: www.affymetrix.com; Illumina, Inc: www.illumina.com)

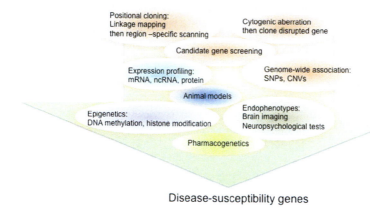

Fig. 2 Strategies to dissect the underlying genetic basis for mood disorders. Genome-wide association studies have grown exponentially very recently. All strategies are discussed in the text

has been declining, it is still beyond the reach of most laboratories. To address this problem, a public–private effort was launched in the United States. The Genetic Association Information Network (GAIN; http://www.fnih.org/GAIN2/home_new.shtml) (2007), a partnership composed of the National Institutes of Health (NIH) Foundation, NIH and Pfizer Global Research and Development was formed to help unravel "the genetics of common disease through whole genome association studies". GAIN is supporting GWAS of several complex diseases, including bipolar disorder and major depression. Independently, in the U.K., the Wellcome Trust Foundation has funded GWAS screens of bipolar disorder and six other common diseases (The Wellcome Trust Case-Control Consortium 2007). Genotyping of bipolar disorder and major depression samples by GAIN has been completed. Both initiatives are making data available to the research community. Easy access by more researchers should hasten the discovery of disease-predisposing genetic factors and stimulate investigations into pathways involved in pathogenesis.

2 Bipolar Disorder

2.1 Clinical Presentation and Epidemiology

Bipolar disorder, also known as manic-depressive illness, is characterized by disabling episodic and recurrent swings of severe elation and depression (Table 1). In rare cases, patients show only recurrent manic episodes with no history of depression ("unipolar mania"). Patients with severe mania referred to as bipolar I (BPI) have markedly impaired social and occupational functioning and often need hospitalization. A milder form of mania, hypomania, is categorized as bipolar II (BPII).

Between 34 and 79% of cases manifest mania as the first episode (Goodwin and Jamison 1990). The cycling patterns in bipolar disorder vary, with episodes occurring irregularly or linked together in a mania–depression–euthymia or a depression–mania–euthymia pattern. Rapid cycling, as defined in DSM-IV, indicates that a patient experiences four or more mood swings or episodes within a 12-month period. An episode can consist of depression, mania, hypomania or even a mixed state. Rapid cycling has been reported in 10–30% of bipolar patients. Rapid cycling has been shown to be more prevalent in women, 1.2:1 female/male ratio, by the National Institute of Mental Health Epidemiologic Catchment Area (ECA) study (Weissman et al. 1988). The term "ultra-rapid cycling" may be applied to those who cycle through episodes within a month or less. If this pattern occurs within a 24-h period, the person's diagnosis could possibly be termed "ultra-ultra-rapid cycling" or "ultraradian". Ultra-rapid cycling is more prevalent in males. As the number of episodes increases, the cycle length (i.e., the interval from the start of one episode to the start of the next) tends to decrease. There is also a tendency for later-onset bipolar disorder to manifest shorter cycle lengths. Later-onset rapid cycling is sometimes evoked by antidepressant medications.

Studies have estimated the risk for bipolar disorder among first degree relatives to be between 2.2 and 15.5 (Tsuang and Faraone 1990). Estimates for lifetime prevalence across several countries ranged from 0.3 to 1% with equal rates of occurrence between females and males (Weissman et al. 1996). An ECA study showed that the median age at onset of the disease is 18 years in men and 20 years in women (Weissman et al. 1988), but a recent study reported a later age at onset of 29.7 years (Bellantuono et al. 2007). A notable early Danish study calculated the concordance rate in monozygotic and dizygotic twins to be 67 and 20%, respectively (Bertelsen et al. 1977; Tsuang and Faraone 1990). Further, heritability in bipolar disorder has been estimated to be >80% (e.g., Bertelsen et al. 1977; McGuffin et al. 2003) indicating that a major proportion of the phenotypic variance is due to genetic factors with a modest but important contribution from environmental entities. Also, McGuffin and colleagues reported a genetic correlation of 65% between mania and depression, but 71% of the genetic variance for mania was distinct from genetic risk for depression.

2.2 Genetic Analysis

2.2.1 Genetic Linkage

Until very recently, linkage mapping was the prevailing strategy for the genetic analysis of bipolar disorder. Linkage findings, however, have been characterized by multiple loci with low signals in broad chromosomal regions. Linkage reports have been summarized in various reviews (e.g., Detera-Wadleigh and Goldin 2002; Hayden and Nurnberger 2006; Craddock and Forty 2006; Serretti and Mandelli 2008). Serretti and Mandelli (2008) have presented an extensive catalog of linkage studies published so far, listing potential positional candidate genes with some relevance to brain function in regions with support in more than one study. So far, these studies reveal mostly suggestive findings with limited consistency across studies. Aside from the possible occurrence of false positives, several factors may account for this predicament. Many linkage mapping and replication studies have been performed on underpowered samples. Substantial locus heterogeneity coupled with allelic heterogeneity elevates difficulty in risk gene detection. Phenotype heterogeneity within a given collection of cases or subsets of families may exist, adding another layer of complexity. Additionally, problems may arise from ascertainment bias, existence of phenocopies and variability in age-at-onset, penetrance and expression. As a diagnostic category, bipolar disorder may be a hodgepodge of miscellaneous traits characterized by an observable outcome that may be underpinned by independent etiologic pathways. Similar challenges exist for major depression (see following section).

In an effort to enlarge sample size and find common disease vulnerability loci, meta-analytic studies on genome scans for bipolar disorder have been conducted (Badner & Gershon 2002; Segurado et al. 2003; McQueen et al. 2005). A summary of the top regions (Table 2) reveals some overlaps but disparate top findings. Nevertheless, it is interesting that these studies showed some support for regions

Table 2 Top hits in meta analysis of genome scans for bipolar disorder

Meta-analysis	Number of independent studies	Number of pedigrees	Top regions (centimorgan, cM)
Badner and Gershon (2002)	11	353	13q (79 cM)
			22q (36 cM)
			4p (16 cM)
			18p (41 cM)
			18q (126 cM)
Segurado et al. (2003)	18	617	9p22.3-p21.1
			10q11.21-q22.1
			14q24.1-q32.12
			14q13.1-q24.1
			18p11-q12.3
			18q12.3-q22.1
			1p32.1-q31.1
			2q22.1-q23.3
			8pter-p22
			8q24.21-qter
			19q13.33-qter
McQueen et al. (2005)	11	1,067	6q (115 cM)
			8q (152 cM)
			9p (46 cM)
			20p (12 cM)
			18q (70 cM)
			17q (98 cM)

previously implicated in individual scans such as 6q, 8q, 9q, 13q, 14q, 18p-q and 22q (see Table 2, and reviews cited above).

Proposed Susceptibility Genes in Linkage Regions: Positional Cloning

Modest lod scores, if supported in other studies, deserve serious consideration because the earliest best replicated genetic finding in complex disease is the association of ApoE4 with late onset Alzheimer's disease. Association was revealed by follow-up analysis of the 19p region that previously showed only suggestive evidence of linkage (Pericak-Vance et al. 1991; e.g., see review by Strittmatter and Roses 1996). For bipolar disorder, the genes that have garnered the best support are those derived by positional cloning, although these findings have not received consistent support. Furthermore, it is interesting that association in these genes has been shown in both bipolar disorder and schizophrenia implying a shared etiologic pathway (Wildenauer et al. 1999; Berrettini 2000). We will present two genes identified using this approach.

G72 (*DAOA*) on 13q33.2

In 1998, a schizophrenia whole genome scan highlighted chromosome 13q32-q33 as one of two top regions that showed significant evidence for linkage (Blouin et al. 1998) supporting an earlier report that implicated the same region (Lin et al. 1997). In a bipolar disorder genome scan, the strongest evidence for linkage centered at approximately the identical 13q region (Detera-Wadleigh et al. 1999). These findings found support in some subsequent linkage studies in independent datasets but not in others (reviewed in Detera-Wadleigh and McMahon 2006).

Exploring the 13q32-q33 region for risk variants for schizophrenia via a SNP screen showed allelic association in a novel primate-specific locus that was given the name G72 (Chumakov et al. 2002). Support for association in bipolar disorder at some of the same G72 SNPs was reported later (Hattori et al. 2003). These studies were followed by a flurry of attempts at replication in both bipolar disorder and schizophrenia that yielded support in some but not in others (Detera-Wadleigh and McMahon 2006). While most of the evidence for association centered at SNP M23 (rs3918342) located ~40 kb downstream of the putative last exon, some studies detected association within the gene, either in an intron or at a potential nonsynonymous SNP in the predicted exon 2. Other reports have implicated the alternative allele(s) adding to the difficulty in interpreting association at this locus. A recent meta-analysis found evidence for allelic association in G72 SNPs, rs947267 and rs778293, with schizophrenia (Shi et al. 2008a). In contrast, five SNPs showed no association with bipolar disorder which may suggest that G72 has no role in bipolar disorder or that more samples may need to be examined and that substantial allelic heterogeneity exists.

G72 has been assigned the acronym DAOA because of its demonstrated in vitro activation of D-amino acid oxidase (DAO) which catalyzes the oxidation of D-serine, an NMDA receptor agonist (Chumakov et al. 2002). A recent study failed to confirm this finding (Kvajo et al. 2007), calling into question whether "DAOA" is an appropriate designation for G72 and indicating a need to confirm either report. Kvajo and co-workers (2007) also presented evidence for a possible G72 function in the mitochondria. Notably, this study has demonstrated the existence in rat amygdala of the elusive native G72 protein although the number of alternate transcripts for the G72 locus is still not known. G72 overlaps with G30, a predicted gene transcribed in the reverse direction. If both genes are active cross-regulation could occur and thus would impact on transcript levels of either gene and contribute to disease risk.

The association at G72 may not fully account for linkage on 13q32-q33, thus potentially several other loci in the region probably contribute to risk, e.g., the RhoGTPase activator, *DOCK9* (Detera-Wadleigh et al. 2007) and *NALCN* (VGCNL1) (Baum et al. 2008)). The case for G72 is complex and unresolved, but there is accumulating evidence that variation in this gene may increase the risk for schizophrenia and mood disorders.

Disrupted-in-Schizophrenia 1 (*DISC1*) on 1q42.2

The discovery of a disease-cosegregating chromosomal aberration facilitated the localization of *DISC1* that otherwise would have been much more formidable as genetic association at this gene seems to involve only a limited group of families. Identification of a balanced (t1,11)(q42.1;q14.3) translocation in an extended Scottish family exhibiting several psychiatric diagnoses (St. Clair et al. 1990) and a strong linkage to 1q42 (Blackwood et al. 2001) led to the discovery of two disrupted genes mapping to this region, *DISC1* and *DISC2* (Millar et al. 2000). Linkage analysis showed that when schizophrenia alone was scored as affected the family displayed LOD 3.6 which increased to 7 when mood disorders were included in the affected category. At that time no transcripts mapped to the translocation breakpoint on 11q; however, recent annotations of the human genomic sequence show ESTs and transcripts in the region (reviewed in Chubb et al. 2008). SNP screening of Finnish samples showed evidence of association of haplotypes on the 3′ end of *DISC1* with bipolar spectrum disorder, and the haplotype on the 5′ end showed over-transmission to males with schizophrenia (Hennah et al. 2003) and psychotic disorder (Palo et al. 2007). As summarized in a recent review, genetic data in some families are consistent with linkage and association in the *DISC1* region in schizophrenia (Chubb et al. 2008).

Of the candidate genes that have been proposed for psychiatric illness, *DISC1* has been one of the most appealing targets for functional studies. The interaction of DISC1 with protein components of multiple pathways involved in neuronal activity has defined the function of DISC1 in vivo and opened a window into how variation in this gene could increase risk for schizophrenia and mood disorders (Chubb et al. 2008). For example, DISC1 interacts with phosphodiesterase 4B

(PDE4B) and other proteins that are important in neuronal signaling, development and morphology (Millar et al. 2005; reviewed in, e.g., Ishizuka et al. 2006; Sawamura and Sawa 2006; Mackie et al. 2007; Chubb et al. 2008). The vital role of DISC1 in neuronal integration in adult rat hippocampus has been recently demonstrated, including the effect of *DISC1* expression knockdown that facilitates an accelerated maturation of neurons (Duan et al. 2007). Reminiscent of the G72/G30 relationship, *DISC1* is overlapped by *DISC2*, an oppositely-transcribed gene that may not give rise to protein but may have a palpable effect on the availability of DISC1 (Chubb et al. 2008).

2.2.2 Genome-Wide Association Studies (GWAS)

In the past year, whole genome association for complex diseases has grown rapidly, transitioning from being in the wish list of geneticists to a facile, fast and efficient approach for fine-scale global search for disease-predisposing variants. Key developments that spurred this progress include: (1) availability of large case-control samples; (2) availability of millions of validated SNPs due to the efforts of the International HapMap consortium (http://www.hapmap.org/); and (3) technological advance enabling the commercial production of chips that can interrogate up to a million SNPs on thousands of samples at a speedy rate. Chips arrayed with SNP-specific oligos from Affymetrix and Illumina can be processed easily in both large and small laboratories. Although the cost of the entire process has become less prohibitive, it still requires an ample budget which is why the support by GAIN and The Wellcome Trust Foundation are truly vital for moving the project expeditiously forward. GAIN is supporting GWAS for mood disorder samples collected by the NIMH Genetics Initiative. Both initiatives provide public access to generated data that could hasten the discovery of genetic determinants and pathways involved in pathogenesis.

In bipolar disorder, three GWAS have been published since 2007 [Wellcome Trust Case Control Consortium (WTCCC) 2007; Baum et al. 2008a; Sklar et al. 2008 advanced on-line pub] (Table 3). The WTCCC U.K. samples consisting of 1,900 bipolar disorder cases and 3,000 controls, and the 1,461 BPI combined cases from the Systematic Treatment Enhancement Program for Bipolar Disorder (STEP-BD) and University College London (UCL) collection. The 2,008 NIMH and UCL controls in the Sklar et al. (2008) study were individually genotyped on the Affymetrix GeneChip Human Mapping 500 K chips. Baum et al. (2008a) genotyped several independent duplicate pools of 461 BPI cases from the NIMH Genetics Initiative and another set of pools consisting of NIMH controls. A similar pooling regimen was done for 772 cases and 876 controls from Germany. Both NIMH and German pools were genotyped using the Illumina 550 K bead array.

A stringent and thorough GWAS quality control prior to and during data analysis is implemented in order to exclude various confounding factors. A detailed account of the process of data clean-up has been described (Wellcome Trust Case Control Consortium 2007; Baum et al. 2008a; Sklar et al. 2008).

Decoding the Genetics and Underlying Mechanisms of Mood Disorders

Table 3 Top highlighted loci in three GWAS for bipolar disorder.[a]

GWAS	Top regions	Genes or SNPs &/or number of SNPs
WTCCC (2007)	16p	rs420259
	2p25	one SNP each (see WTCCC 2007)
	2q12	
	2q37	
	3p23	
	3q27	
	6p21	
	8p21	
	9q32	
	14q22	
	14q32	
	1612	
	2013	
Baum et al. (2008)	13q14.11	DGKH/3 SNPs
	4p16	SORCS2/3 SNPs
	9q32	DFNB31/rs942518
	13q33.1	VGCNL1/rs9513877
	16p13.3	A2BP1/3 SNPs
	3q24	PLSCR4/rs3762685
	17p13.3	NXN/rs2360111
Sklar et al. (2008)	18q21.1	MYO5B/rs4939921
	12q21.1	TSPAN8/rs1705236
	7p11.2	EGFR/2 SNPs
	9q32	C9orf91/rs16931058
	18q22.1	CDH7/6 SNPs

[a]For other highlighted regions, refer to cited publications

Despite some of its attendant drawbacks, analysis of pooled samples for association has the advantage of dramatically cutting the cost of genotyping. By employing this strategy on two sets of cases and controls from the NIMH Genetics Initiative and Germany, 88 SNPs in 80 genes met the criterion of association in both series (Baum et al. 2008a) (Table 3). This finding was verified by analyzing samples individually on a different genotyping platform, i.e., the TaqMan allele discrimination assay. Genome-wide significance was detected with rs1012053 in intron 1 of *DGKH*, a gene located at 13q14.11 (Baum et al. 2008a) (Table 3), a region first highlighted in a linkage scan of Australian pedigrees (Badenhop et al. 2001). DGKH is one of 10 members of the diacylglycerol kinase (DGK) family that catalyzes the conversion of diacylglycerol (DAG), a potent second messenger, into phosphatidic acid (PA) (Sakane et al. 2007). Cellular constraints on DAG concentration could alter the level of enzymatically active protein kinase C (PKC) and perturb the plethora of PKC-responsive downstream events. Notably, PKC is a key enzyme that connects the lithium-responsive phosphoinositide and Wnt signaling pathways and the link of DGKH to these pathways shines on its potential role in bipolar disorder.

In the WTCCC (2007) bipolar analysis, the SNP that had the strongest evidence for association and met criteria for genome-wide significance mapped to 16p12 (Table 3), a region previously shown to have modest linkage signal in bipolar disorder families (Ekholm et al. 2003; Cheng et al. 2006). There are a number of genes in this region but no specific gene was highlighted in this report (WTCCC 2007). The association signals on 14q22 and 14q32 raise interest because this region showed suggestive evidence for linkage in a meta-analysis of bipolar disorder whole genome linkage scans (Segurado et al. 2003).

An outright comparison of the data derived from the Affymetrix 500 K and Illumina 550 K chips is constrained by the fact that of the 500,000 SNPs approximately only ~46,000 are common to both platforms. Fortunately, WTCCC has generated imputed genotype data (2007) that can be used for meta-analysis. In silico genotyping is a powerful and inexpensive tool for generating enormous numbers of genotypes, an example of this program is called MACH that has been developed by Abecasis (http://www.sph.umich.edu/csg/abecasis/MACH/tour/). Imputation is immensely useful as it adds tremendous amount of data through computer work alone, skipping the laborious, time-consuming and tedious wet lab work.

A meta-analysis performed on the Baum et al. (2008a) individually-determined genotypes and the WTCCC imputed genotypes revealed the best overlaps at rs10791345 on *JAM3* on 11q25 and rs4806874 at *SLC39A3* (ZIP3) on 19p13.3 with $p = 5 \times 10^{-6}$ and 9×10^{-6}, respectively, which were short of genome-wide significance (Gershon et al. 2008). Another meta-analysis on both datasets yielded similar association results including additional SNPs on other genes (Baum et al. 2008b). It is interesting to note that suggestive linkage evidence at 19p13 to schizoaffective disorder has been reported previously (Hamshere et al. 2005).

In the study by Sklar et al. (2008), the strongest evidence for association was displayed by a single SNP on *MYO5B* on 18q21.1, and haplotype analysis highlighted *MYO5B*, *TSPAN8* on 12q21.1 and *EGFR* on 7p11.2 (Table 3). Analysis of the combined STEP-BD-UCL and WTCCC data which both used the Affymetrix GeneMapping 500 K SNP Array generated convergence at *CACNA1C* that maps to 12p13.33 (Sklar et al. 2008). Weak evidence of overlap at *DFNB31* on 9q32 between the Baum et al. (2008a) and Sklar et al.'s (2008) studies was also reported (Sklar et al. 2008).

In summary, the three GWAS highlighted diverse sets of associated genetic variants with minimal overlaps in the top findings (Table 3). This lack of consistency seems to mirror reports on linkage and candidate gene associations. It is also interesting that none of the candidate genes that have been the subject of many investigations have appeared in the list of top hits. A number of factors could account for the discrepancy, prominent among which are extensive genetic (locus and allelic) heterogeneity, rare susceptibility alleles, and/or most of the risk variants have very low effect sizes. To detect weak effect variants that are well-replicated might require very large sample sizes, perhaps in the tens of thousands, as shown by the compelling association findings in Type II diabetes (see meta-analysis by Zeggini et al. 2008). The use of chips designed to interrogate rare SNPs is another plausible strategy. Targeting population isolates

Decoding the Genetics and Underlying Mechanisms of Mood Disorders

(Venken and Del-Favero 2007) in the search for infrequent vulnerability variants may yield some success, and extended families would be useful in tracking co-segregation of affection and variation. In addition, searches could be focused on suggestive linkage regions that have prior support in at least two large studies through fine-scale SNP screening combined with deep re-sequencing of exons, splice junctions and promoter regions to identify private susceptibility variants.

2.2.3 Candidate Gene Screening

Brain-Derived Neurotrophic Factor (*BDNF*) on 11p14.1

We selected *BDNF* as an example of a gene that has been repeatedly targeted for analysis in mood disorders. *BDNF* is located at 11p14.1, which is either at or near a region that has shown suggestive linkage for bipolar disorder in Old Order Amish (Egeland et al. 1987) and other family series (McInnes et al. 1996; Detera-Wadleigh et al. 1999). Studies have established the importance of BDNF in neurogenesis, neuronal development and survival (Jones et al. 1994; see recent review by Post 2007 and discussion in the Depression section below). A conditional knockout mouse model specifically targeting brain *bdnf* displayed weight gain and increased locomotor activity, the latter simulating a manic phenotype (Rios et al. 2001). Functionally therefore BDNF is an appealing candidate gene for bipolar disorder, major depression and schizophrenia, and many studies investigating this possibility attest to the interest in this gene.

Genetic studies on *BDNF* in bipolar disorder have generated conflicting results. Most studies have focused on the G > A nonsynonymous SNP that gives rise to Val66Met substitution that has been shown to alter function. In a study by Egan et al. (2003), met-met homozygotes were found to have poorer episodic memory. The same study showed that, unlike val-BDNF expression plasmid, the met-BDNF transfected construct showed an abnormal subcellular distribution in primary hippocampal neurons. Results from studies in bipolar disorder involving samples of European ancestry are generally weakly supportive of association of the G allele (Val of Val66Met) with bipolar disorder (Sklar et al. 2002; Neves-Pereira et al. 2002; Lohoff et al. 2005). Samples of Asian origin, however, have not shown evidence of association (see meta-analysis by Kanazawa et al. 2007). In a large study of ~1,000 bipolar cases and >2,000 controls, association at rs6265 (Val66Met) was found only in the subset of 131 rapid cyclers suggesting a role for the val allele in a specific phenotype (Green et al. 2006). Similar findings on BDNF and rapid cycling have been reported in another sample (Müller et al. 2006).

Haplotype analysis on rs6265 and a dinucleotide $(GT)_n$ repeat has yielded evidence for modest association with bipolar disorder (Neves-Pereira et al. 2002; Schumacher et al. 2005). But a recently published meta-analysis failed to support this finding (Kanazawa et al. 2007), although it is not clear whether the combined samples had sufficient power to detect minor effect. It is also possible that sequence changes in the *BDNF* elevate risk for specific bipolar disorder subphenotypes (Green et al. 2006; Müller et al. 2006).

A recent study in primary rat cortical neurons depolarized by KCl has shown that lithium and valproate induced activation of the BDNF promoter IV (Yasuda et al. 2007). A similar effect was found by inhibiting either GSK3 or HDAC suggesting a new and important step in mood stabilizer action involving BDNF.

2.2.4 Other Genetic Mechanisms

Copy Number Variation (CNV)

Accumulating genomic data indicates that CNV is widespread in the human genome. Historically, de novo genomic structural changes have been demonstrated in sporadic genomic disorders and CNVs have been reported in Mendelian diseases (reviewed in McCarroll and Altshuler 2007). The number of structural variation in each and every chromosome continues to burgeon and is now cataloged in a website (http://projects.tcag.ca/variation/). A recent analysis indicates that CNVs in the database are actually smaller than what is currently recorded (Perry et al. 2008). This study also found that smaller CNVs are embedded within larger ones adding complexity to its genomic architecture, a salient point to consider when searching for disease-predisposing variation.

A genome-wide survey of SNPs and CNVs in 29 populations worldwide illuminates the usefulness of CNVs in population and disease genetics (Jakobsson et al. 2008). The contribution of CNVs to psychiatric disease has been recognized, and studies published so far have been mostly focused on autistic spectrum disorders (ASD). Sebat and colleagues (2007) reported that de novo CNVs were significantly associated with autism. In addition, they found that CNVs were small and heterogeneous, and structural changes were found within genes. Another study that used array comparative genomic hybridization (aCGH) on 397 ASD patients presented 51 autism-specific CNVs in 46 patients (Christian et al. 2008). Microdeletions and microduplications were seen in 272 genes and the authors speculated that some of these could be candidate genes for ASD. A combination of SNP array and karyotyping on ASD cases and families showed association of a CNV at 16p11.2 (Marshall et al. 2008). Identification of novel loci and further support for the involvement of *SHANK3-NLGN4-NRXN1* genes in susceptibility were also reported.

Exploration for CNVs in bipolar disorder and schizophrenia has been scant, and to our knowledge only a few studies have been published. To search for bipolar disorder- and schizophrenia-associated CNVs DNA extracted from postmortem brains were subjected to aCGH (Wilson et al. 2006). CNVs found exclusively in cases were detected in *EFNA5*, *GLUR7*, *CACNG2* and *AKAP5*, all of which have been shown to have neuronal function. Findings were validated by quantitative real time PCR. A recent study showed that a CNV (duplication) on GSK3β on 3q13.3 was overrepresented in bipolar patients compared to controls (Lachman et al.

2007). It is unclear whether these findings are replicable, but we expect that more investigations on CNV in mood disorders will be reported in the near future.

MicroRNA

MicroRNAs (miRNA) are short (~22 nt), non-coding sequences that regulate gene expression and mRNA decay by binding to 3'untranslated regions (3'UTRs) of target transcripts. Since their discovery in *C. elegans* (Lee et al. 1993), approximately 500 mammalian miRNA are either predicted or known to exist, half of which are expressed from non-protein coding transcripts and the other half from introns transcribed as 3–4 kb transcripts (Saini et al. 2007). miRNAs have been shown to exhibit tissue and organ-specific expression and brain expression varies during development and neuronal differentiation (e.g., Krichevsky et al. 2003; Sempere et al. 2004; Miska et al. 2004; Smirnova et al. 2005; Kapsimali et al. 2007). A recent report shows that miR-134 which is brain-specific exerts a negative effect on the size of dendritic spines in rat hippocampal neurons (Schratt et al. 2006). In Drosophila, mutant miR-8 leads to increased atrophin activity, elevated apoptosis in brain, and behavioral defects (Karres et al. 2007).

In psychiatric disease, expression analysis on 264 microRNAs has been performed on postmortem prefrontal cortex from a few patients with schizophrenia, schizoaffective disorder, and controls, and it led to the detection of 12 differentially expressed miRNAs including miR-24, miR-29 and miR-7 that were validated by quantitative PCR (Perkins et al. 2007). Hansen et al. (2007) have taken a different strategy by first identifying SNPs located ± 100 bp from brain-expressed miRNAs and genotyped each of them to test for possible association with schizophrenia in Scandinavian samples. Two SNPs that displayed weak signals were in the vicinity of miR-206 and miR-198, miRNAs predicted to target multiple genes. Microarray analysis of miRNA in Alzheimer's brains showed a decline in expression of miR-107 at the early stages of the disease (Wang et al. 2008b). The regulatory function of miRNA is varied and wide, but most of its function is unknown. The role of microRNAs in mood disorder pathobiology remains to be elucidated.

2.3 Alternative and Complementary Strategies

2.3.1 Endophenotypes

Endophenotypes are intermediate phenotypes in the pathogenic pathway and are thought to be less complex than the categorical phenotype. Employing endophenotypes in genetic analysis could enable stratification of samples toward greater homogeneity and thus implicate fewer genes. Recent reviews have discussed the utility and difficulties associated with the application of endophenotypes in psychiatry (Gottesman and Gould 2003; Hasler et al. 2006); Bearden and Freimer 2006; Flint and Munafo 2007).

Several criteria that have been proposed for useful endophenotypes (Gottesman and Gould 2003; Bearden and Freimer 2006)) are listed below:

1. Associated with causes rather than effects of disease
2. Heritable
3. State-independent
4. Co-segregate within families
5. Present in unaffected members of families that display the endophenotype at a higher rate than in the general population
6. Good reliability

Some examples include brain imaging studies such as functional magnetic resonance imaging (fMRI) that focus on specific candidate genes, but the need for expanded sample size is important. Neurological deficits in mood disorders may be reflected in neuropsychological tests designed to generate quantitative trait measures for cognition, neuroanatomy, and temperament (Bearden and Freimer 2006; Flint and Munafo 2007). Neuropsychological tests have used small sample sizes and they need to be extended to examining families. A recent example of these tests on a small number (52) of Finnish families with bipolar disorder found significant heritability of certain variables involved in general intellectual functioning, attention and working memory, and executive functions (Antila et al. 2007). Bearden and Freimer (2006) have suggested also that nonhuman primates and gene expression profiles may be exploited in the search for reliable endophenotypes.

A critique of the endophenotype concept concluded that an endophenotype may be just as complex as the categorical phenotype that, so far, has not proven that it simplifies the genetic complexity in psychiatric disorders (Flint and Munafo 2007). Meta-analysis on various endophenotypic studies including correlation of *COMT* Val158/108Met polymorphism with results from various tests such as Wisconsin Card Sorting Task (WCST) failed to show either large effects or a significant association. Perhaps analysis of much larger samples would lend validity to some of the proposed endophenotypes. However, a review of the behavioral, biochemical and immunological phenotypes, as well as gene expression profiles, in mice reveals minor effect size of the phenotypes and transcript levels, which may be instructive as endophenotypes in psychiatric diseases are examined (Flint and Munafo 2007).

2.3.2 Epigenetic Mechanisms: DNA Methylation and Histone Methylation and Acetylation

DNA nucleotide sequence-independent processes have been shown to elicit profound effects on neural development and disease. Epigenetic mechanisms that alter chromatin structure, such as the DNA methylation on CpG islands in promoters and histone modification, i.e., methylation and acetylation (reviewed in

Tsankova et al. 2007), regulate gene activity variably in a spatial and temporal manner. Maintaining the delicate balance of the level of methylation has been shown to be critical during the neuroblast stage when mitosis still occurs but not in postmitotic neurons. This was demonstrated in a mouse model created by conditional knockout of DNA methyltransferase (*Dnmt1*), the enzyme that transfers a methyl group to cytosine (Fan et al. 2001). Embryos with Dnmt1 deletion that carried 95% hypomethylated brain cells died after birth, but those with 30% hypomethylation survived into adulthood and mutant brain cells were eliminated after 3 weeks of birth. Deletion of another protein that has an important role in methylation, the methyl-CpG binding protein 1 (MBD1), produced mice that appeared normal but which exhibited defects in neuronal differentiation and adult neurogenesis (Zhao et al. 2003). These animals also showed reduction in long term potentiation and impairment in spatial learning.

DNA methylation has been shown to exert a role in memory formation and consolidation (Miller and Sweatt 2007). Changes in histone H4 acetylation in rat hippocampus at the promoters for glutamate receptor 2 and BDNF upon pilocarpine-induced seizure have been reported (Huang et al. 2002), suggesting that histone modification may be an early event preceding seizure.

A recent report on epigenomic profiling using postmortem frontal cortex from schizophrenia and bipolar disorder patients demonstrated alteration in the methylation status of multiple genes including those involved in the glutamate, gamma-amino butyric acid (GABA), and Wnt neuronal signaling systems (Mill et al. 2008). Methylation at certain loci also appeared to be sex-specific. Investigations into the role of epigenetics in mood disorders are just at the beginning stage and much remains to be uncovered.

Evidence for the convergence of gene-environment interaction (G X E) and epigenetics in behavior has been depicted through a notable approach. High levels of maternal nurturing in rats produced less anxious offspring that elicited increased expression of the glucocorticoid receptor (GR) gene, which contrasts with the behavior of rats that received low levels of maternal care (reviewed in Meaney et al. 2007 and Szyf et al. 2007) (Fig. 3). The rise in GR transcript levels has been traced to the fact that the NGFI-A consensus cis elements on the GR_7 promoter had lower levels of methylation in the highly nurtured animals. The potential relevance to mood disorder pathology is intriguing and still needs to be demonstrated.

2.3.3 Therapeutics and Associated Mechanisms; Pharmacogenetics

Lithium, valproic acid, and carbamazepine are mood stabilizers that prevent relapse into either a depressive or manic phase. Valproic acid and carbamazepine are also used as anti-epileptics. In the process of searching for the molecular targets of mood stabilizers, including their positive effects on neuronal survival and neurite formation, several etiological hypotheses have been proposed and these are discussed in the succeeding sections (also see review, e.g., Shaltiel et al. 2007). Various signaling pathways have been shown to be responsive to lithium, and it is

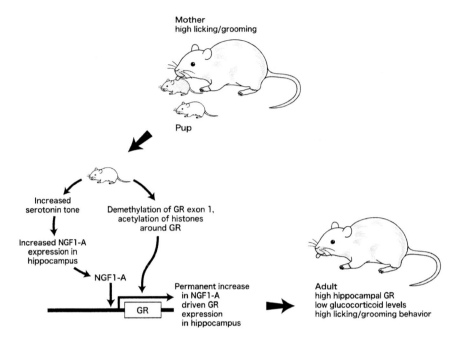

Fig. 3 Epigenetic molecular mechanisms in transmission of traits from mother to offspring. Maternal behavior characterized by high levels of licking and grooming produces two interacting pathways of changes in the pup: (1) Increased serotonin tone in the hippocampus leads to increased expression of the transcription factor NGF1-A. (2) The first exon of the glucocorticoid receptor gene in the hippocampus is demethylated, and the histones surrounding it are acetylated. This makes glucocorticoid receptor gene constitutively activated by NGF1-A, and generates a greater number of glucocorticoid receptor molecules in the hippocampus of the rat when it becomes an adult

possible that mutations in components of these pathways contribute to increase the risk for bipolar disorder (for a review, see Detera-Wadleigh 2001) and are therefore plausible targets for genetic studies. It is also well known that responses to mood stabilizers among patients are variable, hence patients or patient families can be categorized on the basis of response. Moreover, it has been shown that lithium responders have distinctive characteristics such as family history and absence of rapid cycling; this subgroup might have lower genetic heterogeneity (see review by Alda et al. 2005). Turecki et al. (2001) conducted a whole genome linkage scan on lithium-responsive families and reported suggestive signals on 15q, 7q, 6p and 22q. Multipoint analysis failed to support evidence of linkage on 15q and the top findings centered on other chromosomal regions (Alda et al. 2005).

Inositol depletion hypothesis (phosphoinositide cycle/protein kinase C signaling cascade). The phosphoinositide cycle was first implicated in bipolar disorder by the unreplicated finding of changes in plasma inositol levels of affective disorder patients (Barkai et al. 1978). Later findings revealed inhibition of inositol monophosphatase (IMPase: EC 3.1.3.25) by lithium creating a reduction in inositol levels, and this led to the proposal of the inositol depletion hypothesis (Berridge 1989). Supporting evidence from various investigations have been reported, including imaging studies that found reduced inositol in brains of lithium-treated bipolar patients (Davanzo et al. 2001; Moore et al. 1999). Inositol supplementation has been reported to attenuate lithium enhancement of pilocarpine-induced seizures (Patishi et al. 1996) which is augmented by inositol reuptake inhibitors (Wolfson et al. 2000).

Inositol depletion is biologically relevant to the efficacy of not just lithium but also carbamazepine and valproic acid. Firstly, these drugs activate autophagy, a major proteolytic pathway involved in the degradation of bulky protein aggregates (Sarkar et al. 2005). In mammalian cells, autophagy can be induced by rapamycin, a drug that inhibits mTOR, which is a negative regulator of autophagy. The induction of autophagy by lithium is a novel mTOR-independent pathway that enhances the clearance of autophagy substrates, like mutant huntingtin and α-synucleins. The autophagy-enhancing property of lithium has been shown to be mediated through inhibition of IMPase that leads to inositol depletion and a decrease in myo-inositol-1,4,5-triphosphate (IP3) levels. Carbamazepine has been shown to reduce protein aggregates, attenuate polyglutamine toxicity in COS-7 cells, and enhance the clearance of accumulated synuclein. Valproic acid yielded similar results. These data suggest that autophagy is mediated at the level of (or downstream of) IP3, because pharmacologic treatments that increased IP3 abrogated this process (Sarkar et al. 2005).

Secondly, evidence has been reported that all three drugs inhibit the collapse of sensory neuron growth cones and trigger an increase in growth cone areas (Williams et al. 2002). These effects did not involve inhibition of GSK3β or histone deacetylase (HDAC). Note that the action of valproic acid has been linked to both inositol depletion and inhibition of HDAC. Inositol reversed the drug effects on growth cones. To investigate the cellular effects of lithium and valproic acid, Harwood used *Dictyostelium* as it is a simple model organism that is sensitive to both drugs (see review by Harwood 2005). It was found that resistance to both drugs was conferred by deletion of the prolyl oligopeptidase gene that also regulates inositol metabolism. In addition, inhibitors of prolyl oligopeptidase reversed the effects of lithium, valproic acid, and carbamazepine on sensory neuron growth cone area (Williams et al. 2002).

A key enzyme involved in the rate-limiting step for generating inositol in vivo is IMPase. IMPase catalyzes the release of the phosphate group from inositol monophosphate, an important step for the regeneration of free inositol. The first enzyme found to exhibit IMPase activity was referred to as IMPA1 (Diehl et al. 1990; McAllister et al. 1992). The gene maps to 8q21.13. So far, there is no genetic evidence linking *IMPA1* to mental disorders (Sjoholt et al. 2004).

IMPA2 is a close homolog of IMPA1, and the full-length cDNA was cloned in 1997 (Yoshikawa et al. 1997). The newly-identified gene had attracted attention because it is located at 18p11.2, a susceptibility region for bipolar disorder aside from the possibility that the gene product is a major target of lithium. To determine whether variation on *IMPA2* shows association with bipolar disorder, transmission disequilibrium tests were performed on 92 Palestinian Arab trios (Sjoholt et al. 2004). Two SNPs on the 5′ flanking region of *IMPA2* displayed slight overtransmission of alleles. Functional evidence was presented by Ohnishi et al. who detected upregulation of *IMPA2* mRNA in postmortem brains of bipolar patients that harbor risk-associated promoter SNPs (Ohnishi et al. 2007a). Extensive biochemical profiling showed that IMPA2 forms homodimers in vivo but does not heterodimerize with IMPA1 (Fig. 4) (Ohnishi et al. 2007b).

A comparative characterization of IMPA1 and IMPA2 revealed several interesting properties unique to each enzyme isoform. IMPA2 yielded a significantly lower activity towards *myo*-inositol monophosphate than IMPA1, and the former was inhibited by higher concentrations of lithium and restricted magnesium concentrations. *IMPA1* and *IMPA2* displayed differential expression patterns in various tissues including brain suggesting disparate functions in vivo. The crystal structure of human IMPA2 portrays an overall similarity with that of IMPA1, except for the loop regions (Bone et al. 1992; Arai et al. 2007). The wide-open cavity of IMPA2 implies that the physiological substrate may be a larger compound than that for IMPA1. IMPA2 was found to have two Ca^{2+} and one phosphate binding sites. Both isoforms had identical sites for Mn^{2+} and phosphate, suggesting similar mechanisms (Fig. 4).

Wnt/Glycogen synthase kinase-3β (GSK-3β) signaling cascade. The possible involvement of GSK-3β, a key enzyme in the Wnt signaling pathway, in bipolar

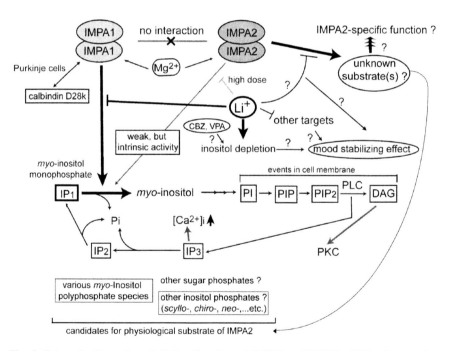

Fig. 4 Schematic illustration of distinct functions of IMPA1 and IMPA2. *CBZ* carbamazepine, *VPA* valproic acid, *IP1 myo*-inositol monophosphate, *IP2 myo*-inositol bisphosphate, *IP3 myo*-inositol 1,4,5-trisphosphate, *PLC* phospholipase C, *DAG* diacylglycerol, *PI* phosphatidylinositol, *PIP* phosphatidylinositol phosphate, *PIP2* phosphatidylinositol 4,5-bisphosphate, *G6P* glucose 6-phosphate. This figure is cited from the paper of Ohnishi et al. (2007)

disorder stems mainly from the fact that lithium (and possibly valproic acid) inhibits its activity (Klein and Melton 1996; Chen et al. 1999a). Furthermore, GSK-3 is a downstream target of monoaminergic systems and growth factor cascades (Rowe et al. 2007). In dopamine transporter knock-out mice, the hyperactivity observed has been shown to be mediated, at least in part, through a GSK-3–dependent mechanism, and that amphetamine administration causes a reduction in the phosphorylation of GSK-3α and GSK-3β in the striatum, an effect opposite to that observed with lithium administration (Beaulieu et al. 2004). Genetic analysis of GSK3β has not revealed a direct role for the gene in bipolar disorder etiology, although this must be considered tentative since underpowered samples may have been used. The potential contribution to susceptibility of other components of the Wnt signaling pathway remains to be investigated.

Neurotrophic cascade. Animal and human studies have provided converging evidence for the role of mood stabilizers in neuroprotection and neuronal survival. Both lithium and valproic acid have been shown to inhibit the proapoptotic action of GSK-3β in the central nervous system and induce an increase in expression in the frontal cortex of B-cell lymphoma protein-2 (bcl-2), a neuroprotective protein (Chen et al. 1999b). Bcl2 is a member of a well-characterized protein family that regulates apoptotic cell death, acting on the mitochondria to stabilize membrane integrity and prevent the release of apoptogenic factors.

Postmortem neuropathologic studies have documented reduced neuronal density in brains of bipolar patients, particularly in the anterior cingulate (Benes et al. 2001). Lithium treatment has been shown to elicit an increase in total gray matter volume in bipolar patients by 3%, on average, after 4 weeks (Moore et al. 2000). Thereafter, larger total gray matter volumes in lithium-treated bipolar patients as compared to both untreated patients and control subjects have been reported (Sassi et al. 2002). In addition, magnetic resonance spectroscopy (MRS) has revealed an increase in cortical N-acetyl-aspartate (NAA), a putative marker of neuronal integrity, in bipolar patients following lithium administration (Silverstone et al. 2003). Recently, Bearden et al. (2007) reported that cortical gray matter density was significantly greater in lithium-treated bipolar patients relative to control subjects. The greatest difference was found in bilateral cingulate and paralimbic cortices, brain regions critical for attentional, motivational, and emotional modulation. Greater gray matter density in the right anterior cingulate was observed among lithium-treated bipolar patients compared to those who were not on lithium. Loss of neurons in bipolar brains seems to be the pattern that emerges from these studies, also that mood stabilizers could, to a certain extent, reverse this process thus promoting neuronal regeneration and survival.

2.3.4 Animal Models: Some Gene Mutants that Produce Bipolar-Like Phenotypes in Animals.

To explore the effect of *IMPA1* and *IMPA2* gene deletions, mouse knockout models were generated. *Impa1* (−/−) deficient mice died in utero between days 9.5 and 10.5 postcoitum demonstrating its importance in early embryonic development (Cryns et al. 2007a). The embryonic lethality can be reversed by *myo*-inositol supplementation to pregnant mothers. Although IMPase activity levels were reduced (up to 65% in hippocampus) in brains of adult *Impa1* (−/−) mice, inositol levels were not altered. These mice showed increased motor activity in open-field test and forced swim test (FST; see Depression section for a brief description) as well as an increased sensitivity to pilocarpine-induced seizures, supporting the idea that Impa1 represents a physiologically pertinent target for lithium. The same group generated *Impa2* deficient mice, *Impa2* (−/−), that exhibited increased rearing in open field, suggesting increased exploratory behavior (Cryns et al. 2007b). Overall, there was no clear effect of *Impa2* gene deletion on depression-like behavior, and the animal model does not reproduce the effects of lithium treatment. To draw clear conclusions on the relationship between IMPase activity and bipolar-like phenotypes, it would be necessary to generate *Impa1* and *Impa2* transgenic mice and evaluate their behaviors.

Dysregulation of the circadian rhythm is a commonly observed feature in mood disorders (see review by McClung 2007), and therefore animal knockout models of genes involved in maintaining the molecular clock are of interest. Disruption of the *Clock* gene in mice revealed mania-like phenotypes such as hyperactivity, decreased sleep, lower anxiety, etc. (Roybal et al. 2007). Chronic administration of lithium reinstated many of the behavioral responses to wild-type levels. In addition, the *Clock* mutants exhibited an increase in dopaminergic activity in the ventral tegmental area, and their behavioral abnormalities were rescued by expressing a functional Clock protein. Suggestive evidence for association of *CLOCK* gene variation with bipolar disorder (Shi et al. 2008b) and insomnia in bipolar disorder (Serretti et al. 2003) has been presented, although validation in larger cohorts will be required.

DBP, a PAR (proline and acidic amino acid-rich basic leucine zipper)-containing transcription factor, whose transcript levels oscillate in the suprachiasmatic nucleus (SCN), exerts an effect on circadian rhythm. DBP deficient mice (*dbp-/-*) have been shown to be less active and display abnormal aspects of sleep–wake cycle (Lopez-Molina et al. 1997; Franken et al. 2000). Recently, this animal model was used to study the effect of chronic stress and mood stabilizer treatment (Le-Niculescu et al. 2008b). Microarray analysis of transcripts from amygdala and blood depicted changes in an array of genes, but whether these genes have relevance to etiology remains to be established.

A mouse GSK-3β knockout has been generated (O'Brien et al. 2004). The GSK-3β heterozygote knockout mice displayed decreased immobility time in FSTs, mimicking a manic phenotype, which does not seem to be consistent with the inhibitory effect of lithium on GSK-3β activity (O'Brien et al. 2004).

Chronic progressive external ophthalmoplegia (CPEO) with comorbid mood disorders has been previously described (Kasahara et al. 2006). In CPEO patients, mitochondrial DNA (mtDNA) deletions accumulate due to mutations in nuclear-encoded genes such as *POLG*. Transgenic mice in which mutant *Polg* was expressed in a neuron-specific manner displayed forebrain-specific defects of mtDNA and altered monoaminergic functions in the brain (Kasahara et al. 2006). The mutant mice exhibited characteristic behavioral phenotypes, a distorted day–night rhythm, and a robust periodic activity pattern associated with the estrous cycle. These behaviors worsened upon tricyclic antidepressant treatment but improved upon lithium administration. Antidepressant-induced mania-like behavior and long-lasting irregularity of activity in some mutant animals caused mood disorder-like phenotypes, similar to treatment responses in bipolar disorder patients.

To determine the molecular repertoire perturbed by lithium treatment in brain, mRNA profiling in treated mice and in untreated controls was performed by using GeneChips arrayed with 39,000 genes (McQuillin et al. 2007). Changes in expression were detected in 121 genes. Three genes that showed the highest increase in expression were alanine-glyoxylate aminotransferase 2-like 1 (*Agxt2l1*), c-mer proto-oncogene tyrosine kinase (*Mertk*), and sulfotransferase family 1A phenol-preferring member 1 (*Sult1a1*). Altered expression was displayed by genes encoding period homolog 2 (*Per2*: upregulated), the metabotropic glutamate receptor (*Grm3*: downregulated), secretogranin II (*Scg2*: upregulated), several myelin-related genes and protein phosphatases. Under lower stringency, significant changes were detected in BDNF (upregulated), β-phosphatidylinositol transfer protein (*Pitpnb*: upregulated), *Impa1* (upregulated), and five genes related to thyroxine metabolism. Altered expression was also found in 45 genes related to phosphatidylinositol metabolism. A similar mRNA profiling study in lithiuim-treated mice highlighted altered expression of a separate set of genes (Chetcuti et al. 2008). The lack of overlap of highlighted genes in these two studies is less than encouraging as RNA was derived from fresh brain tissues, nonetheless it could reflect differences in experimental conditions while many of the observed expression changes may be spurious.

2.3.5 Gene Expression Profiling in Postmortem Brains and Other Tissues

The use of postmortem brains as template for studies carry with it potential confounding factors such as variations in postmortem delays, drug profile, cause of death, health or illness status, etc. Despite these issues several microarray studies have been pursued (e.g., see review by Sequeira and Turecki 2006). It is unclear whether altered expression is an etiological marker or if it is the effect of the overall condition of the individual and the tissue. The hope is that candidate genes that contribute to genetic vulnerability may be identified through this method but, not surprisingly, these studies report different sets of altered transcripts. Two examples are briefly mentioned here.

Transcriptional analysis with Affymetrix GeneChips performed on brains from both bipolar disorder and schizophrenia patients highlighted expression changes in many genes important in mitochondrial function almost exclusively in bipolar disorder, which might support the case for a mitochondrial role in genetic risk (Konradi et al. 2004). Of note, somatostatin (*SST*), one of two genes involved in neurotransmission that showed changes in transcript levels, yielded the highest change in expression.

The expression levels of more than 12,000 genes in the Brodmann's Area (BA) 46 (BA46) (dorsolateral prefrontal cortex) from bipolar I disorder and control samples were profiled on Affymetrix GeneChips (Nakatani et al. 2006). A total of 108 differentially expressed genes were found. Validation studies using quantitative RT-PCR on the two original diagnostic cohorts, plus tissue from schizophrenic subjects, confirmed a bipolar-specific differential expression of eight genes (*RAP1GA1, SST, HLA-DRA, KATNB1, PURA, NDUFV2, STAR* and *PAFAH1B3*). The expression change in *SST* parallels a prior finding by Konradi et al. (2004). The increase in *RAP1GA1* transcript levels was reflected in its protein level in the BA46 region of bipolar brains. Interestingly, network analysis based on the expression data, highlighted cellular growth and nervous system development pathways as potential targets in the molecular pathophysiology of bipolar disorder.

To follow-up on the expression data, the study was extended to family-based association analysis in 229 NIMH bipolar trios utilizing 43 SNPs in nine genes that displayed altered expression. Analysis exposed nominal association and modest empirical haplotypic association ($P = 0.033$) between *SST* and disease (Nakatani et al. 2006).

Profiling gene expression in blood specimens drawn during the two extreme phases of bipolar disorder showed changes in transcript levels of genes involved in myelin formation and growth factor signaling, some of which have been shown to be altered in mood disorder postmortem brains (Le-Niculescu et al. 2008a). The authors suggest that these genes may be candidate biomarkers. As only a modest number of individuals were studied, support from a larger panel of samples would be important.

Cultured fibroblasts from skin biopsies of a small number of bipolar disorder patients and controls were grown and starved of fetal bovine serum, followed by serum shock to follow the oscillation of expression of several genes involved in circadian rhythm regulation (Yang et al. 2008). Apparently, the rhythmic properties of these genes in fibroblast mimic what is observed in the suprachiasmatic nucleus (SCN). A trend toward reduced expression and amplitude of these selected genes as well as diminished phosphorylation of GSK3β in bipolar disorder patient fibroblasts was reported. Whether expression in blood cells or fibroblasts reflects expression in disease-relevant brain structures continues to be an open question.

3 Depression

3.1 Clinical Presentation and Epidemiology

Depression, also called unipolar affective disorder, is one of the most common mental disorders worldwide. Depression is a term with meanings ranging from the transient dips in mood that are characteristic of life itself, to a clinical syndrome of substantial severity, duration, and associated signs and symptoms that are markedly different from normal. Depression is episodic, and periods of normal mood state are interspersed by depressed phases contrasting with schizophrenia that is progressive and chronic, and some mental defects persist even after remission. Major depressive disorder (Table 1) is identified by the presence of one or more major depressive episodes in the absence of a history of mania or hypomania. Accordingly, there are a number of depression subtypes, and there has been much debate about how to most accurately capture and organize the features and subtypes of major depression. There are no definite biological measures for depression, and diagnosis is made based solely on symptoms and their severity.

The clinical features of depression fall into four broad categories (Jefferson and Greist 1994):

(1) Mood (affect): sad, blue, depressed, unhappy, down-in-the dumps, empty, worried, irritable.
(2) Cognition: loss of interest, difficulty concentrating, low self-esteem, negative thoughts, indecisiveness, guilt, suicidal ideation, hallucinations, delusions.
(3) Behavior: psychomotor retardation or agitation, crying, social withdrawal, dependency, suicide.
(4) Somatic (physical): sleep disturbance (insomnia or hypersomnia), fatigue, decreased or increased appetite, weight loss or gain, pain, gastrointestinal upset, decreased libido.

Various studies have shown that heritability for unipolar depression ranges between 40 and 50% (reviewed in Levinson 2006), although a study in twins estimated heritability to be >70% (McGuffin et al. 1996). The ECA study surveyed about 20,000 adults in five U.S. communities, and reported a 6-month prevalence of 2.2% and a lifetime prevalence of 4.4% (Weissman et al. 1988). The mean age at onset was reported to be 27 years in both sexes, but the prevalence of depression in women was about two-fold higher than in men. Following this study, similar epidemiologic studies were done in other countries that generally found equal or even higher prevalence rates.

The ECA study was followed by the National Comorbidity Survey (NCS) in the United States in 1992. A 12-month prevalence rate of 7.7% and a lifetime prevalence of 12.7% (Kessler et al. 1994) were reported, rates that were almost three times higher than those in the ECA study. The NCS was a part of the WHO International Consortium of Psychiatric Epidemiology (WHO-ICPE). The statistics in the other countries that joined the WHO-ICPE showed a 12-month prevalence of 2.0 ~ 5.9% and lifetime prevalence of 6.3 ~ 15.7% (Andrade et al. 2003).

3.2 Genetic Analysis of Depression

Several linkage studies on nonbipolar depression indicate modest evidence and lack of consistency across studies (reviewed in Levinson 2006). These results reflect genetic complexity including weak-effect predisposing variants, but other factors are involved such as the use of underpowered samples and variable definitions of affection phenotypes in different studies. The largest consortium on depression called Genetics of Recurrent Early-Onset Major Depression (GenRED) that analyzed

656 families found suggestive evidence for linkage on 15q25-q26 and more modest signals on 8p and 17p (Holmans et al. 2007). Follow-up finer scale analysis with SNPs increased the linkage evidence (Levinson et al. 2007). Similarly, the 15q linkage region has been reported in Utah families with increased sharing particularly among males (Camp et al. 2005). Support for linkage on 15q and 12q (Abkevich et al. 2003) has been presented in a study on European families that also highlighted the regions 1p36 and 13q31 (McGuffin et al. 2005). Interestingly, the 1p region encodes methylenetetrahydrofolate reductase (*MTHFR*). The TT genotype for 677T/C SNP of *MTHFR* showed significant association with depression in the absence of anxiety in a study on ~6,000 depressives (Bjelland et al. 2003). The modest linkage signal on 13q in McGuffin et al.'s study is upstream of the location of G72 (*DAOA*) on 13q33. However, a recent analysis on samples of 500 major depressives and >1,000 controls found evidence for association at the more distal SNPs (M21 to M24) of G72 with major depression (Rietschel et al. 2008). Association was also observed with underlying neuroticism in these depressives.

No GWAS on depression have been published so far, but it is one of the diseases included in the GAIN study. It is clear that far fewer whole genome analyses have been conducted on depression. On the other hand, many studies have focused on candidate gene analysis including serotonin transporter and BDNF as discussed below.

3.2.1 Serotonin Transporter (SLC6A4, 5-HTT, SERT); Gene x Environment Interaction

Active investigations on 5′HTT vis-à-vis depression and other psychiatric traits continue to proliferate, some of which have yielded fascinating, provocative, but inconsistent findings. We only present a brief overview as studies pertaining to this gene is beyond the scope of this chapter and several recent reviews have been published (e.g., Levinson 2006; Hariri and Holmes 2006; Hahn and Blakely 2007; Canli and Lesch 2007; Murphy and Lesch 2008). The most studied variation on *SLC6A4* is a repeat length polymorphism on the promoter designated as 5-HTTLPR, and the alleles are referred to as short (s) or long (l). Compared to the "l" allele, the "s" allele has shown lower transcriptional activity and has been found to be associated with anxiety and depression related personality traits (Lesch et al. 1996). The latter was the first demonstration that *SLC6A4* promoter polymorphism may be relevant to a psychiatric disease. Sequence homologous to the human HTTLPR is absent in mouse; nevertheless, disruption of the gene in mice produces various behavioral traits that model anxiety, depression (learned helplessness), stress susceptibility, etc. (reviewed in Murphy and Lesch 2008).

Despite its functional link to depression, compelling association of *SLC6A4* variation with the categorical phenotypes of mood disorders has not been established. Various imaging studies such as the use of positron emission tomography (PET) and fMRI have been reported. Imaging a modest number of subjects by fMRI has shown association of the "s" allele with greater reactivity to negative versus neutral stimuli in the amygdala (see review by Hariri and Holmes 2006). PET studies on major depressives have shown a higher binding potential of the SERT ligand ^{11}C-DASB in various serotonergic nerve terminals among a subset of patients with highly dysfunctional negative attitudes, but not among all depressives that were studied compared to controls (Meyer et al. 2004). In a PET study of major depressives and bipolar disorder patients, Cannon and co-workers (2007) have reported that the binding potential of ^{11}C-DASB in the insula, thalamus, and striatum was higher during the depressed phase of the disease suggesting a state-dependent phenomenon.

Studies on life stress and depression in relation to HTTLPR has opened an avenue toward gene-environment interaction (Caspi et al. 2003; see editorial by Zammit and Owen 2006 and review by Uher and McGuffin 2008). A prospective longitudinal study was conducted on >1,000 individuals in Dunedin, New Zealand, who were followed at regular intervals from age 3 to 26 to monitor stressful life events and depression (Caspi et al. 2003). Depression was assessed just prior to age 26. To explore genetic association with variation on *SLC6A4*, genotyping of the HTTLPR was performed. Individuals who were either homozygous or heterozygous for the "s" allele displayed higher incidence of depression, depressive symptoms, and suicidality in relation to stressful events when compared to those that carry the "l" allele. These findings suggest a gene × environment interaction, but the authors also caution that their results could be suggestive of gene–unknown gene interaction. Similar studies have been pursued, but conflicting findings have been reported. A variety of factors could cause these discrepancies (see reviews by Zammit and Owen 2006 and Uher and McGuffin 2008).

3.3 Therapeutics and Associated Mechanisms; Pharmacogenetics

Accumulated pharmacological data so far have established the "monoamine depletion hypothesis of depression" (Bunney and Davis 1965; Schildkraut 1965). "Monoamine" represents neurotransmitters serotonin (5-HT) and norepinephrine (NE). The discovery that the prototypical antidepressants, the tricyclics (TCA) and monoamine oxidase inhibitor (MAOI), acutely increase synaptic levels of monoamines has led to the proposal of the monoamine hypothesis. TCAs have been shown to inhibit monoamine transporters that are responsible for the reuptake of monoamine neurotransmitters into the presynaptic cleft leading to their removal and inactivation (Duman 2004a). The monoamine oxidase inhibitors block one of the primary enzymes responsible for the degradation of 5-HT and NE and thereby increase monoamine concentrations. In contrast, early clinical studies demonstrated that administration of reserpine, an antihypertensive drug, can cause depression in some individuals. Reserpine depletes monoamines from presynaptic storage supporting the monoamine hypothesis. Selective 5-HT reuptake inhibitors (SSRIs) and selective 5-HT and NE inhibitors (SNRIs) have been developed that lack many of the side effects of the early tricyclic antidepressants.

However, the relationship between monoamines and depression is not simple. For example, although antidepressants rapidly increase the levels of monoamines in the brain, therapeutic action appears only after chronic administration (i.e., several weeks or even months). In addition, a substantial proportion of depressive patients are refractory to antidepressants, which probably reflects heterogeneity of depression etiology. It has been hypothesized that the delayed time course of antidepressant treatment is due to neural plasticity or adaptation. The mechanisms underlying neuronal adaptation involve receptor-coupled intracellular signal transduction pathways and neuronal gene expression. Sub-chronic administration of antidepressants is known to alter the sensitivity of receptors for monoamines, i.e., downregulation of $\beta 1$ adrenergic and 5-HT_{2A} receptors, and upregulation of 5-HT_{1A} receptor. These delayed changes of the monoamine receptor sensitivities are not sufficient to explain the therapeutic time course/action of all drugs. Downregulation of $\beta 1$ adrenergic and 5-HT_{2A} receptors may be an outcome of the adaptive process in response to sustained activation of receptor function by increased monoamines (Duman 2004a). It is interesting that the largest study, so far, on treatment outcome with the antidepressant citalopram, involving ~2,000 depressed patients, showed a significant allelic association with response at a SNP on the 3'end of *HTR2A*, specifically among Caucasians (McMahon et al. 2006).

Stimulation of monoamine receptors leads to activation of intracellular signaling pathway including the cyclic AMP (cAMP) cascade (Fig. 5). The transcription factor, cAMP response element binding protein (CREB), is a potential site of convergence for chronic antidepressant treatments. The expression of CREB mRNA and protein in rat hippocampus is increased by chronic treatment with antidepressants (both SSRI and selective NE reuptake inhibitors) (Nibuya et al. 1996). Upregulation of CREB implies that genes with cAMP response element (CRE) in their promoters may be targets for antidepressants. Gene expression is regulated in a complicated region-specific manner, and is dependent upon the function and interaction of multiple transcription factors and cofactors, that either enhance or inhibit transcriptional activity. Hence, identification of CREB target genes that are important for the efficacy of antidepressants needs region-by-region and gene-by-gene labor-intensive analyses. A published study has predicted that there could be as many as 10,000 CREB-binding sites in the mammalian genome (Euskirchen et al. 2004), and a few years ago ~4,000 had been identified (Zhang et al. 2005). For example, genes including *BDNF* and corticotropin-releasing factor (*CRF*) contain CREs (Guardiola-Diaz et al. 1994; Tao et al. 1988); nonetheless, the presence of CRE elements does not indicate that the gene is either a direct target or CREB mediates its regulation. Many transcription factors, including CREM, ATF-1, ATF-2, ATF-3, and C/EBP, can bind CRE sequences (Hurst 1995), thus further complicating the idea that gene regulation by CREB is based solely on promoter sequence information. In spite of these complexities, a search for CREB target genes may provide insights into the therapeutic action of antidepressants.

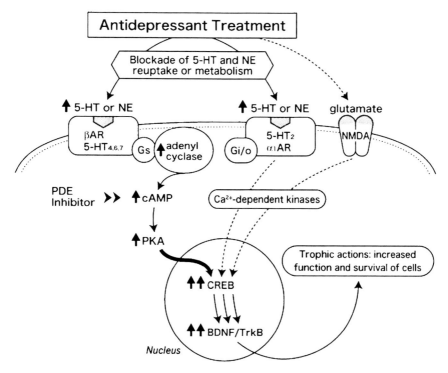

Fig. 5 A model of molecular mechanisms of antidepressant actions.
Gs, *Gi*/o subfamilies of G proteins, *PKA* protein kinase A, *PDE* phosphodiesterase (degrades cAMP), *CREB* cAMP response element binding protein, *NMDA* N-methyl-D-aspartate type glutamate receptor complex, *BDNF* brain-derived neurotrophic factor, *TrkB* receptor for BDNF

3.4 Neurogenesis

For a long time, there have been debates but no definitive evidence on whether new neuronal cells are continually born in the adult brain. In 1983, Goldman and Nottebohm observed that new neurons were added to adult avian brain (Goldman and Nottebohm 1983). A few years thereafter, Erickson and colleagues demonstrated that neurogenesis occurred in adult human brains (Eriksson et al. 1998). They identified new neurons which incorporated the thymidine analog bromodeoxyuridine (BrdU) in the hippocampi of deceased cancer patients who were receiving BrdU to assess tumor progression. Once adult neurogenesis was described in all mammalian species surveyed, investigators began to focus on identifying the physiological role of adult-born neurons. Some clues to the normal function of neurogenesis came from an observation that robust adult neurogenesis is exclusively restricted to the subependymal cells of the ventricular system (SVZ) and the subgranular zone of the dentate gyrus in the hippocampus (SGZ) in rodents. In humans, adult neurogenesis has been demonstrated exclusively in the hippocampus (Eriksson et al. 1998). In the hippocampus, neural progenitor cells are localized to the SGZ where they continue to divide and give rise to new cells (see review by Zhao et al. 2008). These new neurons in the SGZ differentiate and migrate into the granule cell layer and extend dendrites to the molecular layer and axons to the CA3 pyramidal layer, and they attain morphological and physiological characteristics of adult granule cells. However, it is thought that approximately half of the new neurons are lost within 3– 4 weeks of proliferation.

The rate of proliferation and survival of newborn neurons in the hippocampus is dynamically regulated up or down by a variety of stimuli. Stress is one of the most robust negative regulators of adult neurogenesis. Exposure to acute or repeated stressors, including intruder stress, predator odor, maternal separation, or footshock, degrades neurogenesis in the adult hippocampus (Duman 2004b). Most of the stress models that decrease adult neurogenesis concomitantly display a decline in BDNF expression. However, the role of BDNF in regulating neurogenesis is incompletely understood. Infusion of BDNF into the lateral ventricles increases neurogenesis in several brain regions but not in the hippocampus, possibly due to low levels of diffusion to hippocampus (Pencea et al. 2001). Chronic infusion of BDNF directly into the hippocampus does increase neurogenesis, but this paradigm causes seizure activity and greater induction of neurogenesis on the contralateral side, as well as the ipsilateral infusion side. It is difficult to distinguish the direct effects of BDNF on neurogenesis from the effects of seizure activity, and previous studies demonstrate that a single seizure is sufficient to increase neurogenesis (Madsen et al. 2000).

The finding that antidepressants increase neurotrophic factor expression in the adult hippocampus provides the background and rationale for studies of adult neurogenesis. This led to the discovery that antidepressant treatment significantly increases neurogenesis in the adult hippocampus (Malberg et al. 2000; Sapolsky 2001). The upregulation of neurogenesis is observed with chronic, but not acute, administration of different classes of antidepressants, including SSRI, NESRI (norepinephrine selective reuptake inhibitor), MAOI, and ECS, indicating that neurogenesis is a common target of antidepressant medications. It is also notable that other treatments that produce an antidepressant response, including AMPAkines and exercise, also increase neurogenesis (Duman 2004b). Antidepressant treatment blocks or reverses the downregulation of neurogenesis that occurs in response to stress (Duman 2004b). In addition, obstructing neurogenesis blocks the behavioral effects of antidepressants, demonstrating a direct link between neurogenesis and behavioral responses (Santarelli et al. 2003; Dranovsky and Hen 2006).

3.4.1 BDNF

Stress is used as a model to study alterations of brain structure and function because mood disorders are often precipitated or exacerbated by acute or chronic stressful life events. Stress can cause damage and atrophy of neurons in certain brain structures, most notably the hippocampus, which expresses high levels of GR. Alterations of hippocampal structure and function in response to stress provided the rationale for the analysis of neurotrophic factors. Various types of acute and chronic stress paradigms are known to decrease *Bdnf* expression in the hippocampus (Duman and Monteggia 2006). In addition, corticosterone administration, which simulates stress, decreases *Bdnf* expression (Schaaf et al. 1998), but adrenalectomy elevates its expression (Chao et al. 1998).

The requirement for long-term, chronic antidepressant treatment has led to the hypothesis that alterations in functional and structural plasticity are required to elicit a therapeutic response (see review by Nestler et al. 2002) (Fig. 5). Because of the known effect of neurotrophic factors during development, neuronal survival and function in the adult brain, the nerve growth factor family has been thought of as potential targets for antidepressants. Consistent with this hypothesis, chronic treatment with different classes of antidepressants including SSRIs, NESRI, MAOI, atypical antidepressants, and electroconvulsive seizures (ECS), has been shown to significantly increase *Bdnf* expression in the major subfields of the hippocampus, e.g., the granule cell layer and the CA1 and CA3 pyramidal cell layers (Nibuya et al. 1995).

Analysis of postmortem hippocampus revealed a decrease in *BDNF* expression in depressed suicide patients and an increase in patients who were on antidepressant medication at the time of death (Chen et al. 2001; Karege et al. 2005). Serum levels of BDNF have been reported to be reduced in depressed patients (Karege et al. 2002; Shimizu et al. 2003). In addition, several studies have demonstrated that antidepressant treatment can reverse this effect (for review see Duman and Monteggia 2006). It is unclear whether peripheral BDNF crosses the blood–brain barrier. However, there is evidence that other peripheral growth factors, including IGF-I and VEGF, gain access to the brain (Duman and Monteggia 2006). Thus, peripheral BDNF may also influence central nervous system function. Brain imaging studies have demonstrated a reduction in the volume of the hippocampus of depressed subjects. In addition, the reduction of hippocampal volume is reversed by antidepressant treatment (Duman and Monteggia 2006). A reduction in hippocampal volume is also observed in posttraumatic stress disorder (PTSD) patients (Bremner et al. 1995).

3.5 The Endocrine System (hypothalamic-pituitary-adrenal (HPA) axis)

It is well known that endocrine disorders like Cushing syndrome and thyroid dysfunction often elicit psychotic features including depression. Abnormality in HPA axis could be a contributing element to a hormonal pathology in depression. The HPA axis is central to stress response and forms the final common pathway. Corticotropin-releasing hormone (CRH) and arginine vasopressin (AVP)-containing neurons in paraventricular nucleus (PVN) of the hypothalamus project to the median eminence leading to the release of CRH and AVP to hypophysial portal vessels, which connect to the capillary vessels of the anterior pituitary. CRH binds its receptors that couple with cAMP synthesis in the anterior pituitary and stimulates the release of adrenocorticotropic hormone (ACTH). AVP can also elicit the release of ACTH through V3 or V1b receptors. ACTH is carried to the adrenal cortex via systemic blood flow and primes the release of cortisol. Cortisol binds to two receptors, at high affinity to the mineralocorticoid receptor (MR) and at low affinity to the more abundant glucocorticoid receptor (GR) (de Kloet et al. 1998). GR and MR are nuclear receptors acting as transcription factors either as homodimers (GR-GR, MR-MR) or as heterodimers (GR-MR) to activate or suppress the activity of a manifold of genes including POMC (proopiomelanocortin: adrenocorticotropin/ β-lipotropin/ (α-melanocyte stimulating hormone/ β-melanocyte stimulating hormone/ β-endorphin), AVP and CRH (Trapp et al. 1994). In turn, cortisol sends negative feedback signals to the pituitary, hypothalamus and further upstream to the hippocampus through binding to GR/MR.

Impaired regulation of the HPA system during an acute episode is the most consistent laboratory finding in depression (Holsboer 2000). Depressed patients show increased levels of plasma and urinary free cortisol and insufficient suppression of cortisol following a low dose of dexamethasone (DEX) (Carroll et al. 1981), a synthetic corticosteroid acting almost entirely at GR. Depression seems to be associated with increased concentrations of CRH in the cerebrospinal fluid (Nemeroff et al. 1984). Most of these changes have been shown to gradually normalize after successful antidepressant treatment (Holsboer and Barden 1996); therefore, dysregulation of the HPA system is deemed to be state-dependent, not reflecting a trait marker. Postmortem studies in suicide victims and depressed patients have shown elevated expression of hypothalamic and frontal CRH peptide, together with a lower CRH receptor binding capacity and reduced CRH receptor 1 transcript in the frontopolar cortex (Merali et al. 2004). These alterations have been construed as consequences of a disturbed balance between signaling of glucocorticoid and mineralocorticoid receptors, resulting in a blunted negative feedback inhibition of the HPA system in the hypothalamic PVN and pituitary (de Kloet et al. 2005).

The combined DEX/CRH test is a more sensitive test for detecting altered HPA system regulation. In this test, the stimulating effects of CRH on ACTH and cortisol are examined under the suppressive action of dexamethasone. Impaired glucocorticoid receptor (GR) signaling leads to increased secretion of the hypothalamic CRH and vasopressin peptides and an escape from the suppressive effects of dexamethasone at the pituitary. Both factors, elevated levels of hypothalamic neuropeptides and attenuated suppression at the pituitary, result in an increased ACTH and cortisol response to the combined DEX/CRH test. Elevated cortisol responses to the combined DEX/CRH test and superior sensitivity of the combined DEX/CRH test compared with the regular dexamethasone suppression test have been consistently observed in patients suffering from an acute major depressive episode, bipolar disorder, and chronic depression (Ising et al. 2005; Kunugi et al. 2006; Rybakowski and Twardowska 1999; Watson et al. 2002). During antidepressant treatment, neuroendocrine response to the DEX/ CRH test ameliorates, irrespective of the primary pharmacological mode of action of the drugs possibly due to the singular effect of antidepressants on GR gene expression, restoring GR sensitivity (Holsboer 2000; Holsboer and Barden 1996).

3.6 Animal Studies

Three of the most common stress-based models of depression-like behavior that are responsive to antidepressants are the forced swim test (FST), the tail suspension test (TST), and the learned helplessness (LH) models (Fig. 6). In addition, a fourth test, novelty suppressed feeding (NSF), has recently been used. In the FST, rodents alternate between active responses and immobility after placement in a beaker of water. In this paradigm, the more time an animal spends in an inactive or immobile state versus active state is interpreted as a measure of depression-like behavior (Porsolt et al. 1977). Most importantly, this model also reliably predicts antidepressant efficacy as antidepressant treatment decreases immobility time compared with a vehicle-treated animal. As such, the FST has been interpreted as model of "behavioral despair" and has been used to examine depression-like and antidepressant-like behavioral responses in numerous genetic and pharmacological models in rodents. In the TST, mice spend periods of activity and immobility after being suspended by their tails (Steru et al. 1985). This paradigm is similar to the FST in that the time an animal spends immobile is interpreted as a measure of depression-like behavior. The TST is also able to predict antidepressant compounds, as animals administered antidepressants prior to the test display more active escape responses than those administered vehicle. While the FST and TST are similar, in that they assess depression-like behavior by increased immobility or "giving up," these tests are not necessarily interchangeable. For example, a recent study has demonstrated that GABA B receptor knockout mice display less immobility in the FST but not in the TST (Mombereau et al. 2004). In addition, differential antidepressant responses have been shown in the FST and the TST as not all antidepressants work equally well in both paradigms. In particular, the SSRI antidepressants produce a reliable response in the TST, which is generally not observed in the FST (Cryan et al. 2002). The LH paradigm has also been interpreted as a model of

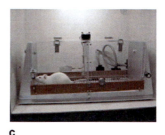

Fig. 6 Behavioral paradigms for animal models for depression. **a** forced swim test, **b** tail suspension test, **c** learned helplessness test

behavioral despair and is responsive to antidepressant treatment in both rats and mice. In LH, animals that are exposed to inescapable shock subsequently fail to escape from a situation in which escape is possible (Overmier and Seligman 1967; Telner and Singhal 1984). In the LH paradigm, animals are rated as helpless if they exhibit a decreased number of escapes or increased escape latencies following inescapable shock relative to control animals. This test also reliably predicts antidepressant efficacy, measured as increased escapes as well as shorter latencies to escape following antidepressant treatment. In addition to the escape deficits, helpless animals exhibit a variety of other behavioral and physiological changes that have been compared with depressive symptoms in humans, including decreased motor activity, loss of appetite and weight, reduced performance in self-stimulation paradigms, and immunosuppression (Willner 1990). Although this paradigm is responsive to all classes of antidepressants, some antidepressants have not yet been validated in this model. The NSF test measures anxiety-like behavior but is also responsive to chronic, and not acute, antidepressant treatment (Bodnoff et al. 1988). This paradigm measures the time, or latency, for a food-deprived animal to move into a brightly lit open field to consume food. The latency to eat is considered a measure of anxiety-related behavior, and chronic, but not acute, antidepressant administration decreases the latency. Although this test provides an important prerequisite (chronic treatment) for understanding the therapeutic action of antidepressants, it is also responsive to acute anxiolytics and thereby lacks pharmacological specificity. These tests have been used to demonstrate abnormalities in mutant mouse models where genes potentially involved in antidepressant action are disrupted: these genes include those belonging to HPA system (for review see Müller and Holsboer 2006) and those for neurotrophins and their receptors (for a review see Duman and Monteggia 2006) among others. Recently, mutant model mice have also been generated from N-ethyl-N-nitrosourea (ENU) mutagenesis (Clapcote et al. 2007).

Rodent models of depression have been used to identify the genetic underpinnings and molecular pathways implicated in disease. It is known that there are inter-strain differences in baseline performances of the FST and the TST in inbred mice. Yoshikawa and colleagues assumed that the baseline immobility time can suitably depict an innate vulnerability to stressors and a predisposition to despair under distress (Yoshikawa et al. 2002). They analyzed 560 F2 mice from an intercross between the C57BL/6 (B6) strain that displayed the longest immobility time among all strains that were examined and the C3H/He (C3) strain that showed the shortest immobility time. Composite interval mapping revealed five major loci (suggestive and significant linkage) affecting immobility in the FST, and four loci by TST. Quantitative trait loci (QTL) on chromosomes 8 and 11 showed overlap between the two behavioral measures. Genome-wide interaction analysis, which was developed to identify locus pairs that may contribute epistatically to a phenotype, detected two pairs of chromosomal loci for the TST. The QTL on chromosome 11 and its associated epistatic TST-QTL on chromosome X encode GABA A receptor subunits as candidates. Sequence and expression analyses of these genes from the two parental strains revealed a significantly lower expression of the alpha 1 subunit (*Gabra1*) gene in the frontal cortex of B6 mice compared to C3 mice. Subsequently, the same group reported that polymorphisms on *GABRA1* and

GABRA6 genes displayed significant associations with mood disorders in female patients (Yamada et al. 2003). A study on a different ethnic group supported association between *GABRA1* and mood disorders (Horiuchi et al. 2004). A large-scale analysis of gene expression in the LH rat model of human depression, using DNA microarrays, has been reported (Nakatani et al. 2004). Gene expression in the frontal cortex (FC) and hippocampus (HPC) of untreated controls and LH rats treated with either saline (LH-S) or imipramine or fluoxetine was examined. A total of 34 and 48 transcripts were differentially expressed in the FC and HPC, respectively, between control and LH-S groups. Unexpectedly, the only genes with altered expression in both FC and HPC were NADH dehydrogenase and zinc transporter, suggesting a limited overlap of molecular processes in specific areas of the brain. Principal component analysis revealed that sets of genes for upregulated metabolic enzymes in the FC and downregulated genes for signal transduction in the HPC can distinguish clearly between depressed and control animals, as well as explain the responsiveness to antidepressants. *Limk1* (LIM domain kinase 1) was the gene that displayed the most dramatic decrease in the frontal cortex of LH rats. Limk1 inactivates ADF/cofilins (Cfl1) in mammals by phosphorylating the Ser3 residue, thereby controlling the cycling of actin monomers through the polymerized network.

Nakatani et al. (2007) performed an integrative study by combining mouse QTL information and genome-wide expression data on mouse GeneChips, and found that adenylate cyclase-associated protein 1 (Cap1), which has a central role in actin turnover, is a compelling player for dysregulation in mood disturbances (Nakatani et al. 2007). They extended the search to other 'core' actin network elements, and observed upregulation of three transcripts including *Cfl1* and downregulation of gene for a Rho-family GTPase member (Pak1) in the frontal cortex of B6 animals. It is worth noting that, when the direction of expressional changes of these transcripts is taken into account, the balance of actin dynamics in depression-prone B6 displays a tendency toward actin depolymerization and fragmentation of filamentous actin (F-actin). The gene expression study on human bipolar prefrontal cortex showed a trend of CAP1 reduction, concordant with results on mouse brains. In addition, the postmortem brain examination detected a trend of diminished Arp2/3 complex formation. This may lead to a decrease in actin fiber tributary, thus reducing the sites of actin polymerization/ elongation. The 'disturbed cytoskeletal theory of mood disorders' has other supporting evidence. Millar et al. (2000) reported that *DISC1* is a susceptibility gene for major psychiatric illnesses including schizophrenia, bipolar disorder, and recurrent major depression in a large Scottish family where affected members harbor a balanced chromosome (1;11) (q42.1;q14.3) translocation (Millar et al. 2000). Subsequent studies suggest that DISC1 plays a major role in the dynamic organization of the cytoskeleton (Ishizuka et al. 2006). The 'core' actin network also plays a crucial role in neurogenesis (Haendel et al. 1996; Laplante et al. 2004), one of mechanisms that antidepressants are thought to target, as described earlier. A high-speed optical imaging technique [voltage-sensitive dye imaging (VSDI)] permitted measurement of cellular activity with millisecond resolution in hippocampal slices from chronic mild stress-induced depression rat model (Airan et al. 2007). Changes in the circuitry between the dentate gyrus and CA1 were recorded in hippocampal slices from antidepressant-treated rats. The study led to the proposal that depression may depend on changes in organizing the flow of information through active networks.

3.7 Postmortem Brain Studies

Oligonucleotide microarray analysis on postmortem prefrontal cortices from patients with either major depression, bipolar disorder or schizophrenia and control subjects from the Stanley Foundation Brain Collection revealed that each disease had its own pattern of altered gene expression, suggesting the molecular distinctiveness of these mental disorders (Iwamoto et al. 2004). Upregulation of genes encoding proteins involved in transcription or translation was a feature shared by these disorders. Changes in expression of genes validated by quantitative RT-PCR included *AQP4* (a gene for aquaporin 4) (in bipolar disorder) and *PDLIM5* (in both schizophrenia and bipolar disorder). AQP4 is associated with dystrophin-associated protein complex (DPC) by interacting with syntrophin, a component of DPC, and is involved in the water permeability across the blood–brain barrier and cerebrospinal fluid–brain interface. *PDLIM5* encodes an adaptor protein linking PKC epsilon and an N-type calcium channel. Its altered function may lead to altered calcium signaling in brains of psychiatric patients including depression.

In another transcriptional profiling study, downregulation of oligodendrocyte-related genes in the temporal cortex has been detected in major depression, which was similar to changes reported in schizophrenia and bipolar disorder (Aston et al. 2004). Evans and others found altered expression of fibroblast growth factor (FGF)-related genes in major depression by comparing 21 FGF genes systematically (Evans et al. 2004). Downregulation of two glial high-affinity glutamate transporters (*SLC1A2* and *SLC1A3*), as well as glutamate-ammonia ligase, which are predominantly expressed in astroglia, has been reported (Choudary et al. 2005). Neuronally-expressed GABA A receptors were shown also to be upregulated, and the authors posited that astroglial dysfunction has a role in major depression. On the other hand, Sibille and others found no altered gene expression in the prefrontal cortex of patients with major depression who committed suicide (Sibille et al. 2004).

PET scanning has shown a decrease in the activity of the subgenual prefrontal cortex of major depressives with familial history compared to controls (Drevets et al. 1997). A recent gene expression analysis was performed using mRNA from punched-out sections of the dorsolateral prefrontal cortex (DLPFC) taken from 15 pairs of major depressives and control brains (Kang et al. 2007). The most prominently altered expression were those of genes encoding the neuropeptide stresscopin and forkhead box D3 transcription factor. This study confirmed prior findings on altered expression of FGF genes (Evans et al. 2004). It is a vexing reality that various expression studies in postmortem brains have not coalesced into a common set of altered transcripts. As in bipolar disorder, the role of genes that display disease-associated expression changes remains to be established.

4 Perspectives

Advances in genetics and technology have provided new knowledge and tools to facilitate identification of genetic factors that contribute to the overall risk for mood disorders. Most prominently, whole genome association on large samples has permitted a rapid search for disease-predisposing variants. Several new candidate susceptibility genes have been highlighted, although convergence of results between studies is still lacking owing primarily to weak effect size risk variants, locus heterogeneity, allelic heterogeneity, phenotype heterogeneity, and not-large enough sample size. Sufficient power to identify replicable minor effect variants may be achieved with tens of thousands of samples. Meta-analysis of large studies may yield new gene variants that were not in the list of top findings in individual GWAS, and these would be the more compelling candidates for functional studies.

So far, screening has been focused on common variants; there is a need to extend the search to identify rare, private risk alleles that exert greater effect on vulnerability. GWAS have focused on single nucleotide changes through SNPs; scanning for CNVs and microRNAs is important and are being pursued in some laboratories.

We expect that more studies will be directed outside the DNA nucleotide sequence as epigenetics is another area of interest in psychiatric disease. DNA methylation and histone acetylation has been shown to exert a vital role in other diseases. Association of brain imaging patterns and neuropsychological test results with risk genotypes could provide important information on brain function for a specific genetic background. The growth of transcriptional profiling and animal modeling has yielded a panoply of genes that display altered expression, but whether these are etiologic or outcome factors still remains to be determined. Possibly, proteome analysis will be embarked on and may have similar problems as mRNA profiles.

Analysis of gene–gene interaction, for which some insight may be gained from animal studies, is an area that needs to be developed further. In mood disorders, evidence has been presented for gene × environment interaction that relates life stress, depression and a serotonin transporter variation. The work on the effect of maternal nurturing on offspring behavior and its relation to the methylation status of the glucocorticoid receptor is of great interest and saliency in mood disorders.

Additionally, we did not elaborate on evidence that show genetic overlaps between bipolar disorder and schizophrenia as gleaned from linkage and a few association studies. As greater numbers of associated variation are identified, more genes common to the pathogenesis of various psychiatric disorders may be revealed. These shared risk factors may serve as anchor points of pathways leading to specific clinical presentations that might elicit parsimony behind the complexity of mental illness.

How many and what genetic variation are necessary and sufficient to trigger mood disorders is an open question. Extensive locus and allelic heterogeneity would predict that the number and type of etiologic determinants for mood disorders will vary across subsets of patients, with each group sharing some common features. As larger samples become available, analysis can be directed at case subsets stratified on the basis of specific phenotypes that might have lower heterogeneity compared to the entire sample.

The study of genetic determinants in psychiatry will benefit from advances in neuroscience as we embark on unraveling function and mechanisms associated with risk genes. As more susceptibility genes are validated, a coherent and unified interacting molecular network could be generated that could illuminate the pathogenic mechanism of mood regulation. Possibly, some components of this network can be exploited in the development of fast-acting, better-tolerated and symptom-specific drugs. An array of variants may be useful to improve diagnostic precision and be utilized as prognostic factors. A huge amount of information on the biology and genetics of organisms across the evolutionary scale and advanced tools are available to the geneticist, who is now probably better equipped than ever to address the challenges that remain.

Acknowledgments The authors would like to extend their sincerest thanks to Dr. Tetsuo Onishi for his very generous and kind assistance and also Apoorv Gupta for his help.

Appendix: Addendum After this chapter was written a report on the combined analysis of two published GWAS in bipolar disorder (WTCCC 2007; Sklar et al. 2008) that interrogated ~2 million variants in 4,237 cases and 6,209 controls has shown a compelling evidence for association at a SNP in *ANK3* and additional support for association in *CACNA1C* (Ferreira MAR, O'Donovan MC, Meng YA, Jones IR, Ruderfer DM, Jones L, Fan J et al. (2008) Collaborative genome-wide association analysis supports role for *ANK3* and *CACNA1C* in bipolar disorder. Nature Genet 40(9):1056–1058). Also, it is noteworthy that this recent study adds support to a previous finding that detected association in *ANK3* with bipolar disorder in the NIMH and German samples (Baum et al. 2008a).

References

Abkevich V, Camp NJ, Hensel CH, Neff CD, Russell DL, Hughes DC, Plenk AM et al (2003) Predisposition locus for major depression at chromosome 12q22–12q23.2. Am J Hum Genet 73:1271–1281

Airan RD, Meltzer LA, Roy M, Gong Y, Chen H, Karl Deisseroth K (2007) High-speed imaging reveals neurophysiological links to behavior in an animal model of depression. Science 317(5839):819–823

Alda M, Grof P, Rouleau GA, Turecki G, Young LT (2005) Investigating responders to lithium prophylaxis as a strategy for mapping susceptibility genes for bipolar disorder. Prog Neuropsychopharmacol Biol Psychiatry 29(6):1038–1045

American Psychiatric Association (1994) Diagnostic and statistical manual of mental disorders, 4th edn. American Psychiatric Association, Washington, DC

Andrade L, Caraveo-Anduaga JJ, Berglund P, Bijl RV, De Graaf R, Vollebergh W, Dragomirecka E, Kohn R, Keller M, Kessler RC, Kawakami N, Kilic C, Offord D, Ustun TB, Wittchen HU (2003) The epidemiology of major depressive episodes: results from the International Consortium of Psychiatric Epidemiology (ICPE) Surveys. Int J Methods Psychiatr Res 12:3–21

Antila M, Tuulio-Henriksson A, Kieseppä T, Soronen P, Palo OM, Paunio T, Haukka J et al (2007) Heritability of cognitive functions in families with bipolar disorder. Am J Med Genet B Neuropsychiatr Genet 144(6):802–808

Arai R, Ito K, Ohnishi T, Ohba H, Akasaka R, Bessho Y, Hanawa-Suetsugu K, Yoshikawa T, Shirouzu M, Yokoyama S (2007) Crystal structure of human myo-inositol monophosphatase 2 (IMPA2), the product of the putative susceptibility gene for bipolar disorder, schizophrenia and febrile seizures. PROTEINS: Struct, Funct, Bioinform 67:732–742

Aston C, Jiang L, Sokolov BP (2004) Microarray analysis of postmortem temporal cortex from patients with schizophrenia. J Neurosci Res 77:858–866

Badenhop RF, Moses MJ, Scimone A, Mitchell PB, Ewen KR, Rosso A, Donald JA (2001) A genome screen of a large bipolar affective disorder pedigree supports evidence for a susceptibility locus on chromosome 13q. Mol Psychiatry 6(4):396–403

Badner JA, Gershon ES (2002) Meta analysis of whole-genome linkage scans of bipolar disorder and schizophrenia. Mol Psychiatry 7(4):405–411

Barkai AI, Dunner DL, Gross HA, Mayo P, Fieve RR (1978) Reduced myo-inositol levels in cerebrospinal fluid from patients with affective disorder. Biol Psychiatry 13:65–72

Baum AE, Akula N, Cabanero M, Cardona I, Corona W, Klemens B, Schulze TG et al (2008a) A genome-wide association study implicates diacylglycerol kinase eta (DGKH) and several other genes in the etiology of bipolar disorder. Mol Psychiatry 13(2):197–207

Baum AE, Hamshere M, Green E, Cichon S, Rietschel M, Noethen MM, Craddock N, McMahon FJ (2008b) Meta-analysis of two genome-wide association studies of bipolar disorder reveals important points of agreement. Mol Psychiatry 13(5):466–467

Bearden CE, Freimer NB (2006) Endophenotypes for psychiatric disorders: ready for primetime? Trends Genet 22(6):306–313

Bearden CE, Thompson PM, Dalwani M, Hayashi KM, Lee AD, Nicoletti M, Trakhtenbroit M (2007) Greater cortical gray matter density in lithium-treated patients with bipolar disorder. Biol Psychiatry 62(1):7–16

Beaulieu JM, Sotnikova TD, Yao WD, Kockeritz L, Woodgett JR, Gainetdinov RR, Caron MG (2004) Lithium antagonizes dopamine-dependent behaviors mediated by an AKT/glycogen synthase kinase 3 signaling cascade. Proc Natl Acad Sci USA 101:5099–5104

Bellantuono C, Barraco A, Rossi A, Goetz I (2007) The Management of Bipolar Mania: a national survey of Baseline data from tha EMBLEM study in Italy. BMC Psychiatry 7:33 [Epub ahead of print]

Benes FM, Vincent SL, Todtenkopf M (2001) The density of pyramidal and nonpyramidal neurons in anterior cingulate cortex of schizophrenic and bipolar subjects. Biol Psychiatry 50:395–406.

Berrettini WH (2000) Susceptibility loci for bipolar disorder: overlap with inherited vulnerability to schizophrenia. Biol Psychiatry 47(3):245–251

Berridge MJ (1989) The Albert Lasker Medical Awards. Inositol trisphosphate, calcium, lithium, and cell signaling. JAMA 262:1834–1841

Bertelsen A, Hauge M, Harvald B (1977) A Danish twin study of manic-depressive disorders. Br J Psychiatry 130:330–351

Bjelland I, Tell GS, Vollset SE, Refsum H, Ueland PM (2003) Folate, vitamin B12, homocysteine, and the MTHFR 677C- > T polymorphism in anxiety and depression: the Hordaland Homocysteine Study. Arch Gen Psychiatry 60(6):618–626

Blackwood DH, Fordyce A, Walker MT, St Clair DM, Porteous DJ, Muir WJ (2001) Schizophrenia and affective disorders-co-segregation with a translocation at 1q42 that directly disrupts brain-expressed genes: clinical and P300 findings in a family. Am J Hum Genet 69:428–433

Blouin JL, Dombroski BA, Nath SK, Lasseter VK, Wolyniec PS, Nestadt G, Thornquist M et al (1998) Schizophrenia susceptibility loci on chromosomes 13q32 and 8p21. Nat Genet 20(1):70–73

Bodnoff SR, Suranyi-Cadotte B, Aitken DH, Quirion R, Meaney MJ (1988) The effects of chronic antidepressant treatment in an animal model of anxiety. Psychopharmacology 95:298–302

Bone R, Springer JP, Atack JR (1992) Structure of inositol monophosphatase, the putative target of lithium therapy. Proc Natl Acad Sci USA 89:10031–10035

Bremner JD, Randall P, Scott TM, Bronen RA, Seibyl JP, Southwick SM, Delaney RC, McCarthy G, Charney DS, Innis RB (1995) MRI-based measurement of hippocampal volume in patients with combat-related posttraumatic stress disorder. Am J Psychiatry 152:973–981

Bunney WE, Davis JM (1965) Norepinephrine in depressive reactions. A review. Arch Gen Psychiatry 13:483–494

Camp NJ, Lowry MR, Richards RL, Plenk AM, Carter C, Hensel CH et al (2005) Genome-wide linkage analyses of extended Utah pedigrees identifies loci that influence recurrent, early onset major depression and anxiety disorders. Am J Med Genet B Neuropsychiatr Genet 135:85–93

Canli T, Lesch KP (2007) Long story short: the serotonin transporter in emotion regulation and social cognition. Nat Neurosci 10(9):1103–1109

Cannon DM, Ichise M, Rollis D, Klaver JM, Gandhi SK, Charney DS, Manji HK, Drevets WC (2007) Elevated serotonin transporter binding in major depressive disorder assessed using positron emission tomography and [11C]DASB; comparison with bipolar disorder. Biol Psychiatry 62(8):870–877

Carroll BJ, Feinberg M, Greden JF, Tarika J, Albala AA, Haskett RF, James NM, Kronfol Z, Lohr N, Steiner M, de Vigne JP, Young E (1981) A specific laboratory test for the diagnosis of melancholia Standardization, validation, and clinical utility. Arch Gen Psychiatry 38:15–22

Caspi A, Sugden K, Moffitt TE, Taylor A, Craig IW, Harrington H, McClay J et al (2003) Influence of life stress on depression: moderation by a polymorphism in the 5-HTT gene. Science 301(5631):386–389

Chao H, Sakai RR, Ma LY, McEwen BS (1998) Adrenal steroid regulation of neurotrophic factor expression in the rat hippocampus. Endocrinology 139:3112–3118

Chen G, Huang LD, Jiang YM, Manji HK (1999a) The mood-stabilizing agent valproate inhibits the activity of glycogen synthase kinase-3. J Neurochem 72:1327–1330

Chen G, Zeng WZ, Yuan PX, Huang LD, Jiang YM, Zhao ZH, Manji HK (1999b) The mood-stabilizing agents lithium and valproate robustly increase the levels of the neuroprotective protein bcl-2 in the CNS. J Neurochem 72:879–882

Chen B, Dowlatshahi D, MacQueen GM, Wang J-F, Young LT (2001) Increased hippocampal BDNF immunoreactivity in subjects treated with antidepressant medication. Biol Psychiatry 50:260–265

Cheng R, Juo SH, Loth JE, Nee J, Iossifov I, Blumenthal R, Sharpe L et al (2006) Genome-wide linkage scan in a large bipolar disorder sample from the National Institute of Mental Health genetics initiative suggests putative loci for bipolar disorder, psychosis, suicide, and panic disorder. Mol Psychiatry 11(3):252–260

Chetcuti A, Adams LJ, Mitchell PB, Schofield PR (2008) Microarray gene expression profiling of mouse brain mRNA in a model of lithium treatment. Psy Genet 18(2):64–72

Choudary PV, Molnar M, Evans SJ, Tomita H, Li JZ, Vawter MP, Myers RM et al (2005) Altered cortical glutamatergic and GABAergic signal transmission with glial involvement in depression. Proc Natl Acad Sci 102(43):15653–15658.

Christian SL, Brune CW, Sudi J, Kumar RA, Liu S, Karamohamed S, Badner JA, Matsui S et al (2008) Novel submicroscopic chromosomal abnormalities detected in autism spectrum disorder. Biol Psychiatry 63(12):1111–1117

Chubb JE, Bradshaw NJ, Soares DC, Porteous DJ, Millar JK (2008) The DISC locus in psychiatric illness. Mol Psychiatry 13(1):36–64

Chumakov I, Blumenfeld M, Guerassimenko O, Cavarec L, Palicio M, Abderrahim H, Bougueleret L et al (2002) Genetic and physiological data implicating the new human gene G72 and the gene for D-amino acid oxidase in schizophrenia. Proc Natl Acad Sci 99(21):13675–13680

Clapcote SJ, Lipina TV, Millar JK, Mackie S, Christie S, Ogawa F, Lerch JP, Trimble K, Uchiyama M, Sakuraba Y, Kaneda H, Shiroishi T, Houslay MD, Henkelman RM, Sled JG, Gondo Y, Porteous DJ, Roder JC (2007) Behavioral phenotypes of Disc1 missense mutations in mice. Neuron 54:387–402

Cohen JC, Kiss RS, Pertsemlidis A, Marcel YL, McPherson R, Hobbs HH (2004) Multiple rare alleles contribute to low plasma levels of HDL cholesterol. Science 305(5685):869–872

Craddock N, Forty L (2006) Genetics of affective (mood) disorders. Eur J Hum Genet 14(6):660–668

Cryan J, Markou A, Lucki I (2002) Assessing antidepressant activity in rodents: recent developments and future needs. Trends Pharmacol Sci 23:238–245

Cryns K, Shamir A, Van Acker N, Levi I, Daneels G, Goris I, Bouwknecht JA, Andries L, Kass S, Agam G, Belmaker H, Bersudsky Y, Steckler T, Moechars D (2007a) IMPA1 is essential for embryonic development and lithium-like pilocarpine sensitivity. Neuropsychopharmacology 33:674–684

Cryns K, Shamir A, Shapiro J, Daneels G, Goris I, Van Craenendonck H, Straetemans R, Belmaker RH, Agam G, Moechars D, Steckler T (2007b) Lack of lithium-like behavioral and molecular effects in IMPA2 knockout mice. Neuropsychopharmacology 32:881–891

Davanzo P, Thomas MA, Yue K, Oshiro T, Belin T, Strober M, McCracken J (2001) Decreased anterior cingulate myo-inositol/creatine spectroscopy resonance with lithium treatment in children with bipolar disorder. Neuropsychopharmacology 24:359–369

De Kloet ER, Vreugdenhil E, Oitzl MS, Joels M (1998) Brain corticosteroid receptor balance in health and disease. Endocr Rev 19:269–301

De Kloet ER, Joels M, Holsboer F (2005) Stress and the brain: from adaptation to disease. Nat Rev Neurosci 6:463–475

Detera-Wadleigh SD, Badner JA, Berrettini WH, Yoshikawa T, Goldin LR, Turner G, Rollins DY et al (1999) A high-density genome scan detects evidence for a bipolar-disorder susceptibility locus on 13q32 and other potential loci on 1q32 and 18p11.2. Proc Natl Acad Sci 96(10):5604–5609

Detera-Wadleigh SD (2001) Lithium-related genetics of bipolar disorder. Ann Med 33(4):272–285

Detera-Wadleigh SD, Goldin LR (2002) Affective disorders. In: King RA, Rotter JI, Motulsky AG (eds) The genetic basis of common diseases, 2nd ed. Oxford University Press, London, pp. 831–849

Detera-Wadleigh SD, McMahon FJ (2006) G72/G30 in schizophrenia and bipolar disorder: review and meta-analysis. Biol Psychiatry 60(2):106–114

Detera-Wadleigh SD, Liu CY, Maheshwari M, Cardona I, Corona W, Akula N, Steele CJ et al (2007) Sequence variation in DOCK9 and heterogeneity in bipolar disorder. Psychiatr Genet 17(5):274–286

Diehl RE, Whiting P, Potter J, Gee N, Ragan CI, Linemeyer D, Schoepfer R, Bennett C, Dixon RA (1990) Cloning and expression of bovine brain inositol monophosphatase. J Biol Chem 265:5946–5949

Dranovsky A, Hen R (2006) Hippocampal neurogenesis regulation by stress and antidepressants. Biol Psych 59:1136–1143

Drevets WC, Price JL, Simpson JR Jr, Todd RD, Reich T, Vannier M, Raichle ME (1997) Subgenual prefrontal cortex abnormalities in mood disorders. Nature 386(6627):824–827

Duan X, Chang J, Ge S, Faulkner R, Kim J, Kitabatake Y, Lu X et al (2007) Disrupted-in-schizophrenia 1 regulates integration of newly generated neurons in the adult brain. Cell 130 (6):1146–1158

Duman RS (2004a) The neurochemistry of depressive disorders: preclinical studies. In: Charney DS, Nestler EJ (eds) Neurobiology of mental illness, 2nd edn. Oxford University Press, New York, pp 421–439

Duman R (2004b) Depression: a case of neuronal life and death? Biol Psychiatry 56:140–145

Duman RS, Monteggia LM (2006) A neurotrophic model for stress-related mood disorders. Biol Psychiatry 59:1116–1127

Egan MF, Kojima M, Callicott JH, Goldberg TE, Kolachana BS, Bertolino A, Zaitsev E et al (2003) The BDNF val66met polymorphism affects activity-dependent secretion of BDNF and human memory and hippocampal function. Cell 112(2):257–269

Egeland JA, Gerhard DS, Pauls DL, Sussex JN, Kidd KK, Allen CR, Hostetter AM et al (1987) Bipolar affective disorders linked to DNA markers on chromosome 11. Nature 325(6107):783–787

Ekholm JM, Kieseppä T, Hiekkalinna T, Partonen T, Paunio T, Perola M, Ekelund J et al (2003) Evidence of susceptibility loci on 4q32 and 16p12 for bipolar disorder. Hum Mol Genet 12(15):1907–1915

Eriksson PS, Perfilieva E, Bjork-Eriksson T, Albornb AM, Nordborg C, Peterson DA, Gage FH (1998) Neurogenesis in the adult human hippocampus. Nat Med 4:1313–1317

Euskirchen G, Royce TE, Bertone P, Martone R, Rinn JL, Nelson FK, Sayward F, Luscombe NM, Miller P, Gerstein M, Weissman S, Snyder M (2004) CREB binds to multiple loci on human chromosome 22. Mol Cell Biol 24:3804–3814

Evans SJ, Choudary PV, Neal CR, Li JZ, Vawter MP, Tomita H, Lopez JF, Thompson RC, Meng F, Stead JD, Walsh DM, Myers RM, Bunney WE, Watson SJ, Jones EG, Akil H (2004) Dysregulation of the fibroblast growth factor system in major depression. Proc Natl Acad Sci USA 101:15506–15511

Fan G, Beard C, Chen RZ, Csankovszki G, Sun Y, Siniaia M, Biniszkiewicz D et al (2001) DNA hypomethylation perturbs the function and survival of CNS neurons in postnatal animals. J Neurosci 21(3):788–797

Flint J, Munafò MR (2007) The endophenotype concept in psychiatric genetics. Psychol Med 37(2):163–180

Franken P, Lopez-Molina L, Marcacci L, Schibler U, Tafti M (2000) The transcription factor DBP affects circadian sleep consolidation and rhythmic EEG activity. J Neurosci 20(2):617–625

Gershon ES, Liu C, Badner JA (2008) Genome-wide association in bipolar. Mol Psychiatry 13(1):1–2

Goldman SA, Nottebohm F (1983) Neuronal production, migration, and differentiation in a vocal control nucleus of the adult female canary brain. Proc Natl Acad Sci USA 80:2390–2394

Goodwin FK, Jamison KR (1990) Manic-depressive illness. Oxford University Press, New York

Gorlov IP, Gorlova OY, Sunyaev SR, Spitz MR, Amos CI (2008) Shifting paradigm of association studies: value of rare single-nucleotide polymorphisms. Am J Hum Genet 82(1):100–112

Gottesman II, Gould TD (2003) The endophenotype concept in psychiatry: etymology and strategic intentions. Am J Psychiatry 160(4):636–645

Green EK, Raybould R, Macgregor S, Hyde S, Young AH, O'Donovan MC, Owen MJ et al (2006) Genetic variation of brain-derived neurotrophic factor (BDNF) in bipolar disorder: case-control study of over 3000 individuals from the UK. Br J Psychiatry 188:21–25

Guardiola-Diaz HM, Boswell C, Seasholtz AF (1994) The cAMP-responsive element in the corticotropin-releasing hormone gene mediates transcriptional regulation by depolarization. J Biol Chem 269:14784–14791

Haendel MA, Bollinger KE, Baas PW (1996) Cytoskeletal changes during neurogenesis in cultures of avain neural crest cells. J Neurocytol 25:289–301

Hahn MK, Blakely RD (2007) The functional impact of SLC6 transporter genetic variation. Annu Rev Pharmacol Toxicol 47:401–441

Hamshere ML, Bennett P, Williams N, Segurado R, Cardno A, Norton N, Lambert D et al (2005) Genomewide linkage scan in schizoaffective disorder: significant evidence for linkage at 1q42 close to DISC1, and suggestive evidence at 22q11 and 19p13. Arch Gen Psychiatry 62(10):1081–1088

Hansen T, Olsen L, Lindow M, Jakobsen KD, Ullum H, Jonsson E, Andreassen OA et al (2007) Brain Expressed microRNAs Implicated in Schizophrenia Etiology. PLoS One 2(9):e873

Hariri AR, Holmes A (2006) Genetics of emotional regulation: the role of the serotonin transporter in neural function. Trends Cogn Sci 10(4):182–191

Harwood AJ (2005) Lithium and bipolar mood disorder: the inositol-depletion hypothesis revisited. Mol Psychiatry 10(1):117–126

Hasler G, Drevets WC, Gould TD, Gottesman II, Manji HK (2006) Toward constructing an endophenotype strategy for bipolar disorders. Biol Psychiatry 60(2):93–105

Hattori F, Liu C, Badner JA, Bonner TI, Christian SL, Maheshwari M, Detera-Wadleigh SD et al (2003) Polymorphisms at the G72/G30 gene locus, on 13q33, are associated with bipolar disorder in two independent pedigree series. Am J Hum Genet 72(5):1131–1140

Hayden EP, Nurnberger JI Jr (2006) Molecular genetics of bipolar disorder. Genes Brain Behav 5(1):85–95

Hennah W, Varilo T, Kestilä M, Paunio T, Arajärvi R, Haukka J, Parker A et al (2003) Haplotype transmission analysis provides evidence of association for DISC1 to schizophrenia and suggests sex-dependent effects. Hum Mol Genet 12(23):3151–3159

Holmans P, Weissman M, Zubenko G, Sheftner WA, Crowe RR, DePaulo JR, Knowles JA (2007) Genetics of recurrent early-onset major depression (genred): final genome scan report. Am J Psychiatry 164:248–258

Holsboer F, Barden N (1996) Antidepressants and hypothalamic-pituitary-adrenocortical regulation. Endocr Rev 17:187–205

Holsboer F (2000) The corticosteroid receptor hypothesis of depression. Neuropsychopharmacol 23:477–501

Horiuchi Y, Nakayama J, Ishiguro H, Ohtsuki T, Detera-Wadleigh SD, Toyota T, Yamada K, Nankai M, Shibuya H, Yoshikawa T, Arinami T (2004) Possible association between a haplotype of the GABA-A receptor alpha 1 subunit gene (GABRA1) and mood disorders. Biol Psychiatry 55:40–45

Huang Y, Doherty JJ, Dingledine R (2002) Altered histone acetylation at glutamate receptor 2 and brain-derived neurotrophic factor genes is an early event triggered by status epilepticus. J Neurosci 22(19):8422–8428

Hurst HC (1995) Transcription factors. Protein Profiles 2:105–168

Hyman SE (2007) Can neuroscience be integrated into the DSM-V? Nat Rev Neurosci 8(9):725–732

Ishizuka K, Paek M, Kamiya A, Sawa A (2006) A review of Disrupted-In-Schizophrenia-1 (DISC1): neurodevelopment, cognition, and mental conditions. Biol Psychiatry 59:1189–1197

Ising M, Künzel HE, Binder EB, Nickel T, Modell S, Holsboer F (2005) The combined dexamethasone/CRH test as a potential surrogate marker in depression. Prog Neuropsychopharmacol Biol Psychiatry 29(6):1085–1093

Iwamoto K, Kakiuchi C, Bundo M, Ikeda K, Kato T (2004) Molecular characterization of bipolar disorder by comparing gene expression profiles of postmortem brains of major mental disorders. Mol Psychiatry 9(4):406–416

Jakobsson M, Scholz SW, Scheet P, Gibbs JR, VanLiere JM, Fung HC, Szpiech ZA et al (2008) Genotype, haplotype and copy-number variation in worldwide human populations. Nature 451(7181):998–1003

Jefferson JW, Greist JH (1994) Mood disorders. In: Hales RE, Yudofsky SC, Talbott JA (eds) Textbook of psychiatry, 2nd edn. American Psychiatric Press, Washington, DC, pp 465–494

Jones KR, Fariñas I, Backus C, Reichardt LF (1994) Targeted disruption of the BDNF gene perturbs brain and sensory neuron development but not motor neuron development. Cell 76(6):989–999

Kanazawa T, Glatt SJ, Kia-Keating B, Yoneda H, Tsuang MT (2007) Meta-analysis reveals no association of the Val66Met polymorphism of brain-derived neurotrophic factor with either schizophrenia or bipolar disorder. Psychiatr Genet 17(3):165–170

Kang HJ, Adams DH, Simen A, Simen BB, Rajkowska G, Stockmeier CA, Overholser JC et al (2007) Gene expression profiling in postmortem prefrontal cortex of major depressive disorder. J Neurosci 27(48):13329–13340

Kapsimali M, Kloosterman WP, de Bruijn E, Rosa F, Plasterk RH, Wilson SW (2007) MicroRNAs show a wide diversity of expression profiles in the developing and mature central nervous system. Genome Biol 8:R173

Karege F, Perret H, Bondolfi G, Schwald M, Bertschv G, Aubrey JM (2002) Decreased serum brain-derived neurotrophic factor levels in major depressed patients. Psychol Res 109:143–148

Karege F, Vaudan G, Schwald M, Perroud N, La Harpe R (2005) Neurotrophin levels in postmortem brains of suicide victims and the effects of antemortem diagnosis and psychotropic drugs. Brain Res Mol Brain Res 136:29–37

Karres JS, Hilgers V, Carrera I, Treisman J, Cohen SM (2007) The conserved microRNA MiR-8 tunes atrophin levels to prevent neurodegeneration in Drosophila. Cell 131(1):136–145

Kasahara T, Kubota M, Miyauchi T, Noda Y, Mouri A, Nabeshima T, Kato T (2006) Mice with neuron-specific accumulation of mitochondrial DNA mutations show mood disorder-like phenotypes. Mol Psychiatry 11:577–593

Kessler RC, McGonagle KA, Zhao S, Nelson CB, Hughes M, Eshleman S, Wittchen HU, Kendler KS (1994) Lifetime and 12-month prevalence of DSM-III-R psychiatric disorders in the United States. Results from the National Comorbidity Survey. Arch Gen Psychiatry 51:8–19

Kessler RC, Angermeyer M, Anthony JC, DE Graaf R, Demyttenaere K, Gasquet I, DE Girolamo G et al (2007) Lifetime prevalence and age-of-onset distributions of mental disorders in the World Health Organization's World Mental Health Survey Initiative. World Psych 6(3):168–176

Klein PS, Melton DA (1996) A molecular mechanism for the effect of lithium on development. Proc Natl Acad Sci 93(16):8455–8459

Konradi C, Eaton M, MacDonald ML, Walsh J, Benes FM, Heckers S (2004) Molecular evidence for mitochondrial dysfunction in bipolar disorder. Arch Gen Psychiatry 61(3):300–308

Krichevsky AM, King KS, Donahue CP, Khrapko K, Kosik KS (2003) A microRNA array reveals extensive regulation of microRNAs during brain development. RNA 9:1274–1281

Kunugi H, Ida I, Owashi T, Kimura M, Inoue Y, Nakagawa S, Yabana T et al (2006) Assessment of the dexamethasone/CRH test as a state-dependent marker for hypothalamic-pituitary-adrenal (HPA) axis abnormalities in major depressive episode: a Multicenter Study. Neuropsychopharmacology 31(1):212–220

Kvajo M, Dhilla A, Swor DE, Karayiorgou M, Gogos JA (2007) Evidence implicating the candidate schizophrenia/bipolar disorder susceptibility gene G72 in mitochondrial function. Mol Psychiatry 13:685–696

Lee RC, Feinbaum RL, Ambros V (1993) The C. elegans heterochronic gene lin-4 encodes small RNAs with antisense complementarity to lin-14. Cell 75(5):843–854

Lachman HM, Pedrosa E, Petruolo OA, Cockerham M, Papolos A, Novak T, Papolos DF et al (2007) Increase in GSK3beta gene copy number variation in bipolar disorder. Am J Med Genet B Neuropsychiatr Genet 144(3):259–265

Laplante I, Beliveau R, Paquin J (2004) RhoA/ROCK and Cdc42 regulate cell-cell contact and N-cadherin protein level during neurodetermination of P19 embryonal stem cells. J Neurobiol 60:289–307

Le-Niculescu H, Kurian SM, Yehyawi N, Dike C, Patel SD, Edenberg HJ, Tsuang MT et al (2008a) Identifying blood biomarkers for mood disorders using convergent functional genomics. Mol Psychiatry. Epub, February 26, 2008

Le-Niculescu H, McFarland MJ, Ogden CA, Balaraman Y, Patel S, Tan J, Rodd ZA et al (2008b) Phenomic, convergent functional genomic, and biomarker studies in a stress-reactive genetic animal model of bipolar disorder and co-morbid alcoholism. Am J Med Genet B Neuropsychiatr Genet 147(2):134–166

Lesch KP, Bengel D, Heils A, Sabol SZ, Greenberg BD, Petri S, Benjamin J et al (1996) Association of anxiety-related traits with a polymorphism in the serotonin transporter gene regulatory region. Science 274(5292):1527–1531

Levinson DF (2006) The genetics of depression: a review. Biol Psychiatry 60(2):84–92

Levinson DF, Evgrafov OV, Knowles JA, Potash JB, Weissman MM, Scheftner WA, Depaulo JR Jr et al (2007) Genetics of recurrent early-onset major depression (GenRED): significant linkage on chromosome 15q25-q26 after fine mapping with single nucleotide polymorphism markers. Am J Psychiatry 164(2):259–264

Lin MW, Sham P, Hwu HG, Collier D, Murray R, Powell JF (1997) Suggestive evidence for linkage of schizophrenia to markers on chromosome 13 in Caucasian but not Oriental populations. Hum Genet 99(3):417–420

Lohoff FW, Sander T, Ferraro TN, Dahl JP, Gallinat J, Berrettini WH (2005) Confirmation of association between the Val66Met polymorphism in the brain-derived neurotrophic factor (BDNF) gene and bipolar I disorder. Am J Med Genet B Neuropsychiatr Genet 139B(1):51–53

Lopez-Molina L, Conquet F, Dubois-Dauphin M, Schibler U (1997) The DBP gene is expressed according to a circadian rhythm in the suprachiasmatic nucleus and influences circadian behavior. EMBO J 16(22):6762–6771

Mackie S, Millar JK, Porteous DJ (2007) Role of DISC1 in neural development and schizophrenia. Curr Opin Neurobiol 17(1):95–102

Madsen TM, Treschow A, Bengzon J, Bolwig TG, Lindrall O, Tingstrom A (2000) Increased neurogenesis in a model of electro convulsive therapy. Biol Psychiatry 47:1043–1049

Malberg J, Eisch AJ, Nestler EJ, Duman RS (2000) Chronic antidepressant treatment increases neurogenesis in adult hippocampus. J Neurosci 20:9104–9110

Marshall CR, Noor A, Vincent JB, Lionel AC, Feuk L, Skaug J, Shago M et al (2008) Structural variation of chromosomes in autism spectrum disorder. Am J Hum Genet 82(2):477–488

Mathers CD, Loncar D (2006) Projections of global mortality and burden of disease from 2002 to 2030. plos medicine 3:e442

McAllister G, Whiting P, Hammond EA, Knowles MR, Atack JR, Bailey FJ, Maigetter R, Ragan CI (1992) cDNA cloning of human and rat brain myo-inositol monophosphatase. Expression and characterization of the human recombinant enzyme. Biochem J 284:749–754

McCarroll SA, Altshuler DM (2007) Copy-number variation and association studies of human disease. Nat Genet 39(7 Suppl):S37–S42

McClellan JM, Susser E, King MC (2007) Schizophrenia: a common disease caused by multiple rare alleles. Br J Psychiatry 190:194–199

McClung (2007) Circadian genes, rhythms and the biology of mood disorders. Pharmacol Therap 114:222–232

McGuffin P, Katz R, Watkins S, Rutherford J (1996) A hospital-based twin register of the heritability of DSM-IV unipolar depression. Arch Gen Psychiatry 53(2):129–136

McGuffin P, Rijsdijk F, Andrew M, Sham P, Katz R, Cardno A (2003) The heritability of bipolar affective disorder and the genetic relationship to unipolar depression. Arch Gen Psychiatry 60(5):497–502

McGuffin P, Knight J, Breen G, Brewster S, Boyd PR, Craddock N, Gill M et al (2005) Whole genome linkage scan of recurrent depressive disorder from the depression network study. Hum Mol Genet 14(22):3337–3345

McInnes LA, Escamilla MA, Service SK, Reus VI, Leon P, Silva S, Rojas E et al (1996) A complete genome screen for genes predisposing to severe bipolar disorder in two Costa Rican pedigrees. Proc Natl Acad Sci USA 93(23):13060–13065

McMahon FJ, Buervenich S, Charney D, Lipsky R, Rush AJ, Wilson AF, Sorant AJM et al (2006) Variation in the gene encoding the serotonin 2A receptor is associated with outcome of antidepressant treatment. Am J Hum Genet 76:804–814

McQueen MB, Devlin B, Faraone SV, Nimgaonkar VL, Sklar P, Smoller JW, Abou Jamra R et al (2005) Combined analysis from eleven linkage studies of bipolar disorder provides strong evidence of susceptibility loci on chromosomes 6q and 8q. Am J Hum Genet 77(4):582–595

McQuillin A, Rizig M, Gurling HM (2007) A microarray gene expression study of the molecular pharmacology of lithium carbonate on mouse brain mRNA to understand the neurobiology of mood stabilization and treatment of bipolar affective disorder. Pharmacogenet Gen 17(8):605–617

Meaney MJ, Szyf M, Seckl JR (2007) Epigenetic mechanisms of perinatal programming of hypothalamic-pituitary-adrenal function and health. Trends Mol Med 13(7):269–277

Merali Z, Du L, Hrdina P, Palkovits M, Faludi G, Poulter MO, Anisman H (2004) Dysregulation in the suicide brain: mRNA expression of corticotropin-releasing hormone receptors and GABA(A) receptor subunits in frontal cortical brain region. J Neurosci 24:1478–1485

Melchior M, Caspi A, Milne BJ, Danese A, Poulton R, Moffitt TE (2007) Work stress precipitates depression and anxiety in young, working women and men. Psychol Med 37(8):1119–1129

Meyer JH, Houle S, Sagrati S, Carella A, Hussey DF, Ginovart N, Goulding V, Kennedy J, Wilson AA (2004) Brain serotonin transporter binding potential measured with carbon 11-labeled DASB positron emission tomography: effects of major depressive episodes and severity of dysfunctional attitudes. Arch Gen Psychiatry 61(12):1271–1279

Mill J, Tang T, Kaminsky Z, Khare T, Yazdanpanah S, Bouchard L, Jia P et al (2008) Epigenomic profiling reveals DNA-methylation changes associated with major psychosis. Am J Hum Genet 82(3):696–711

Millar JK, Wilson-Annan JC, Anderson S, Christie S, Taylor MS, Semple CA, Devon RS, Clair DM, Muir WJ, Blackwood DH, Porteous DJ (2000) Disruption of two novel genes by a translocation co-segregating with schizophrenia. Hum Mol Genet 9:1415–1423

Millar JK, Pickard BS, Mackie S, James R, Christie S, Buchanan SR, Malloy MP et al (2005) DISC1 and PDE4B are interacting genetic factors in schizophrenia that regulate cAMP signaling. Science 310(5751):1187–1191

Miller CA, Sweatt JD (2007) Covalent modification of DNA regulates memory formation. Neuron 53(6):857–869

Miska EA, Alvarez-Saavedra E, Townsend M, Yoshii A, Sestan N, Rakic P, Constantine-Paton M et al (2004) Microarray analysis of microRNA expression in the developing mammalian brain. Genome Biol 5(9):R68

Mombereau C, Kaupmann K, Froestl W, Sansig G, van der Putten H, Cryan JF (2004) Genetic and pharmacological evidence of a role for GABA(B) receptors in the modulation of anxiety- and antidepressant-like behavior. Neuropsychopharmacology 29:1050–1062

Moore GJ, Bebchuk JM, Parrish JK, Faulk MW, Arfken CL, Strahl-Bevacqua J, Manji HK (1999) Temporal dissociation between lithium-induced changes in frontal lobe myo-inositol and clinical response in manic-depressive illness. Am J Psychiatry 156:1902–1808

Moore GJ, Bebchuk JM, Wilds IB, Chen G, Manji HK (2000) Lithium-induced increase in human brain grey matter. Lancet 356:1241–1242

Müller MB, Holsboer F (2006) Mice with mutations in the HPA-system as models for symptoms of depression. Biol Psychiatry 59:1104–1115

Müller DJ, De Luca V, Sicard T, King N, Strauss K, Kennedy JL (2006) Brain-derived neurotrophic factor (BDNF) gene and rapid-cycling bipolar disorder. Br J Psychiatry 189:317–323

Murphy DL, Lesch KP (2008) Targeting the murine serotonin transporter: insights into human neurobiology. Nat Rev Neurosci 9(2):85–96

Nakatani N, Aburatani H, Nishimura K, Semba J, Yoshikawa T (2004) Comprehensive expression analysis of a rat depression model. Pharmacogenomics J 4:114–126

Nakatani N, Hattori E, Ohnishi T, Dean B, Iwayama Y, Matsumoto I, Kato T, Osumi N, Higuchi T, Niwa S, Yoshikawa T (2006) Genome-wide expression analysis detects eight genes with robust alterations specific to bipolar I disorder: relevance to neuronal network perturbation. Hum Mol Genet 15:1949–1962

Nakatani N, Ohnishi T, Iwamoto K, Watanabe A, Iwayama Y, Yamashita S, Ishitsuka Y, Moriyama K, Nakajima M, Tatebayashi Y, Akiyama H, Higuchi T, Kato T, Yoshikawa T (2007) Expression analysis of actin-related genes as an underlying mechanism for mood disorders. Biochem Biophys Res Commun 352:780–786

Nemeroff CB, Widerlov E, Bissette G, Walleus H, Karlsson I, Eklund K, Kilts CD, Loosen PT, Vale W (1984) Elevated concentrations of CSF corticotropin-releasing factor-like immunoreactivity in depressed patients. Science 226:1342–1344

Nestler EJ, Barrot M, DiLeone RJ, Eisch AJ, Gold SJ, Monteggia LM (2002) Neurobiology of depression. Neuron 34(1):13–25

Neves-Pereira M, Mundo E, Muglia P, King N, Macciardi F, Kennedy JL (2002) The brain-derived neurotrophic factor gene confers susceptibility to bipolar disorder: evidence from a family-based association study. Am J Hum Genet 71(3):651–655

Nibuya M, Morinobu S, Duman RS (1995) Regulation of BDNF and trkB mRNA in rat brain by chronic electroconvulsive seizure and antidepressant drug treatments. J Neurosci 15:7539–7547

Nibuya M, Nestler EJ, Duman RS (1996) Chronic antidepressant administration increases the expression of cAMP response element binding protein (CREB) in rat hippocampus. J Neurosci 16:2365–2372

O'Brien WT, Harper AD, Jove F, Woodgett JR, Maretto S, Piccolo S, Klein PS (2004) Glycogen synthase kinase-3beta haploinsufficiency mimics the behavioral and molecular effects of lithium. J Neurosci 24:6791–649

Ohnishi T, Yamada T, Ohba H, Iwayama Y, Toyota T, Hattori E, Inada T, Kunugi H, Tatsumi M, Ozaki N, Iwata N, Sakamoto K, Iijima Y, Iwata Y, Tsuchiya KJ, Sugihara G, Nanko S, Osumi N, Detera-Wadleigh SD, Kato T, Yoshikawa T (2007a) A promoter haplotype of the inositol monophosphatase 2 gene (IMPA2) at 18p11.2 confers a possible risk for bipolar disorder by enhancing transcription. Neuropsychopharmacology 32:1727–1737

Ohnishi T, Ohba H, Seo K-C, Im J, Sato Y, Iwayama Y, Furuichi T, Chung S-K, Yoshikawa T (2007b) Spatial expression patterns and biochemical properties distinguish a second myo-inositol monophosphatase, IMPA2 from IMPA1. J Biol Chem 282:637–646

Overmier JB, Seligman ME (1967) Effects of inescapable shock upon subsequent escape and avoidance responding. J Comp Physiol Psychol 63:28–33

Palo OM, Antila M, Silander K, Hennah W, Kilpinen H, Soronen P, Tuulio-Henriksson A et al (2007) Association of distinct allelic haplotypes of DISC1 with psychotic and bipolar spectrum disorders and with underlying cognitive impairments. Hum Mol Genet 16(20):2517–2528

Patishi Y, Belmaker RH, Bersudsky Y, Kofman O (1996) A comparison of the ability of myo-inositol and epi-inositol to attenuate lithium-pilocarpine seizures in rats. Biol Psychiatry 39:829–832

Pencea V, Bingaman KD, Wiegand SJ, Luskin MB (2001) Infusion of brain-derived neurotrophic factor into the lateral ventricle of the adult rat leads to new neurons in the parenchyma of the striatum, sseptum, thalamus, and hypothalamus. J Neurosci 21:6706–6717

Pericak-Vance MA, Bebout JL, Gaskell PC Jr, Yamaoka LH, Hung WY, Alberts MJ, Walker AP et al (1991) Linkage studies in familial Alzheimer disease: evidence for chromosome 19 linkage. Am J Hum Genet 48(6):1034–1050

Perkins DO, Jeffries CD, Jarskog LF, Thomson JM, Woods K, Newman MA, Parker JS et al (2007) microRNA expression in the prefrontal cortex of individuals with schizophrenia and schizoaffective disorder.Genome Biol 8(2):R27

Perry GH, Ben-Dor A, Tsalenko A, Sampas N, Rodriguez-Revenga L, Tran CW, Scheffer A et al (2008) The fine-scale and complex architecture of human copy-number variation. Am J Hum Genet 82(3):685–695

Porsolt R, Le Pichon, Jalfre M (1997) Depression: a new animal model sensitive to antidepressant treatments. Nature 266:730–732

Post RM (2007) Role of BDNF in bipolar and unipolar disorder: clinical and theoretical implications. J Psychiatr Res 41(12):979–990

Pritchard JK (2001) Are rare variants responsible for susceptibility to complex diseases? Am J Hum Genet 69 (1):124–137

Reich DE, Lander ES (2001) On the allelic spectrum of human disease. Trends Genet 17(9):502–510

Rietschel M, Beckmann L, Strohmaier J, Georgi A, Karpushova A, Schirmbeck F, Boesshenz KV et al (2008) G72 and its association with major depression and neuroticism in large population-based groups from Germany. Am J Psychiatry 165:753–762

Rios M, Fan G, Fekete C, Kelly J, Bates B, Kuehn R, Lechan RM, Jaenisch R (2001) Conditional deletion of brain-derived neurotrophic factor in the postnatal brain leads to obesity and hyperactivity. Mol Endocrinol 15(10):1748–1757

Rowe MK, Wiest C, Chuang DM (2007) GSK-3 is a viable potential target for therapeutic intervention in bipolar disorder. Neurosci Biobehav Rev 31(6):920–931

Roybal K, Theobold D, Graham A, DiNieri JA, Russo SJ, Krishnan V, Chakravarty S (2007) Mania-like behavior induced by disruption of CLOCK. Proc Natl Acad Sci USA 104(15):6406–6411

Rybakowski JK, Twardowska K (1999) The dexamethasone/corticotropin-releasing hormone test in depression in bipolar and unipolar affective illness. J Psychiatr Res 33:363–370

Saini HK, Griffiths-Jones S, Enright AJ (2007) Genomic analysis of human microRNA transcripts. Proc Natl Acad Sci USA 104(45):17719–17724

Sakane F, Imai S, Kai M, Yasuda S, Kanoh H (2007) Diacylglycerol kinases: why so many of them? Biochim Biophys Acta 1771(7):793–806

Santarelli L, Saxe M, Gross C, Surget A, Battaglia F, Dulawa S, Weisstaub N et al (2003) Requirement of hippocampal neurogenesis for the behavioral effects of antidepressants. Science 301(5634):805–809

Sapolsky RM (2001) Depression, antidepressants, and the shrinking hippocampus. Proc Natl Acad Sci 98(22):12320–12322

Sarkar S, Floto RA, Berger Z, Imarisio S, Cordenier A, Pasco M, Cook LJ, David C, Rubinsztein DC (2005) Lithium induces autophagy by inhibiting inositol monophosphatase. J Cell Biol 170:1101–1111

Sassi RB, Nicoletti M, Brambilla P, Mallinger AG, Frank E, Kupfer DJ, Keshavan MS, Soares JC (2002) Increased gray matter volume in lithium-treated bipolar disorder patients. Neurosci Lett 329:243–245

Sawamura N, Sawa A (2006) Disrupted-in-schizophrenia-1 (DISC1): a key susceptibility factor for major mental illnesses. Ann NY Acad Sci 1086:126–133

Schaaf M, de Jong J, de Kloet ER, Vreugdenhil E (1998) Downregulation of BDNF mRNA and protein in the rat hippocamppus by corticosterone. Brain Res 813:112–120

Schildkraut JJ (1965) The catecholamine hypothesis of affective disorders: a review of supporting evidence. Am J Psychiatry 122:509–522

Schratt GM, Tuebing F, Nigh EA, Kane CG, Sabatini ME, Kiebler M, Greenberg ME (2006). A brain-specific microRNA regulates dendritic spine development. Nature 439:283–289

Schumacher J, Jamra RA, Becker T, Ohlraun S, Klopp N, Binder EB, Schulze TG et al (2005) Evidence for a relationship between genetic variants at the brain-derived neurotrophic factor (BDNF) locus and major depression. Biol Psychiatry 58(4):307–314

Sebat J, Lakshmi B, Malhotra D, Troge J, Lese-Martin C, Walsh T, Yamrom B et al (2007) Strong association of de novo copy number mutations with autism. Science 316(5823):445–449

Segurado R, Detera-Wadleigh SD, Levinson DF, Lewis CM, Gill M, Nurnberger JI Jr, Craddock N et al (2003) Genome scan meta-analysis of schizophrenia and bipolar disorder, part III: Bipolar disorder. Am J Hum Genet 73(1):49–62

Sempere LF, Freemantle S, Pitha-Rowe I, Moss E, Dmitrovsky E, Ambros V (2004) Expression profiling of mammalian microRNAs uncovers a subset of brain-expressed microRNAs with possible roles in murine and human neuronal differentiation. Genome Biol 5:R13

Sequeira A, Turecki G (2006) Genome wide gene expression studies in mood disorders. OMICS 10(4):444–454

Serretti A, Benedetti F, Mandelli L, Lorenzi C, Pirovano A, Colombo C, Smeraldi E (2003) Genetic dissection of psychopathological symptoms: insomnia in mood disorders and CLOCK gene polymorphism. Am J Med Genet B Neuropsychiatr Genet 121(1):35–38

Serretti A, Mandelli L (2008) The genetics of bipolar disorder: genome 'hot regions,' genes, new potential candidates and future directions. Mol Psychiatry 13(8):742–771

Shaltiel G, Chen G, Manji HK (2007) Neurotrophic signaling cascades in the pathophysiology and treatment of bipolar disorder. Curr Opin Pharmacol 7(1):22–26

Shi J, Badner JA, Gershon ES, Liu C (2008a) Allelic association of G72/G30 with schizophrenia and bipolar disorder: a comprehensive meta-analysis. Schizophr Res 98(1–3):89–97

Shi J, Wittke-Thompson JK, Badner JA, Hattori E, Potash JB, Willour VL, McMahon FJ et al (2008b) Clock genes may influence bipolar disorder susceptibility and dysfunctional circadian rhythm. Am J Med Genet B Neuropsychiatr Genet 147B(7):1047–1055

Shimizu E, Hashimoto K, Olamura N, Koike K, Komatsu N, Kumakiri C, Nakazato M, Watanabe H, Shinoda N, Okada S, Iyo M (2003) Alterations of serum levels of brain-derived neurotrophic factor (BDNF) in depressed patients with or without antidepressants. Biol Psychiatry 54:70–75

Sibille E, Arango V, Galfalvy HC, Pavlidis P, Erraji-Benchekroun L, Ellis SP, John Mann J (2004) Gene expression profiling of depression and suicide in human prefrontal cortex. Neuropsychopharmacology 29:351–361

Silverstone PH, Wu RH, O'Donnell T, Ulrich M, Asghar SJ, Hanstock CC (2003) Chronic treatment with lithium, but not sodium valproate, increases cortical N-acetyl-aspartate concentrations in euthymic bipolar patients. Int Clin Psychopharmacol 18:73–79

Sjoholt G, Ebstein RP, Lie RT, Berle JO, Mallet J, Deleuze JF, Levinson DF, Laurent C, Mujahed M, Bannoura I, Murad I, Molven A, Steen VM (2004) Examination of IMPA1 and IMPA2 genes in manic-depressive patients: association between IMPA2 promoter polymorphisms and bipolar disorder. Mol Psychiatry 9:621–629

Sklar P, Gabriel SB, McInnis MG, Bennett P, Lim YM, Tsan G, Schaffner S et al (2002) Family-based association study of 76 candidate genes in bipolar disorder: BDNF is a potential risk locus. Brain-derived neutrophic factor. Mol Psychiatry 7(6):579–593

Sklar P, Smoller JW, Fan J, Ferreira MA, Perlis RH, Chambert K, Nimgaonkar VL et al (2008) Whole-genome association study of bipolar disorder. Mol Psychiatry 13(6):558–569

Smirnova L, Gräfe A, Seiler A, Schumacher S, Nitsch R, Wulczyn FG (2005) Regulation of miRNA expression during neural cell specification. Euro J Neurosci 21 (6):1469–1477

Decoding the Genetics and Underlying Mechanisms of Mood Disorders

St. Clair D, Blackwood D, Muir W, Carothers A, Walker M, Spowart G et al (1990) Association within a family of a balanced translocation with major mental illness. Lancet 336:13–16

Steru L, Chermat R, Thierry B, Simon P (1985) The tail suspension test: a new method for screening antidepressants in mice. Psychopharmacology 85:367–370

Strittmatter WJ, Roses AD (1996) Apolipoprotein E and Alzheimer's disease. Annu Rev Neurosci 19:53–77

Szyf M, Weaver I, Meaney M (2007) Maternal care, the epigenome and phenotypic differences in behavior. Reprod Toxicol 24(1):9–19

Tao X, Finkbeiner S, Arnold DB, Shaywitz AJ, Greenberg ME (1988) Ca2+ influx regulates BDNF transcription by a CREB family transcription factor-dependent mechanism. Neuron 20:709–726

Telner JI, Singhal RL (1984) Psychiatric progress. The learned helplessness model of depression. J Psychiat Res 18:207–215

Trapp T, Rupprecht R, Castren M, Reul JM, Holsboer F (1994) Heterodimerization between mineralocorticoid and glucocorticoid receptor a new principle of glucocorticoid action in the CNS. Neuron 13:1457–1462

Tsankova N, Renthal W, Kumar A, Nestler EJ (2007) Epigenetic regulation in psychiatric disorders. Nat Rev Neurosci 8(5):355–367

Tsuang MT, Faraone SV (1990) The genetics of mood disorders. The Johns Hopkins University Press, Baltimore

Turecki G, Grof P, Grof E, D'Souza V, Lebuis L, Marineau C, Cavazzoni P et al (2001) Mapping susceptibility genes for bipolar disorder: a pharmacogenetic approach based on excellent response to lithium. Mol Psychiatry 6(5):570–578

Uher R, McGuffin P (2008) The moderation by the serotonin transporter gene of environmental adversity in the aetiology of mental illness: review and methodological analysis. Mol Psychiatry 13(2):131–146

Venken T, Del-Favero J (2007) Chasing genes for mood disorders and schizophrenia in genetically isolated populations. Hum Mutat 28(12):1156–1170

Wang SS, Kamphuis W, Huitinga I, Zhou JN, Swaab DF (2008a) Gene expression analysis in the human hypothalamus in depression by laser microdissection and real-time PCR: the presence of multiple receptor imbalances. Mol Psychiatry 13(8):786–799

Wang WX, Rajeev BW, Stromberg AJ, Ren N, Tang G, Huang Q, Rigoutsos I et al (2008b) The expression of microRNA miR-107 decreases early in Alzheimer's Disease and may accelerate disease progression through regulation of β-site amyloid precursor protein-cleaving enzyme 1. J Neurosci 28(5):1213–1223

Watson S, Gallagher P, Del-Estal D, Hearn A, Ferrier IN, Young AH (2002) Hypothalamic-pituitary-adrenal axis function in patients with chronic depression. Psychol Med 32:1021–1028

Weissman MM, Leaf PJ, Tischler GL, Blazer DG, Karno M, Bruce ML, Florio LP (1988) Affective disorders in five United States communities. Psychol Med 18:141–153

Weissman MM, Bland RC, Canino GJ, Faravelli C, Greenwald S, Hwu HG, Joyce PR et al (1996) Cross-national epidemiology of major depression and bipolar disorder. JAMA 276(4):293–299

Wellcome Trust Case Control Consortium (WTCCC) (2007) Genome-wide association study of 14,000 cases of seven common diseases and 3,000 shared controls. Nature 447(7145):661–678

Wildenauer DB, Schwab SG, Maier W, Detera-Wadleigh SD (1999) Do schizophrenia and affective disorder share susceptibility genes? Schizophr Res 39(2):107–111

Williams RSB, Cheng L, Mudge AW, Harwood AJ (2002) A common mechanism of action for three mood-stabilizing drugs. Nature 417:292–295

Willner P (1990) Animal models of depression: an overview. Pharmacol Ther 45:425–455

Wilson GM, Flibotte S, Chopra V, Melnyk BL, Honer WG, Holt RA (2006) DNA copy-number analysis in bipolar disorder and schizophrenia reveals aberrations in genes involved in glutamate signaling. Hum Mol Genet 15(5):743–749

Wolfson M, Einat H, Bersudsky Y, Berkin V, Belmaker RH, Hertz L (2000) Nordidemnin potently inhibits inositol uptake in cultured astrocytes and dose-dependently augments lithium's proconvulsant effect in vivo. J Neurosci Res 60:116–121

Yamada K, Watanabe A, Iwayama-Shigeno Y, Yoshikawa T (2003) Evidence of association between gamma-aminobutyric acid type A receptor genes located on 5q34 and female patients with mood disorders. Neurosci Lett 349:9–12

Yang S, Van Dongen HP, Wang K, Berrettini W, Bućan M (2008) Assessment of circadian function in fibroblasts of patients with bipolar disorder. Mol Psychiatry. Epub, February 26, 2008

Yasuda S, Liang MH, Marinova Z, Yahyavi A, Chuang DM (2007) The mood stabilizers lithium and valproate selectively activate the promoter IV of brain-derived neurotrophic factor in neurons. Mol Psychiatry. Epub, October 9, 2007

Yoshikawa T, Turner G, Esterling LE, Sanders AR, Detera-Wadleigh SD (1997) A novel human myo-inositol monophosphatase gene, IMP.18p, maps to a susceptibility region for bipolar disorder. Mol Psychiatry 2:393–397

Yoshikawa T, Watanabe A, Ishitsuka Y, Nakaya A, Nakatani N (2002) Identification of multiple genetic loci linked to the propensity for "behavioral despair" in mice. Genome Res 12:357–366

Zammit S, Owen MJ (2006) Stressful life events, 5-HTT genotype and risk of depression. Br J Psychiatry 188:199–201

Zeggini E, Scott LJ, Saxena R, Voight BF, Marchini JL, Hu T, de Bakker PI et al (2008) Meta-analysis of genome-wide association data and large-scale replication identifies additional susceptibility loci for type 2 diabetes. Nat Genet 40:638–645

Zhang X, Odom DT, Koo SH, Conkright MD, Canettieri G, Best J, Chen H et al (2005) Genome-wide analysis of cAMP-response element binding protein occupancy, phosphorylation, and target gene activation in human tissues. Proc Natl Acad Sci 102 (12):4459–4464

Zhao X, Ueba T, Christie BR, Barkho B, McConnell MJ, Nakashima K, Lein ES et al (2003) Mice lacking methyl-CpG binding protein 1 have deficits in adult neurogenesis and hippocampal function. Proc Natl Acad Sci 100(11):6777–6782

Zhao C, Deng W, Gage FH (2008) Mechanisms and functional implications of adult neurogenesis. Cell 132(4):645–660

Dissecting the Molecular Causes of Schizophrenia

Dieter B. Wildenauer, Diah Mutiara B. Wildenauer, and Sibylle G. Schwab

Contents

1   Introduction 52
2   Diagnosis 52
3   Therapy 52
4   Incidence, Prevalence 53
5   The Search for Biological Causes 53
6   Epidemiology 54
7   Neuropathology 56
8   Neurotransmitter Hypotheses 57
9   Molecular Genetics 58
    9.1   Localization of Susceptibility Genes by Linkage Analysis 58
    9.2   Genetic Association 59
10   Gene Expression 69
    10.1   mRNA Expression 70
    10.2   miRNA 70
    10.3   Protein Expression 71
    10.4   Epigenetics 71
11   Animal Models 72
12   Conclusion 73
References   73

Abstract Schizophrenia is a complex genetic disorder with largely unknown aetiology. The search for molecular causes has been the focus for many years studying biochemistry, neurobiology, molecular biology, and molecular genetics of the disorder. Our intention is to provide an overview of approaches with focus on new developments in molecular biology and genetics, as well as describe selected findings, eventually leading to the dissection of molecular causes for this devastating mental disorder.

D.B. Wildenauer (✉)
Center for Research in Neuropsychiatry (CRN), University of Western Australia
Crawley, WA 6009 Australia,
School of Psychiatry and Clinical Neuropsychiatry, University of Western Australia,
Crawley, WA 6009 Australia
Dieter.Wildenauer@uwa.edu.au

D.B. Wildenauer (ed.), *Molecular Biology of Neuropsychiatric Disorders*,
Nucleic Acids and Molecular Biology, © Springer-Verlag Berlin Heidelberg 2009

1 Introduction

Schizophrenia is a devastating mental disorder affecting around 1% of the population. The associated long-lasting social dysfunction, high rates of unemployment and hospitalization, high treatment costs, and the need to provide lifelong care and support in many cases, imposes a severe burden on individuals and their families. People with schizophrenia have an increased mortality rate, which is mainly due to suicide (around 10%).

The disorder is defined as a syndrome composed of several characteristic symptoms, as characterized by the DSM-IV criteria (American Psychiatric Association 1994:

delusions
hallucinations
disorganized speech (frequent derailment or incoherence)
grossly disorganized or catatonic behavior
negative symptoms (affective flattening, alogia, avolition).

2 Diagnosis

The diagnosis schizophrenia requires the presence of at least two of these symptoms during a 1-month period. Further inclusion criteria include social/occupational dysfunction as well as the overall duration of disturbance persisting for at least 6 months. Exclusion criteria include a diagnosis of schizoaffective or mood disorder, as well as schizophrenia-like presentations secondary to drug abuse, medication or general medical condition. The disorder has been further refined into nosological subtypes, e.g., paranoid type, disorganized type, etc. (American Psychiatric Association 1994). The course of schizophrenia can be episodic, i.e. one (22%) or more episodes, without inter-episode residual symptoms (35%). Approximately 43% of patients exhibit symptoms for schizophrenia throughout all or most of the course of their illness (no return to normality after one or between several episodes). The age of onset of schizophrenia peaks at 20–25 years and then steadily declines with age reaching a minimum by 45–50 years (http://www.schizophrenia.com/szfacts.htm).

The diagnosis of schizophrenia relies on assessment by psychiatrists, which includes disorder and family history. Consensus diagnosis of two or more psychiatrists may be necessary. Unlike common disorders, such as diabetes and hypertension, there is no objective biological marker or test for schizophrenia. The difficulty of assessment and the exact definition of the phenotype appear to be major obstacles in the search for molecular causes of schizophrenia.

3 Therapy

Before introduction of the classical neuroleptics by Delay et al. (1952), the pharmacological treatment of schizophrenia was ineffective. Neuroleptic drugs, e.g., chlorpromazine, haloperidol, and a number of related compounds, which treat the symptoms of schizophrenia, have drastically reduced the time spent in hospital, but are not ideal because of severe side effects like dyskinesia, and also because they have little effect on negative symptoms. The more recently introduced, so-called atypical neuroleptics, i.e., clozapine, risperidone, are thought to be superior in treating negative symptoms. However, these drugs have other serious side effects like agranulocytosis, obesity and metabolic syndrome. Identifying schizophrenia-specific molecular targets for development of pharmacological agents is one of key goals driving worldwide research on the molecular causes for schizophrenia.

4 Incidence, Prevalence

The 1986 WHO Collaborative Study on determinants of outcome of severe mental disorders found the *incidence* of schizophrenia (number of new cases diagnosed each year) is comparable across ten countries, ranging from 16 to 42 per 100,000 (Jablensky et al. 1992; Sartorius et al. 1986) for the broad definition of the phenotype. While this report has been widely cited as evidence that the incidence of schizophrenia is relatively uniform across the world, a more recent review (McGrath et al. 2004) that included data from 158 studies and 32 countries reported an incidence range from 7.7 to 43.0 per 100,000. This fivefold difference suggests a noticeable worldwide variation in the incidence of schizophrenia.

Prevalence rates (number of cases at any given time/time interval) also appear to be different in different populations (Saha et al. 2005). Based on 188 studies from 46 countries, the lifetime morbid risk was 0.31–2.71%, resulting in a medium value of 0.72%. This study also provided evidence for a difference in prevalence rates between developed and developing countries, being lower in the later.

5 The Search for Biological Causes

After many years of debate around the putative biological basis for schizophrenia, it is now agreed that biological causes do contribute to the development of the disorder. Evidence is mainly circumstantial and based on the action of neuroleptics, some pathological findings, and genetic transmission studies.

The search for biological causes of schizophrenia was initiated more than 100 years ago with the work of E. Kraepelin and collaborators. Despite this, the evidence for specific biological abnormalities remains inferential rather than definite. In addition to phenotype definition, the complex nature of the disorder, i.e., a contribution of environment and genes, which is shared by other common, multifactorial, polygenic diseases like arteriosclerosis, diabetes, coronary heart disease, and many others, renders the dissection of the molecular basis for schizophrenia extremely difficult. After many years of epidemiological research and controversial discussion ("nature *versus* nurture"), there is now agreement that environment and genes together contribute to development of the disorder ("nature *and* nurture"). However, the number, composition, and mode of interaction of environmental and genetic factors are still unknown.

Research into molecular biology of schizophrenia is mainly based on findings in epidemiology, pharmacology, pathology, and cytology. These findings have been helpful in generating hypotheses. Consequently, we will briefly describe findings in epidemiology, as well as anatomical, cellular and biochemical–pharmacological studies, which have been used to generate hypotheses for research into the molecular biology of schizophrenic disorders.

6 Epidemiology

Epidemiology has revealed a number of risk factors, which are summarized in Fig. 1 (Sullivan 2005). The aetiological contributions of some of these factors are not clear, e.g., season of birth, (Mortensen et al. 1999). Others may implicate stress reactions precipitating the illness, e.g., urban/rural differences (Mortensen et al. 1999), and migration (Cantor-Graae and Selten 2005), or implicate altered brain function in response to activation of the immune system, e.g., virus infection (Buka et al. 2001). Obstetric complications (Cannon et al. 2002) may lead to enlarged ventricles and reduced brain volume, which has been detected in individuals with schizophrenia (McNeil et al. 2000).

One of the best documented and repeatedly confirmed findings in the epidemiology of schizophrenia is evidence for genetic transmission. It has been shown in numerous studies that schizophrenia runs in families. For example, if we look at the average increase in risk reported in past studies (Gottesman 1991), the lifetime risk for developing schizophrenia in children with one parent with schizophrenia or in siblings of an individual with schizophrenia is about 15 times higher as compared to the general population. In second degree relatives, like cousins,

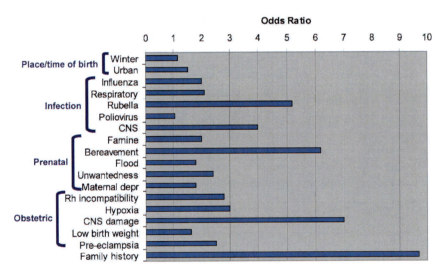

Fig. 1 Selected risk factors for schizophrenia. The figure was prepared and provided by Dr Patrick F Sullivan and first published in PLOS Medicine, July 2005 (Sullivan 2005)

uncles, aunts, and grandparents, the risk is approximately 3 times higher. The lifetime risk for an identical co-twin of an individual with schizophrenia is about 50 times higher as compared to the general population, whilst for a fraternal twin with schizophrenia the risk is with about 15 times in the same range as for siblings (Fig. 2). The heritability rate estimated from these studies is approximately 80%. The concordance rate for identical twins is around 50%, which is clearly lower than the expected 100%, if the disorder was entirely genetic. This provides also evidence for the involvement of environmental factors in development of the illness. Additional evidence for genetic transmission comes from adoption studies. (1) Children of parents with schizophrenia, who were adopted by parents without the illness, had the same disease risk as those brought up by their natural "schizophrenic" parents (Rosenthal et al. 1971). (2) In another design, the risk for biological parents of an offspring who became schizophrenic while raised by adoptive parents was studied. The biological parents had the increased risk as expected for first degree relatives of schizophrenics, whilst the adoptive parents had the risk as seen in the general population (Kety et al. 1971). (3) The risk for children from unaffected natural parents who were adopted from initially unaffected parents, but where one adoptive parent later developed schizophrenia, was studied ("exposure to schizophrenic parenting"). The child's risk was not different from that for the general population (Wender et al. 1974).

In summary, family, twin and adoption studies clearly show that genetic factors are important for the development of schizophrenia, but they show also that there is no Mendelian mode of transmission as in monogenic disorders like Huntingtons' disease or early onset familial Alzheimer's disease.

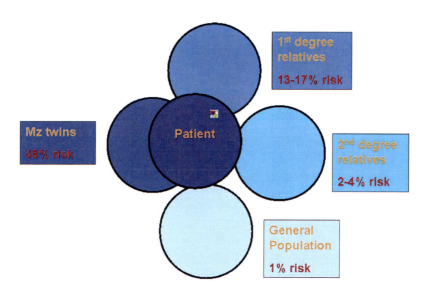

Fig. 2 Risk for schizophrenia in first, second degree relatives, and in identical twins in comparison to the population risk. The risk data are taken from Gottesman (1991)

Studies of the mode of transmission in families does not reveal evidence for dominant or recessive single gene transmission. There was also no evidence that any subtype of the disorder shows Mendelian type inheritance like early onset familial Alzheimer's disease. Genetic factors in schizophrenia appear to be a composition of several genes each presumably making a minor contribution (oligo- or polygenic inheritance). The genes involved in the disorder confer risk (susceptibility), but a single gene may not be sufficient for development of the disorder. As discussed later, this has profound consequences for gene identification by genetic linkage and association studies.

7 Neuropathology

There have been numerous studies published (Weinberger 1995a) on pathological changes in different brain areas, beginning with postmortem studies and followed by computerized axial tomography (CAT) and magnetic resonance imaging (MRI) (Weinberger 1995b; 1996). A consistent finding appears to be increased ventricle size and loss of cortical volume with unequal distribution in favor of the left hemisphere (Crow et al. 1989). These abnormalities are seen in first diagnosis cases and do not progress during the course of the illness, which would be expected if due to a neurodegenerative process. Together with evidence for cytoarchitectural abnormalities (Jakob and Beckmann 1986) and an increased rate of obstetric complications associated with schizophrenia (Cannon et al. 2002), the morphometric findings provide major support for the *neurodevelomental hypothesis* of schizophrenia, which is the rationale for molecular biological investigations of a number of candidate genes, i.e., Neuregulin, Reelin.

Studies on the phenotypical (i.e., cellular and biochemical) level are usually hampered by the relative inaccessibility of the brain in living patients. Numerous investigations have been performed using postmortem brain tissue from individuals with schizophrenia. However, these studies are usually confounded by the difficulty in obtaining material from unaffected individuals and the need to match samples by cause of death, medication use, drug abuse, additional illness, physical condition, postmortem delay, and other parameters, which may cause differential changes not related to schizophrenia. Nevertheless, collections of post mortem brain tissues are now available (i.e., the Stanley Medical Research Institute brain-array collection), which control, to some extent, for these confounds.

Limited access to the environment of the living brain is provided via the cerebrospinal fluid (CSF), which, in the brain, is located in the subarachnoid space and the ventricular system and can be obtained by spinal lumbar puncture.

In addition, the search for surrogate or peripheral tissues has included blood cells or fibroblasts. These can be taken from patients and analyzed for molecular changes that may reflect pathological alterations of the brain responsible for development of the disorder.

8 Neurotransmitter Hypotheses

Since there is a consensus that schizophrenia is a brain disorder, the putative role of neurotransmitter systems in aetiology of the illness has been of considerable interest. Studies of the pharmacology and biochemistry of neuroleptics and psychostimulants lead to the *dopamine hypothesis*. This hypothesis is based on several converging lines of evidence implicating overactivity of dopaminergic pathways (Abi-Dargham 2006): (1) Neuroleptics bind to dopamine D2 receptors in concentrations that correspond to therapeutically active plasma concentrations of these drugs (Creese et al. 1976; Seeman et al. 1975; Seeman and Lee 1975). (2) Psychostimulants, like amphetamine, which stimulate release of dopamine from vesicles and also block reuptake from the synaptic cleft, increase dopamine concentrations in the synaptic cleft and produce psychotic symptoms (Angrist and Gershon 1970). (3) L-dopa, a drug used to stimulate dopamine synthesis in Parkinson's disease, can cause psychotic symptoms in very high doses. (4) Postmortem (Owen et al. 1978) and brain imaging (Positron Emission Tomography, Farde et al. 1992) studies have consistently found evidence for enhanced D2 receptor binding sites in individuals with schizophrenia compared to controls. However, these studies remain controversial due to the very considerable difficulties in matching cases and controls on key confounding factors (see discussion above). (5) There is evidence from functional brain imaging studies for a deficit in DA transmission at D1 receptors in the prefrontal cortex (PFC), which might be implicated in cognitive impairments and negative symptoms of schizophrenia (Abi-Dargham et al. 2002). This leads to the current view that *hyper*activity of subcortical (striatal) transmission at D2 receptors is responsible for the positive symptoms (hallucinations, delusions, etc.), whilst *hypo*-dopaminergia in the dorsolateral prefrontal cortex (DLPFC) at D1 receptors contributes to the pathophysiology of negative (cognitive deficit) symptoms (Abi-Dargham 2006; Abi-Dargham et al. 2002).

The dopamine hypothesis for schizophrenia has stimulated numerous studies analyzing variation at the DNA and RNA level in various components (receptors, transporters, enzymes) of dopamine neurotransmission. In most cases, the association with schizophrenia was found either to be negative or a positive association could not be replicated or confirmed in other studies. Given the various interactions between neurotransmitter systems within the neuronal network, the dopaminergic pathway may not necessarily be the origin of the molecular changes but, instead, may be the endpoint of molecular pathways which finally leads to the symptoms seen in schizophrenia (Abi-Dargham 2006).

Glutamatergic neurotransmission is among the systems interacting with the dopaminergic system and involved in upstream regulation. Glutamate is the major excitatory neurotransmitter in the brain interacting with two types of receptors, the ionotropic receptors with NMDA, Kainate, and AMPA receptor subtypes connected to or representing ion channels, and the metabotropic receptors, which activate G-protein coupled signal transduction. The *glutamate hypothesis* for schizophrenia (Moghaddam 2005) was first proposed by Kim et al. (1980) based on findings of reduced glutamate levels in the cerebrospinal fluid of patients with schizophrenia. Major support for the hypothesis comes from the observation that NMDA receptor agonists like ketamine and phencyclidine (PCP) exacerbate schizophrenia-like symptoms in healthy individuals (Javitt and Zukin 1991). Additional evidence comes from studies on postmortem brain tissues from individuals with schizophrenia which reveal changes in receptor binding, as well as in mRNA and protein expression (Clinton and Meador-Woodruff 2004), and from the fact that most of the potential susceptibility genes discovered by genetic association can be connected to glutamate neurotransmission (Harrison and Owen 2003; Moghaddam 2003).

The excitatory glutaminergic system is also interconnected with the GABA system (Coyle 2006). The *GABA hypothesis* which postulates a hypofunction of GABAergic neurons is supported by postmortem findings showing a reduction in the activity of (1) the enzyme glutamic acid decarboxylase (Bird et al. 1978), (2) the GABA transporter (Reynolds et al. 1990), and (3) the GABA A receptor (Hanada et al. 1987)

9 Molecular Genetics

Whilst a hypothesis should be the basis for studies at the phenotypical level, a major constraint for these studies is the difficulty in differentiating between effects caused by the disease (state dependent effects) and those causing the disease. Molecular genetic methods, however, have made the comprehensive screening of the genome for loci in linkage or association with a disorder possible without needing to invoke a causal hypothesis. This approach is made possible by the availability of genome-wide narrowly spaced DNA sequence polymorphisms (Short Tandem Repeats, Single Nucleotide Polymorphism), which can be efficiently and economically genotyped, as well as by the development of suitable biostatistical methods.

9.1 Localization of Susceptibility Genes by Linkage Analysis

In genetic linkage analysis, it is estimated whether a known polymorphic marker is, in families, more often transmitted together with a genetic trait than one would expect by chance. Detection of linkage has proven to be a valuable first step in gaining information about the localization of genes in monogenic disorders. However, due to multifactorial, polygenic inheritance, linkage analysis in complex disorders like schizophrenia is less informative. Nevertheless, linkage studies may be a starting point for gene identification by providing evidence for chromosomal regions.

Classical LOD-score linkage analysis requires the specification of parameters like mode of inheritance, penetrance, disease allele frequency, as well as the availability of large multigenerational families. Thus, model-free methods, in particular the affected sib-pair method, are preferred. Here, the ideal requirements for the

Dissecting the Molecular Causes of Schizophrenia 59

samples are families with two or more affected siblings and preferably both parents available for genotyping. Allele sharing between affected siblings is analyzed by estimating deviations from the 50% value, which is seen when disease and marker locus are transmitted independently.

More than 30 genome-wide scans for genetic linkage with schizophrenia have been performed using family samples with multiple individuals affected with schizophrenia (Fig. 3 and reviewed in Lewis et al. 2003; Riley 2004; Sullivan 2005). Only a few loci that met the criteria for genome-wide significance were detected in these genome-wide linkage scans: 1q21–22 (Brzustowicz et al. 2000), 1p36 (Abecasis et al. 2004), 6q23 (Lerer et al. 2003), 10p14 (DeLisi et al. 2002), 10q25.3–26.3 (Williams et al. 2003), 13q32 (Blouin et al. 1998). The majority of linkage studies produced LOD scores below the level of genome-wide significance, but may be classified as suggestive evidence (Lander and Kruglyak 1995). These findings are scattered over the whole genome (Fig. 3, Sullivan 2005).

There are two meta-analyses available. Using different approaches, Badner and Gershon (2002) obtained support for loci on 8p, 13q, and 22q, while Lewis et al. (2003) reported the highest evidence for loci on chromosome 2q, 5q, 3p, 11q, 6p, 1q, 22q, 8p, 20q, and 14p.

For some of these regions, gene finding approaches have been applied using methods of association/linkage disequilibrium analysis employing DNA sequence variants (SNPs). Examples include neuregulin on chromosome 8p (Stefansson et al. 2002; 2003), and dysbindin on chromosome 6p (Lewis et al. 2003; Schwab et al. 2003; Straub et al. 2002).

9.2 Genetic Association

In its simplest form, an association study compares allele frequencies of a polymorphism in a sample of individuals affected by a disease with those in a sample of unaffected individuals (controls). Association with the disease phenotype is present when the polymorphism is functionally involved in the expression of the phenotype (direct association), but association can also be caused by linkage disequilibrium between polymorphism and disease locus (indirect association). Linkage disequilibrium appears, when marker and disease locus are located close together, not being separated by recombination over many generations. Studies of association by linkage disequilibrium have expanded enormously since the discovery of SNP's and the development of efficient and reliable SNP detection methods, i.e., allele specific amplification using Taqman- or Amplifluor probes and microarray technology. In 2008, close to 4 million SNPs, validated and tested in different populations, are listed in the HapMap database (http://www.hapmap.org/). Continuous stretches of linkage disequilibrium (haplotypes) have been identified, which can be tagged by single SNPs. Thus, the number of SNPs necessary to cover a region of interest or the whole genome can be reduced by the use of these tag-SNPs.

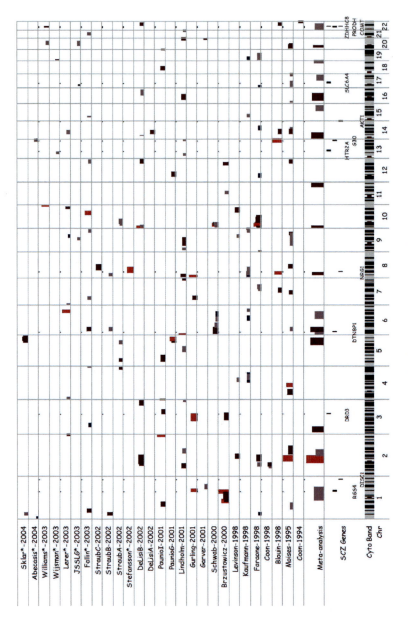

Fig. 3 Genomewide linkage studies of schizophrenia as of July 2005. Linkage findings for 27 samples along with the results of one metaanalysis (Lewis et al. 2003) are summarized in the graph provided by Dr. Patrick F. Sullivan (Sullivan 2005). The height and color of the bars are proportional to the $-\log_{10}$ (P-value). Chromosomal location is given on the bottom together with the position of some selected candidate genes.

Association studies are critically dependent on the quantity and quality of the samples. The number of cases needed are primarily determined by the effect size, which is expected to be small for candidate genes in complex disorders. Consequently samples should contain several hundred individuals for the cases, as well as equal numbers for the controls. Furthermore, a well-defined disease phenotype is essential for the case sample, whilst for individuals of the control sample the disease phenotype should preferentially be absent. In addition, since the allele frequencies of the selected polymorphisms are different in different populations, it is important that case and control samples come from the same ethnical background, in order to avoid detection of population association as opposed to association with the disorder. There are now genetic methods available to identify individuals according to ethnicity. Alternatively, family-based studies may be designed, where the non-transmitted alleles of a polymorphism, detected by genotyping the parents, may serve as internal controls. The Transmission Disequilibrium Test (TDT; Spielman and Ewens 1996) compares the number of alleles transmitted from heterozygous parents with the number of un-transmitted alleles and tests for deviation from 50% expected by chance distribution. However, this approach requires the ascertainment of parents and therefore recruitment may require more effort, time and costs.

At present, four approaches for the identification of susceptibility genes using association analysis are being pursued:

(1) *Based on specific hypotheses* on the pathophysiology, association with the disorder is investigated in candidate genes, selected on the grounds of possible involvement in the development of the disorder or simply representing genes that are involved in organization or function of the brain. This approach has produced a number of ambiguous findings. We will describe evidence for association as well as additional evidence for a possible functional involvement in the development of the illness for some prominent findings (COMT, the dopamine receptors D2 and D3, AKT1).

(2) *Based on linkage results* in family samples (preferentially in families composed of affected sib-pairs and parents), the area defined by linkage is scanned using polymorphic DNA sequence variants for association of candidate genes with the disorder. Genes identified by this approach include neuregulin, dysbindin, RGS4.

(3) *Chromosomal abnormalities* in schizophrenia have been detected only in very few cases, but have been used as the basis for the identification of potential candidate genes (DISC1). Microdeletions, which are associated with psychosis, were detected on chromosome 22q and have been investigated for the presence of candidate genes.

   More recently microarray-based methods have made the detection of microdeletions (Copy Number Variants, CNVs) on a genome-wide level possible.

(4) *Genome-wide screens for association* using a large number of DNA sequence variants distributed at small intervals; this approach has become possible only recently with the availability of SNP genomic arrays containing up to 1 million SNPs on one chip. First studies in schizophrenia research have been published and will be included in this review. Large projects, using case

control samples, with several thousand individuals with schizophrenia are in progress (May 2008).

Replication of association findings

In single gene disorders, a one-to-one relationship between phenotype and genotype usually provides sufficient evidence for identification of a disease-causing gene. In contrast, in complex genetic disorders, a combination of several genes may lead to the disease phenotype, none of these genes being sufficient or necessary. Consequently, these "susceptibility" genes are also present in unaffected individuals. Thus, detection of susceptibility genes is based primarily on statistical support and critically dependent on variables like sample size, gene frequency and genotype relative risk. As a consequence, replication in independent samples is required, in order to support statistical significance. There are several constraints, which have to be considered before successful replication can be claimed. Firstly, phenotypes should be matched as closely as possible. This is, in particular, difficult in psychiatric disorders, where the boundaries are not easy to determine. Secondly, the sample should be larger in size than the sample for the original finding, since effect size is commonly overestimated in the original findings (Ioannidis et al. 2006). Finally, the DNA sequence variants should be identical and genotyping should be of the highest possible standard with a vigorous control for genotyping errors. Guidelines for replication in genetic association studies have been published in connection with the large Wellcome Trust genome-wide association study (Chanock et al. 2007).

Association has been studied for more than 500 genes, ranging from one study up to 100 studies per single gene. A comprehensive list with important details like sample size, diagnosis, type of study (case/control or family based) etc., can be found at the schizophrenia forum web site (http://www.schizophreniaforum.org/).

Whilst some of these association results are convincing, i.e., observed in more than one sample and backed up by meta-analysis, the direct functional involvement of variants in development of schizophrenia has not, to date, been shown convincingly for any of the currently proposed susceptibility genes. Thus, it is likely some of these candidate genes may eventually turn out to be false positives, while others have yet to be discovered.

In the following, we will describe investigations of some potential susceptibility genes in more detail. This selection is biased since it is based on what, in our opinion, are the most promising candidate genes derived from evidence for association with DNA sequence variants, as well as on evidence for functional involvement. Considering the enormous differences in quantity and quality of the samples under investigation as well as the bias against negative studies, the number of positive and negative studies included for the following candidate genes should be interpreted with caution. Moreover, despite converging evidence in some cases, all potential susceptibility genes discovered up to date have to be regarded as preliminary until

Dissecting the Molecular Causes of Schizophrenia 63

a functional impact on molecular mechanisms leading to the illness has been found and confirmed.

9.2.1 Candidate Genes Based on Hypotheses

Dopamine receptor genes

In view of the dopamine hypothesis, the genes encoding the dopamine receptors have been among the first targets for association studies. Despite the large number of studies, a definitive answer to the question of involvement of these genes in the development of schizophrenia cannot be given. Indeed, even a consensus summary or meta-analysis is hard to justify, since studies vary in sample size (from around 50 to several hundreds) and type (case/control, family-based) and are subjected to publication bias. Overall, a majority of published studies are negative, with 49 of 70 association studies listed on the schizophrenia forum website (http://www.schizophreniaforum.org/) for the DRD2 receptor gene and all 11 studies for the DRD1 gene being negative. Fifty-eight negative and 28 positive studies are given for the DRD4 gene, and 41 negative and 8 positive studies are listed for the DRD3 gene.

COMT (Catechol-O-Methyltransferase)

The enzyme plays a role in the breakdown of dopamine. It is conceivable that reduced activity of this enzyme increases dopaminergic activity by a relative enrichment of dopamine, as seen in schizophrenia. Activity of the enzyme in human has been associated with a Val/Met polymorphism (Bertocci et al. 1991), the Met alleles having a lower activity at physiological temperatures (Lotta et al. 1995). Furthermore, there is evidence that COMT is involved in cognitive function and working memory, the high activity Val-COMT being associated with poorer performance and inefficient prefrontal function (reviewed in Tunbridge et al. 2006). Moreover, an association with schizophrenic disorders has been described for the Val/Met polymorphism, but also for haplotypes not including this variant (Shifman et al. 2002). The Schizophrenia Forum (http://www.schizophreniaforum.org/) lists 77 studies, 19 of which report evidence for association.

AKT1 (protein kinase B)

AKT1, a serin/threonine kinase, plays a key role in regulation of intracellular signaling pathways. AKT1 is induced by growth factors, hormones, neurotransmitters, which leads to activation of transcription, cell metabolism, cell growth and survival, and apoptosis. Some of these pathways are involved in synaptic transmission (Kim et al. 2002; Wang et al. 2003) and neuronal plasticity. Convergent evidence for impairment of AKT1 – glycogen-synthasekinase-3 (GSK3) – signaling has been reported by Emamian et al. (2004) pointing to involvement of AKT1 in the pathophysiology of schizophrenia: (1) AKT1 protein levels and GSK3-phosphorylation were decreased in peripheral lymphocytes as well as in hippocampus and frontal cortex of postmortem brains of individuals with schizophrenia. (2) Phosphorylation of AKT1 and GSK3 was increased after haloperidol treatment of mice. (3) Sensitivity to the sensorimotor gating disruptive effect of amphetamine was greater in AKT1 deficient mice, as measured by prepulse inhibition of a startle response, (4) Nominally significant association of SNPs and haplotypes was detected in a family sample.

Since this first report, two positive and three negative family-based association studies have been published (http://www.schizophreniaforum.org/). In addition, six population-based studies of association were published, four with positive and two with negative results, each consisting of more than 300 cases. One of the two negative studies used a large sample of 1,870 cases with schizophrenia (Sanders et al. 2008).

A connection with the dopaminergic signal transmission is suggested by Beaulieu et al. (2004). These authors found that increased dopaminergic neurotransmission, experimentally induced either by amphetamine or by lack of dopamine transporter, resulted in inactivation of AKT.

9.2.2 Candidate Genes Discovered By a Combined Linkage/Association Approach

Neuregulin (NRG1)

The first successful identification of a potential candidate gene using this approach, NRG1, a neurotrophic factor, in a region with evidence for linkage (8p) was published by Stefansson et al. (2002). Association was reported for a single marker SNP8NRG221533 and for a "high risk" haplotype, consisting of seven polymorphisms (Stefansson et al. 2002). Forty-four studies, 26 with positive results, have been published since the initial finding (http://www.schizophreniaforum.org/). A meta-analysis of 27 studies published before September 2007 found that evidence for association with SNP8NRG221533 was non-significant, whilst evidence in support of association of NRG1 with schizophrenia was provided by haplotype based *P*-values (Munafo et al. 2008; Munafo et al. 2006).

Genetics, gene expression and neurobiology of the human NRG1 gene in view of being a possible susceptibility gene have been reviewed by Harrison and Law (2006) and more recently by Mei and Xiong (2008). The genes spans 1.4 megabases, contains more than 20 exons, encodes more than 15 isoforms and generates six types of protein (I-VI). All of these contain a core EGF domain which serves as binding sites for ErbB receptor tyrosine kinases. After dimerisation and autophosphorylation of tyrosine, docking occurs to adaptor proteins, i.e., Shc, Grb2, and the regulatory subunit of phosphoinositide-3-kinase. These interactions activate the MAP kinase and PI3-kinase-pathways, which modulate protein synthesis and transcriptional activity (reviewed in Carpenter 2003). NRG1 signaling has been implicated in a number of neurodevelopmental processes like neuronal migration, axon guidance, synapse formation, oligodendrocyte development as well as in synaptic plasticity and neuronal survival (Mei and Xiong 2008). Thus, NRG1 and the connected pathways appear to be plausible targets in the search for molecular mechanisms conferring susceptibility to schizophrenia. Following investigation of the association, gene expression has been studied in post-mortem brains providing evidence for increased expression (Hashimoto et al. 2004; Law et al. 2004). There is also support for the association of DNA sequence variants in the ERBB4 gene, which is part of a connected pathway (Benzel et al. 2007; Nicodemus et al. 2006; Norton et al. 2006) as well as evidence for altered gene expression in postmortem tissue (Silberberg et al. 2006).

Dystrobrevin binding protein1 (DTNBP1, Dysbindin)

Linkage to chromosome 6p22.3 was first detected in 265 Irish families Straub et al. (1995) and, at the same time, in a sample from Germany consisting of 54 sib-pair families (Schwab et al. 1995). The linked region was analyzed by Straub et al. (2002) using DNA sequence variants for association analysis and resulted in iden- tification of dysbindin. This was immediately confirmed by Schwab et al. (2003) in two independent samples and subsequently by other groups (Kirov et al. 2004; Van Den Bogaert et al. 2003). There are now (May 2008) altogether 20 studies with positive and 23 studies with negative association results listed on the schizophrenia forum website (http://www.schizophreniaforum.org/). None of the associated SNPs or haplotypes seem to be directly involved in conferring risk for the disorder. In addition, the search for "functional" variants is hampered because different alleles and haplotypes were significant in single studies, suggesting allelic heterogeneity even in samples with similar ethnical background.

Recent gene expression studies indicate that altered dysbindin expression is related to schizophrenia pathophysiology. Talbot et al. (2004) reported the pres- ence of the dysbindin protein in cortical and subcortical regions of normal brain. They found reduced presynaptic dysbindin levels (at mRNA and protein level) at hippocampal formation sites, in more than two-thirds of patients with schizophrenia. Their data also suggest that dysbindin reduction is related to glutamatergic alterations in intrinsic hippocampal formation connections and may contribute to the cognitive deficits common in schizophrenia. In addition, Weickert et al. (2004) reported that patients with schizophrenia showed statistically significant reduced dysbindin mRNA levels in multiple layers of the dorsolateral prefrontal cortex (DLPFC). There is also some support for the influence of sequence variation in the untranslated 3´or 5´ regions on gene expression/dysbindin mRNA levels in human cerebral cortex (HCC) (Bray et al. 2003; Weickert et al. 2004), which suggests cis-acting regulatory influences. Bray et al. (2005) showed that a specific haplotype associated with schizophrenia tags one or more cis-acting variants which results in reduced expression of dysbindin in the HCC.

Little is known about the function of dysbindin in neurons. It is believed that dysbindin plays a role in synaptic plasticity and signal transduction. Numakawa et al. (2004) examined the function of dysbindin protein in primary cortical neuronal culture. They focused on the presynaptic machinery in neuronal transmission, as dysbindin is primarily expressed in axonal terminals of mouse brain. Overexpression of protein induced the expression of two pre-synaptic proteins, SNAP25 (a membrane protein involved in intracellular vesicle trafficking and neurotransmitter release) and synapsin I (a cytoskeletal protein, localized to the synaptic vesicle), increased extracellular basal glutamate levels and release of glutamate evoked by potassium. This suggests that dysbindin might be one of the regulatory proteins in excitatory neurotransmission. Overexpression of dysbindin also increased phosphorylation of Akt protein and protected cortical neurons against neuronal death due to serum deprivation. These effects were blocked by LY294002, a phosphatidylinositol 3-kinase (PI3-kinase) inhibitor. Dysbindin may thus promote neuronal viability through PI3-kinase-Akt signaling (Numakawa et al. 2004). Impaired PI3-kinase-Akt signaling has been reported in schizophrenia.

Dysbindin is a component of two functional protein complexes, the dystrophin protein complex (DPC; Benson et al. 2001), which is mainly located in muscles, and postsynaptic at hippocampal formation sites, and the lysosome-related organelles (BLOC-1) complex (Wei 2006), which is ubiquitously expressed. The BLOC-1 protein complex is made up of at least eight proteins. Recently, an interesting connection with the dopaminergic system has been found by studying the effects of transfection of DTNBP1 siRNA on cell surface levels of dopamine D2 receptor (DRD2) in human neuroblastoma cells and in rat primary cortical neurons (Iizuka et al. 2007). Dysbindin protein was decreased, cell surface DRD2 increased, and dopamine-induced DRD2 internalization was blocked, whilst DRD1 levels or internalization was not changed. More DRD2 receptors on the outside of the cells, which appears to be the situation in schizophrenia, would stimulate dopamine-induced intracellular signaling,

Regulator of G-protein signaling4 (RGS4)

RGS4 has been identified in 1q23.3 close to a region (1q21–22), where evidence for significant linkage was previously detected (Brzustowicz et al. 2000). The protein activates a GTPase and accelerates the hydrolysis of G_α-bound GTP (Berman et al. 1996). An association with schizophrenia was detected for several SNPs and haplotypes in a family sample, and replicated in two additional independent samples (Chowdari et al. 2002). The schizophrenia forum database (http://www.schizophreniaforum.org/) now lists 14 positive studies from a total of 28 reports. Reduced RGS4 mRNA expression has been detected in postmortem brain tissue of people with schizophrenia (Mirnics et al. 2001), and this finding has been replicated in two other studies (Bowden et al. 2007; Erdely et al. 2006). In addition, based on correlations between a significantly associated SNP and fMRI results, Buckholtz et al. (2007) concluded that allelic variation is related to structural and functional connectivity in the brain.

Dissecting the Molecular Causes of Schizophrenia 67

9.2.3 Chromosomal Abnormalities

Disrupted in Schizophrenia 1 (DISC1)

A major cytogenetic abnormality lead to the discovery of DISC1. The gene was identified in the breakpoint of a balanced (1;11) chromosomal translocation region, which was co-segregating with psychosis (LOD score 7.1 for broad diagnosis) in one large Scottish family (Millar et al. 2001). A number of studies have been published with evidence for genetic association with different psychiatric disorders (schizophrenia, bipolar disorder, major depression, autism; Callicott et al. 2005; Cannon et al. 2005). The schizophrenia forum (http://www.schizophreniaforum.org/) lists 21 studies in schizophrenia, 13 of which with evidence for association (May 2008). As for most of the candidate genes, association with the DISC1 gene is shown for different DNA sequence polymorphisms and haplotypes. No functional variant associated with the disorder has yet been discovered. A comprehensive review of the genetic data is given in Chubb et al. (2008). The discovery of DISC1 as a potential susceptibility gene for schizophrenia has initiated extensive studies of its involvement in brain development and neuronal function (reviewed in Ross et al. 2006). DISC1 is expressed in many tissues (Millar et al. 2000) and appears to be increased during neuronal development (Schurov et al. 2004). The investigation of protein interaction partners has revealed at least 16 DISC1 interactors (Chubb et al. 2008), which have different cellular functions (Ross et al. 2006). Of particular interest appears to be the interaction between DISC1, NudEL and Lis1, which is connected to Reelin signaling (Brandon et al. 2004). Lis1 mutations cause lisencephaly, a severe disorder of cortical development. Disruption of this complex by a truncated DISC1 may cause schizophrenia as a milder form of cortical disorder (Ross et al. 2006). Other complexes appear to be critical for neuronal migration (dynein, dynaction), cAMP signaling (PDE4B), regulation of synaptic function and synaptic plasticity (Citron), and modulate neurite outgrowth in PC12 cells and cortical neurons (Miyoshi et al. 2003; Ozeki et al. 2003).

22q deletion syndrome (22q11DS)

A microdeletion (approximately 3 Mb) has been identified by phenotype abnormalities on chromosome 22q11 (Karayiorgou et al. 1995). The 22q deletion syndrome occurs in about 1% of people with schizophrenia, but the risk for individuals with 22q11DS to develop schizophrenia is about 25–30% (Bassett et al. 1998; Murphy et al. 1999). The region is supported by several linkage studies and in two meta-analyses (Badner and Gershon 2002; Lewis et al. 2003). Several candidate genes in this region have been supported by association findings in independent samples: COMT (Shifman et al. 2002), PRODH (Liu et al. 2002), ZDHHC8 (Mukai et al. 2004), but these studies have received little support by replication. Recently, strong evidence for GNB1L, located in the region, was obtained for male-specific genotypic association in three independent case-control samples in a sample with 480 schizophrenia parent-proband trios as well as a significant allelic association with psychosis in males in a sample of 83 subjects with the 22q deletion syndrome (Williams et al. 2008). GNB1L encodes a G-protein beta-subunit-like polypeptide without homology to known proteins. It contains WD40 repeats which are known to facilitate heterotrimeric or multiprotein complexes (Williams et al. 2008).

Copy Number Variants (CNVs)

Microarray-based methods have made the genome-wide analysis of large-scale copy number variations possible. The current estimate is that around 10% of the genome is subjected to microduplications or microdeletions (Beckmann et al. 2007). This has led to a change in the estimate of interindividual genomic variation, which was thought to be around 0.1% based on SNP variation, but has been revised to approximately 1% when CNVs are taken into account. In contrast to SNP association studies, which are designed to detect small effects which increase population risk, microdeletions and microduplications are detected in individuals and represent rare mutations, conferring high risk for individuals. There are now several reports on genome-wide studies of association of CNVs with schizophrenia. Kirov et al. (2008a) analyzed 93 individuals with schizophrenia in comparison to a reference data set of 322 controls. A disruption of neurexin1 (NRXN1) and amyloid precursor-binding protein (APBA2), which play a role in synaptic development and function, have been detected. Both genes have also been detected by CNV analysis in patients with autism and in patients with mental retardation. Walsh et al. (2008) investigated CNVs in a sample of 150 individuals with schizophrenia and 268 controls as well as 92 patients with childhood-onset schizophrenia. Novel microdeletions and microduplications were present in 5% of controls, 15% of cases overall and 20% of young-onset cases. Genes located in these deletions and duplications were significantly overrepresented in pathways for brain development, i.e., neuregulin, ERK/MAP, and glutamate receptor signaling pathways.

Xu et al. (2008) reported a strong association of de novo CNVs with sporadic schizophrenia. A sample of 200 individuals with schizophrenia from the Afrikaner population in South Africa was divided in 152 sporadic cases, i.e., no history of disorder in first or second degree relatives, and 48 familial cases, Fifteen rare, de novo, CNVs (not present in parents) were found to be associated with sporadic cases, nil with familial cases ($P = 0.02$) and only two were detected in 159 controls ($P = 0.00078$). Interestingly, three of the de novo CNVs were in 22q11,21, the region of the 22q11DS microdeletion.

Microdeletions and microduplications are rare, except in the 22q11DS region; each of the reported CNVs occurred only once in the sample. Nevertheless detection of genes in these regions would provide candidates which can be analyzed for their impact on development of schizophrenia in general.

Dissecting the Molecular Causes of Schizophrenia

9.2.4 Genome-Wide Association Studies (GWA)

The GWA approach has been applied to a number of common disorders and has produced highly significant results, mainly by using samples of several thousands patients and controls (The Wellcome Trust Case Control Consortium 2007). While large samples of patients with bipolar disorder have been already included in the Wellcome Trust study, GWA studies with large schizophrenia samples are in preparation (May 2008).

In a first GWA study, DNA preparations of 178 patients with schizophrenia and 144 controls were hybridized with 500,000 polymorphisms using chips technology (Lencz et al. 2007). One SNP (rs4129148) located in the pseudoautosomal region at the X- and Y-chromosome reached genomewide significance ($P = 3.7 \times 10^{-7}$). The polymorphism is in the neighborhood of the colony stimulating factor, receptor 2 alpha (CSF 2 RA).

Two other GWA studies used DNA pooling. In a sample of 600 Ashkenazy Jews with schizophrenia (265 females), Shifman et al. (2008) reported the lowest P-value ($P = 2.9 \times 10^{-5}$) for SNP rs7341475 in the 4 intron of the gene for reelin (restricted to females). Four additional samples were genotyped for replication. The same genotypes and alleles were overrepresented, but statistical significance was obtained in only one of the replication samples. The reelin protein is involved in corticogenesis and is a candidate for the neurodevelopmental hypothesis of schizophrenia. Kirov et al. (2008b) conducted a GWA study with 574 parent-offspring trios using DNA pooling and hybridization on a Illumina HumanHap 550 array. The most promising SNPs (cutoff < 0.001) were individually genotyped. Nominally significant P-values were obtained for SNPs in the CCDC60 gene, a coiled-coil domain gene ($P = 1.2 \times 10^{-6}$) and in RBP1, a cellular retinol binding protein ($P = 1.6 \times 10^{-4}$).

Sullivan et al. (2008) reported a GWA study with 738 cases and 733 matched controls using a 500 K two-chip genotyping platform and a custom fill-in chip. The study has been considered as stage I, preceding independent GWA studies with samples of several thousands of patients and controls, which are currently being developed (May 2008). The highest P-value (1.85×10^{-6}) failed to reach genomewide significance. There was no overlap with the previous GWASs, but some clustering of uncorrected P-values was seen in the area of previously reported candidate genes, i.e., COMT and DISC1.

10 Gene Expression

Genetic association of schizophrenia with DNA sequence variation or microdeletions and microduplications is in general not influenced by non-genetic factors. In contrast, transcription into mRNA and translation into polypeptides may be regulated by mechanisms responding to influences of developmental state, environment, and/or acquired phenotype (education, disease, emotion). Studying gene expression at the RNA or protein level is important for dissecting complex genetic disorders for which quantitative influences are assumed. However, studies on gene expression may be confounded by unspecific environmental influences or state variables, which are not connected to the etiology of the illness

10.1 mRNA Expression

Expression studies on single candidate genes, either based on a pathophysiology-based hypothesis or on evidence for association with DNA sequence variants have been performed using postmortem brain tissue or peripheral cells (lymphoblast or fibroblast cell cultures) as a surrogate. As in genetic association studies, sufficient evidence for single genes has not been attained. The availability of microarrays with a genome-wide coverage has facilitated non-hypothesis-based expression profiling. Usually a large number of up- and downregulated genes are found. These studies have provided evidence for the involvement of pathways, for example, synaptic function (Middleton et al. 2002; Vawter et al. 2002), myelination related genes (Hakak et al. 2001), and oligodendrocyte function (Tkachev et al. 2003). Using chip technology for genome-wide SNP detection as well as for genome-wide gene expression, it has become possible to map DNA sequence polymorphisms involved in control of gene expression (Dixon et al. 2007; Goring et al. 2007; Myers et al. 2007). These maps are likely to prove useful in identification of disease-associated expression quantitative trait loci (eQTL).

10.2 miRNA

Gene expression can be modulated by a recently discovered novel class of small, non-coding RNA molecules, miRNAs (reviewed in He and Hannon 2004). miRNAs bind to specific regions in target mRNAs controlling mRNA stability and translation. There are about 500 miRNAs known and 800 predicted, each controlling up to hundreds of genes. miRNAs are believed to play a critical role in brain development and neuronal plasticity processes involved in the etiology of schizophrenia. Using miRNA microarrays, Perkins et al. (2007) compared the expression of 264 miRNAs from postmortem prefrontal cortex tissue of 15 individuals with schizophrenia with tissue of 21 unaffected individuals as control. Sixteen miRNAs were found to be differentially expressed, with 15 at lower levels. Beveridge et al. (2008) using a similar approach obtained evidence for upregulation of two miRNAs by comparing the expression pattern of 21 individuals with schizophrenia and carefully matched controls in postmortem cortical grey matter from the superior temporal gyrus. The finding has been validated and confirmed by quantitative real-time PCR analysis. The two miRNA regulate targets (calcium sensor gene visinin-like 1 = VSNL1 and ionotropic AMPA glutamate receptor subunit, GRIA2), for which altered mRNA expression has been shown in previous microarray experiments. The finding was further supported by (1) decreased expression of the two genes in the same tissue, (2) both genes were suppressed in cells transfected with the upregulated miRNA, and (3) both genes contained functional miRNA recognition sites as shown by reporter gene assay.

Additional support for the involvement of miRNAs in schizophrenia comes from a case-control association study between schizophrenia and genetic variants in miRNA genes associated with brain expression (Hansen et al. 2007). Two SNPs revealed nominal significance in two of the three samples studied.

10.3 Protein Expression

The final goal in molecular studies is the identification of abnormalities in structure or function of proteins (enzymes, receptors, etc.) involved in development of schizophrenia. Proteins are usually investigated based on hypotheses on pathophysiology or following evidence obtained by molecular genetic studies. Protein expression levels as well as possible structural alterations can be studied by electrophoresis and detection with specific antibodies using Western blotting. Advances in proteomics, in particular the application of standardized 2D electrophoresis techniques in combination with mass spectrometry, have made analysis on the level of proteome possible and may be applied to postmortem tissue, CSF, and peripheral cells as brain surrogate. Considering the large number of protein molecules generated by transcriptional (splicing) and posttranslational modification, which considerably exceed the number of genes, as well as the difficulties in obtaining well-characterized and matched postmortem tissue, it may be difficult to differentiate between alterations involved in development of the disease from those caused by the disease. It is hoped that the approach will lead to the discovery of biomarkers which differentiate between trait and state factors. Biomarkers will be useful for diagnosis and monitoring treatment response (Schwarz and Bahn 2008).

10.4 Epigenetics

Gene expression is mainly regulated by transcription factors, miRNA silencing and RNA splicing, which may be influenced by permanent DNA sequence variants within the genome. However, gene expression may also be affected by epigenetic processes like methylation of Cytosine residues in CpG islands of promoter regions or by chromatin remodeling induced by histone methylation or acetylation (Duman and Newton 2007). Differential DNA methylation has been reported for COMT (Abdolmaleky et al. 2005) and Reelin (Grayson et al. 2005). Reelin is involved in cell–cell interactions and neuronal migration during brain development. However, hypermethylation of the Reelin promoter region could not be confirmed by Tochigi et al. (2008) and Mill et al. (2008). An increase of the methyl donor S-Adenosylmethionine associated with over-expression of DNA Methyltransferase I mRNA in schizophrenia and bipolar disorder is also reported (Guidotti et al. 2007). Recently, a genome-wide epigenomic study has been undertaken (Mill et al. 2008) using CpG island microarrays for detection of the methylation status in the frontal cortex. Tissue of people with schizophrenia and of matched controls was provided by the Stanley Medical Research Institute brain-array collection. Decreased epigenetic modularity was uncovered by methylome network analysis. In addition, a strong correlation between lower DNA methylation in the promoter region of the gene encoding mitogen-activated protein kinase I (MEK1) was detected in males with schizophrenia and lifetime antipsychotic use. However, the possibility cannot be excluded that the observed changes are induced by the illness rather than representing causes for the disorder.

11 Animal Models

The establishment of animal models for schizophrenia would undoubtedly help in understanding neurobiology of the disorder, as well as aid in the development of new pharmacological treatments. However, heterogeneity, lack of knowledge of neuropathology, the fact that some symptoms of schizophrenia (hallucinations, delusions etc.) are exclusively detectable in humans, and most crucial, that diagnosis is dependent on language-based communication, render a "mouse model for schizophrenia" unrealistic. Nevertheless, there are some behavioral aspects of schizophrenia which might be modulated in animals (for a comprehensive review, see Arguello and Gogos 2006). The usefulness of such tests depends on the characterization of the particular phenotype to be modeled in terms of specificity and contribution to the complete schizophrenia phenotype. This characterization should follow basically the definition of endophenotypes as summarized by Gottesman and Gould (2003). Endophenotypes are a function of some features included in the phenotype, reducing its complexity. In schizophrenia, for example, sensory motor gating (measured as prepulse inhibition), eye-tracking dysfunction, various cognitive measurements (working memory, executive cognition) have all been proposed as potential endophenotypes. The requirements for an endophenotype are an association with the illness in populations, heritability, state independence (not influenced by medication, psychosis, other conditions), and co-segregation in families. Whether the presently proposed schizophrenia endophenotypes meet these criteria is still under investigation. A comprehensive updated list of animal models is listed on the schizophrenia research forum website (http://www.schizophreniaforum.org/) and includes assessed phenotypes like dopamine-related behavior (activity), gating (prepulse inhibition), cognitive performance (learning and memory), social behavior, molecular/morphological signature, and response to antipsychotic drugs (APD).

Dissecting the Molecular Causes of Schizophrenia 73

12 Conclusion

Molecular biology and molecular genetics have provided methods and information that have been the basis for expanding our knowledge about the functioning of the human brain. This knowledge has boosted the research into the molecular causes of schizophrenia enormously. A large number of candidates (genes, proteins, regulatory mechanisms, pathways) have been identified, but for none of them is the evidence sufficient to draw definite clues. Given the multifactorial, oligo- or polygenic nature of schizophrenia, this is not surprising. Nevertheless, there is still optimism that the accumulation of existing and future data will lead finally to the discovery of molecular causes for this devastating brain disorder.

Acknowledgments We wish to express our sincere gratitude to Professor Dr Patrick F. Sullivan, Departments of Genetics, University of North Carolina, USA, for kindly providing Figs. 1 and 3. We are thankful to Dr. Daniel Rock for critical reading of the manuscript.

References

Abdolmaleky HM, Cheng KH, Russo A, Smith CL, Faraone SV, Wilcox M, Shafa R, Glatt SJ, Nguyen G, Ponte JF et al (2005). Hypermethylation of the reelin (RELN) promoter in the brain of schizophrenic patients: a preliminary report. Am J Med Genet B Neuropsychiatr Genet 134B:60–66

Abecasis GR, Burt RA, Hall D, Bochum S, Doheny KF, Lundy SL, Torrington M, Roos JL, Gogos JA, Karayiorgou M (2004). Genomewide scan in families with schizophrenia from the founder population of Afrikaners reveals evidence for linkage and uniparental disomy on chromosome 1. Am J Hum Genet 74:403–417

Abi-Dargham A 2006. The dopamine hypothesis of schizophrenia. Schizophrenia research forum Available at: http://wwwschizophreniaforumorg/for/curr/Moghaddam/defaultasp

Abi-Dargham A, Mawlawi O, Lombardo I, Gil R, Martinez D, Huang Y, Hwang DR, Keilp J, Kochan L, Van Heertum R et al (2002). Prefrontal dopamine D1 receptors and working memory in schizophrenia. J Neurosci 22:3708–3719

American Psychiatric Association (1994). Diagnostic and Statistical Manual of Mental Disorders, 4th edition. American Psychiatric Press, Washington, DC

Angrist BM, Gershon S (1970). The phenomenology of experimentally induced amphetamine psychosis–preliminary observations. Biol Psychiatry 2:95–107

Arguello PA, Gogos JA (2006). Modeling madness in mice: one piece at a time. Neuron 52:179–196

Badner JA, Gershon ES (2002). Meta-analysis of whole-genome linkage scans of bipolar disorder and schizophrenia. Mol Psychiatry 7:405–411

Bassett AS, Hodgkinson K, Chow EW, Correia S, Scutt LE, Weksberg R (1998). 22q11 deletion syndrome in adults with schizophrenia. Am J Med Genet 81:328–337

Beaulieu JM, Sotnikova TD, Yao WD, Kockeritz L, Woodgett JR, Gainetdinov RR, Caron MG (2004). Lithium antagonizes dopamine-dependent behaviors mediated by an AKT/glycogen synthase kinase 3 signaling cascade. Proc Natl Acad Sci U S A 101:5099–5104

Beckmann JS, Estivill X, Antonarakis SE (2007). Copy number variants and genetic traits: closer to the resolution of phenotypic to genotypic variability. Nat Rev Genet 8:639–646

Benson MA, Newey SE, Martin-Rendon E, Hawkes R, Blake DJ (2001). Dysbindin, a novel coiled-coil-containing protein that interacts with the dystrobrevins in muscle and brain. J Biol Chem 276:24232–24241

Benzel I, Bansal A, Browning BL, Galwey NW, Maycox PR, McGinnis R, Smart D, St Clair D, Yates P, Purvis I (2007). Interactions among genes in the ErbB-neuregulin signalling network are associated with increased susceptibility to schizophrenia. Behav Brain Funct 3:31

Berman DM, Wilkie TM, Gilman AG (1996). GAIP and RGS4 are GTPase-activating proteins for the Gi subfamily of G protein alpha subunits. Cell 86:445–452

Bertocci B, Miggiano V, Da Prada M, Dembic Z, Lahm HW, Malherbe P (1991). Human catechol-O-methyltransferase: cloning and expression of the membrane-associated form. Proc Natl Acad Sci U S A 88:1416–1420

Beveridge NJ, Tooney PA, Carroll AP, Gardiner E, Bowden N, Scott RJ, Tran N, Dedova I, Cairns MJ (2008). Dysregulation of miRNA 181b in the temporal cortex in schizophrenia. Hum Mol Genet 17:1156–1168

Bird ED, Spokes EG, Barnes J, Mackay AV, Iversen LL, Shepherd M (1978). Glutamic-acid decarboxylase in schizophrenia. Lancet 1:156

Blouin JL, Dombroski BA, Nath SK, Lasseter VK, Wolyniec PS, Nestadt G, Thornquist M, Ullrich G, McGrath J, Kasch L et al (1998). Schizophrenia susceptibility loci on chromosomes 13q32 and 8p21. Nat Genet 20:70–73

Bowden NA, Scott RJ, Tooney PA (2007). Altered expression of regulator of G-protein signalling 4 (RGS4) mRNA in the superior temporal gyrus in schizophrenia. Schizophr Res 89:165–168

Brandon NJ, Handford EJ, Schurov I, Rain JC, Pelling M, Duran-Jimeniz B, Camargo LM, Oliver KR, Beher D, Shearman MS et al (2004). Disrupted in schizophrenia 1 and nudel form a neurodevelopmentally regulated protein complex: implications for schizophrenia and other major neurological disorders. Mol Cell Neurosci 25:42–55

Bray NJ, Buckland PR, Owen MJ, O'Donovan MC (2003). Cis-acting variation in the expression of a high proportion of genes in human brain. Hum Genet 113:149–153

Bray NJ, Preece A, Williams NM, Moskvina V, Buckland PR, Owen MJ, O'Donovan MC (2005). Haplotypes at the dystrobrevin binding protein 1 (DTNBP1) gene locus mediate risk for schizophrenia through reduced DTNBP1 expression. Hum Mol Genet 14:1947–1954

Brzustowicz LM, Hodgkinson KA, Chow EW, Honer WG, Bassett AS (2000). Location of a major susceptibility locus for familial schizophrenia on chromosome 1q21-q22. Science 288:678–682

Buckholtz JW, Meyer-Lindenberg A, Honea RA, Straub RE, Pezawas L, Egan MF, Vakkalanka R, Kolachana B, Verchinski BA, Sust S et al (2007). Allelic variation in RGS4 impacts functional and structural connectivity in the human brain. J Neurosci 27:1584–1593

Buka SL, Tsuang MT, Torrey EF, Klebanoff MA, Bernstein D, Yolken RH (2001). Maternal infections and subsequent psychosis among offspring. Arch Gen Psychiatry 58:1032–1037

Callicott JH, Straub RE, Pezawas L, Egan MF, Mattay VS, Hariri AR, Verchinski BA, Meyer-Lindenberg A, Balkissoon R, Kolachana B et al (2005). Variation in DISC1 affects hippocampal structure and function and increases risk for schizophrenia. Proc Natl Acad Sci U S A 102:8627–8632

Cannon M, Jones PB, Murray RM (2002). Obstetric complications and schizophrenia: historical and meta-analytic review. Am J Psychiatry 159:1080–1092

Cannon TD, Hennah W, van Erp TG, Thompson PM, Lonnqvist J, Huttunen M, Gasperoni T, Tuulio-Henriksson A, Pirkola T, Toga AW et al (2005). Association of DISC1/TRAX haplotypes with schizophrenia, reduced prefrontal gray matter, and impaired short- and long-term memory. Arch Gen Psychiatry 62:1205–1213

Cantor-Graae E, Selten JP (2005). Schizophrenia and migration: a meta-analysis and review. Am J Psychiatry 162:12–24

Carpenter G (2003). ErbB-4: mechanism of action and biology. Exp Cell Res 284:66–77

Chanock SJ, Manolio T, Boehnke M, Boerwinkle E, Hunter DJ, Thomas G, Hirschhorn JN, Abecasis G, Altshuler D, Bailey-Wilson JE et al (2007). Replicating genotype-phenotype associations. Nature 447:655–660

Chowdari KV, Mirnics K, Semwal P, Wood J, Lawrence E, Bhatia T, Deshpande SNBKT, Ferrell RE, Middleton FA et al (2002). Association and linkage analyses of RGS4 polymorphisms in schizophrenia. Hum Mol Genet 11:1373–1380

Dissecting the Molecular Causes of Schizophrenia

Chubb JE, Bradshaw NJ, Soares DC, Porteous DJ, Millar JK (2008). The DISC locus in psychiatric illness. Mol Psychiatry 13:36–64

Clinton SM, Meador-Woodruff JH (2004). Abnormalities of the NMDA receptor and associated intracellular molecules in the thalamus in schizophrenia and bipolar disorder. Neuropsychopharmacology 29:1353–1362

Coyle JT (2006). Glutamate and schizophrenia: beyond the dopamine hypothesis. Cell Mol Neurobiol 26:365–384

Creese I, Burt DR, Snyder SH (1976). Dopamine receptors and average clinical doses. Science 194:546

Crow TJ, Ball J, Bloom SR, Brown R, Bruton CJ, Colter N, Frith CD, Johnstone EC, Owens DG, Roberts GW (1989). Schizophrenia as an anomaly of development of cerebral asymmetry. A postmortem study and a proposal concerning the genetic basis of the disease. Arch Gen Psychiatry 46:1145–1150

Delay J, Deniker P, Harl JM (1952). Therapeutic use in psychiatry of phenothiazine of central elective action (4560 RP). Ann Med Psychol (Paris) 110:112–117

DeLisi LE, Shaw SH, Crow TJ, Shields G, Smith AB, Larach VW, Wellman N, Loftus J, Nanthakumar B, Razi K et al (2002). A genome-wide scan for linkage to chromosomal regions in 382 sibling pairs with schizophrenia or schizoaffective disorder. Am J Psychiatry 159:803–812

Dixon AL, Liang L, Moffatt MF, Chen W, Heath S, Wong KC, Taylor J, Burnett E, Gut I, Farrall M et al (2007). A genome-wide association study of global gene expression. Nat Genet 39:1202–1207

Duman RS, Newton SS (2007). Epigenetic marking and neuronal plasticity. Biol Psychiatry 62:1–3

Emamian ES, Hall D, Birnbaum MJ, Karayiorgou M, Gogos JA (2004). Convergent evidence for impaired AKT1-GSK3beta signaling in schizophrenia. Nat Genet 36:131–137

Erdely HA, Tamminga CA, Roberts RC, Vogel MW (2006). Regional alterations in RGS4 protein in schizophrenia. Synapse 59:472–479

Farde L, Nordstrom AL, Halldin C, Wiesel FA, Sedvall G (1992). PET studies of dopamine receptors in relation to antipsychotic drug treatment. Clin Neuropharmacol 15 Suppl 1 Pt A:468A–469A

Goring HH, Curran JE, Johnson MP, Dyer TD, Charlesworth J, Cole SA, Jowett JB, Abraham LJ, Rainwater DL, Comuzzie AG et al (2007). Discovery of expression QTLs using large-scale transcriptional profiling in human lymphocytes. Nat Genet 39:1208–1216

Gottesman II, Gould TD (2003). The endophenotype concept in psychiatry: etymology and strategic intentions. Am J Psychiatry 160:636–645

Gottesman I (1991). Schizophrenia Genesis: The origins of madness. Freeman, New York

Grayson DR, Jia X, Chen Y, Sharma RP, Mitchell CP, Guidotti A, Costa E (2005). Reelin promoter hypermethylation in schizophrenia. Proc Natl Acad Sci U S A 102:9341–9346

Guidotti A, Ruzicka W, Grayson DR, Veldic M, Pinna G, Davis JM, Costa E (2007). S-adenosyl methionine and DNA methyltransferase-1 mRNA overexpression in psychosis. Neuroreport 18:57–60

Hakak Y, Walker JR, Li C, Wong WH, Davis KL, Buxbaum JD, Haroutunian V, Fienberg AA (2001). Genome-wide expression analysis reveals dysregulation of myelination-related genes in chronic schizophrenia. Proc Natl Acad Sci U S A 98:4746–4751

Hanada S, Mita T, Nishino N, Tanaka C (1987). [3H]muscimol binding sites increased in autopsied brains of chronic schizophrenics. Life Sci 40:259–266

Hansen T, Olsen L, Lindow M, Jakobsen KD, Ullum H, Jonsson E, Andreassen OA, Djurovic S, Melle I, Agartz I et al (2007). Brain expressed microRNAs implicated in schizophrenia etiology. PLoS One 2:e873

Harrison PJ, Law AJ (2006). Neuregulin 1 and schizophrenia: genetics, gene expression, and neurobiology. Biol Psychiatry 60:132–140

Harrison PJ, Owen MJ (2003). Genes for schizophrenia? Recent findings and their pathophysiological implications. Lancet 361:417–419

Hashimoto R, Straub RE, Weickert CS, Hyde TM, Kleinman JE, Weinberger DR (2004). Expression analysis of neuregulin-1 in the dorsolateral prefrontal cortex in schizophrenia. Mol Psychiatry 9:299–307

He L, Hannon GJ (2004). MicroRNAs: small RNAs with a big role in gene regulation. Nat Rev Genet 5:522–531

Iizuka Y, Sei Y, Weinberger DR, Straub RE (2007). Evidence that the BLOC-1 protein dysbindin modulates dopamine D2 receptor internalization and signaling but not D1 internalization. J Neurosci 27:12390–12395

Ioannidis JP, Trikalinos TA, Khoury MJ (2006). Implications of small effect sizes of individual genetic variants on the design and interpretation of genetic association studies of complex diseases. Am J Epidemiol 164:609–614

Jablensky A, Sartorius N, Ernberg G, Anker M, Korten A, Cooper JE, Day R, Bertelsen A (1992). Schizophrenia: manifestations, incidence and course in different cultures. A World Health Organization ten-country study. Psychol Med Monogr Suppl 20:1–97

Jakob H, Beckmann H (1986). Prenatal developmental disturbances in the limbic allocortex in schizophrenics. J Neural Transm 65:303–326

Javitt DC, Zukin SR (1991). Recent advances in the phencyclidine model of schizophrenia. Am J Psychiatry 148:1301–1308

Karayiorgou M, Morris MA, Morrow B, Shprintzen RJ, Goldberg R, Borrow J, Gos A, Nestadt G, Wolyniec PS, Lasseter VK et al (1995). Schizophrenia susceptibility associated with interstitial deletions of chromosome 22q11. Proc Natl Acad Sci U S A 92:7612–7616

Kety SS, Rosenthal D, Wender PH, Schulsinger F (1971). Mental illness in the biological and adoptive families of adpoted schizophrenics. Am J Psychiatry 128:302–306

Kim AH, Yano H, Cho H, Meyer D, Monks B, Margolis B, Birnbaum MJ, Chao MV (2002). Akt1 regulates a JNK scaffold during excitotoxic apoptosis. Neuron 35:697–709

Kim JS, Kornhuber HH, Schmid-Burgk W, Holzmuller B (1980). Low cerebrospinal fluid glutamate in schizophrenic patients and a new hypothesis on schizophrenia. Neurosci Lett 20:379–382

Kirov G, Gumus D, Chen W, Norton N, Georgieva L, Sari M, O'Donovan MC, Erdogan F, Owen MJ, Ropers HH et al (2008a). Comparative genome hybridization suggests a role for NRXN1 and APBA2 in schizophrenia. Hum Mol Genet 17:458–465

Kirov G, Ivanov D, Williams NM, Preece A, Nikolov I, Milev R, Koleva S, Dimitrova A, Toncheva D, O'Donovan MC et al (2004). Strong evidence for association between the dystrobrevin binding protein 1 gene (DTNBP1) and schizophrenia in 488 parent-offspring trios from Bulgaria. Biol Psychiatry 55:971–975

Kirov G, Zaharieva I, Georgieva L, Moskvina V, Nikolov I, Cichon S, Hillmer A, Toncheva D, Owen MJ, O'Donovan MC (2008b). A genome-wide association study in 574 schizophrenia trios using DNA pooling. Mol Psychiatry, epub ahead of print March 11

Lander E, Kruglyak L (1995). Genetic dissection of complex traits: guidelines for interpreting and reporting linkage results. Nat Genet 11:241–247

Law AJ, Shannon Weickert C, Hyde TM, Kleinman JE, Harrison PJ (2004). Neuregulin-1 (NRG-1) mRNA and protein in the adult human brain. Neuroscience 127:125–136

Lencz T, Morgan TV, Athanasiou M, Dain B, Reed CR, Kane JM, Kucherlapati R, Malhotra AK (2007). Converging evidence for a pseudoautosomal cytokine receptor gene locus in schizophrenia. Mol Psychiatry 12:572–580

Lerer B, Segman RH, Hamdan A, Kanyas K, Karni O, Kohn Y, Korner M, Lanktree M, Kaadan M, Turetsky N et al (2003). Genome scan of Arab Israeli families maps a schizophrenia susceptibility gene to chromosome 6q23 and supports a locus at chromosome 10q24. Mol Psychiatry 8:488–498

Lewis CM, Levinson DF, Wise LH, DeLisi LE, Straub RE, Hovatta I, Williams NM, Schwab SG, Pulver AE, Faraone SV et al (2003). Genome scan meta-analysis of schizophrenia and bipolar disorder, part II: Schizophrenia. Am J Hum Genet 73:34–48

Liu H, Heath SC, Sobin C, Roos JL, Galke BL, Blundell ML, Lenane M, Robertson B, Wijsman EM, Rapoport JL et al (2002). Genetic variation at the 22q11 PRODH2/DGCR6 locus presents an unusual pattern and increases susceptibility to schizophrenia. Proc Natl Acad Sci U S A 99:3717–3722

Lotta T, Vidgren J, Tilgmann C, Ulmanen I, Melen K, Julkunen I, Taskinen J (1995). Kinetics of human soluble and membrane-bound catechol O-methyltransferase: a revised mechanism and description of the thermolabile variant of the enzyme. Biochemistry 34:4202–4210

Dissecting the Molecular Causes of Schizophrenia 77

McGrath J, Saha S, Welham J, El Saadi O, MacCauley C, Chant D (2004). A systematic review of the incidence of schizophrenia: the distribution of rates and the influence of sex, urbanicity, migrant status and methodology. BMC Med 2:13

McNeil TF, Cantor-Graae E, Weinberger DR (2000). Relationship of obstetric complications and differences in size of brain structures in monozygotic twin pairs discordant for schizophrenia. Am J Psychiatry 157:203–212

Mei L, Xiong WC (2008). Neuregulin 1 in neural development, synaptic plasticity and schizophrenia. Nat Rev Neurosci 9:437–452

Middleton FA, Mirnics K, Pierri JN, Lewis DA, Levitt P (2002). Gene expression profiling reveals alterations of specific metabolic pathways in schizophrenia. J Neurosci 22:2718–2729

Mill J, Tang T, Kaminsky Z, Khare T, Yazdanpanah S, Bouchard L, Jia P, Assadzadeh A, Flanagan J, Schumacher A et al (2008). Epigenomic profiling reveals DNA-methylation changes associated with major psychosis. Am J Hum Genet 82:696–711

Millar JK, Christie S, Anderson S, Lawson D, Hsiao-Wei Loh D, Devon RS, Arveiler B, Muir WJ, Blackwood DH, Porteous DJ (2001). Genomic structure and localisation within a linkage hotspot of Disrupted In Schizophrenia 1, a gene disrupted by a translocation segregating with schizophrenia. Mol Psychiatry 6:173–178

Millar JK, Wilson-Annan JC, Anderson S, Christie S, Taylor MS, Semple CA, Devon RS, Clair DM, Muir WJ, Blackwood DH et al (2000). Disruption of two novel genes by a translocation co-segregating with schizophrenia. Hum Mol Genet 9:1415–1423

Mirnics K, Middleton FA, Stanwood GD, Lewis DA, Levitt P (2001). Disease-specific changes in regulator of G-protein signaling 4 (RGS4) expression in schizophrenia. Mol Psychiatry 6:293–301

Miyoshi K, Honda A, Baba K, Taniguchi M, Oono K, Fujita T, Kuroda S, Katayama T, Tohyama M (2003). Disrupted-In-Schizophrenia 1, a candidate gene for schizophrenia, participates in neurite outgrowth. Mol Psychiatry 8:685–694

Moghaddam B (2003). Bringing order to the glutamate chaos in schizophrenia. Neuron 40:881–884

Moghaddam B 2005. Glutamate Hypothesis of Schizophrenia In: Forum SR (ed) http: //wwwschizophren iaforumorg/for/curr/Moghaddam/defaultasp.

Mortensen PB, Pedersen CB, Westergaard T, Wohlfahrt J, Ewald H, Mors O, Andersen PK, Melbye M (1999). Effects of family history and place and season of birth on the risk of schizophrenia. N Engl J Med 340:603–608

Mukai J, Liu H, Burt RA, Swor DE, Lai WS, Karayiorgou M, Gogos JA (2004). Evidence that the gene encoding ZDHHC8 contributes to the risk of schizophrenia. Nat Genet 36:725–731

Munafo MR, Attwood AS, Flint J (2008). Neuregulin 1 genotype and schizophrenia. Schizophr Bull 34:9–12

Munafo MR, Thiselton DL, Clark TG, Flint J (2006). Association of the NRG1 gene and schizophrenia: a meta-analysis. Mol Psychiatry 11:539–546

Murphy KC, Jones LA, Owen MJ (1999). High rates of schizophrenia in adults with velo-cardiofacial syndrome. Arch Gen Psychiatry 56:940–945

Myers AJ, Gibbs JR, Webster JA, Rohrer K, Zhao A, Marlowe L, Kaleem M, Leung D, Bryden L, Nath P et al (2007). A survey of genetic human cortical gene expression. Nat Genet 39:1494–1499

Nicodemus KK, Luna A, Vakkalanka R, Goldberg T, Egan M, Straub RE, Weinberger DR (2006). Further evidence for association between ErbB4 and schizophrenia and influence on cognitive intermediate phenotypes in healthy controls. Mol Psychiatry 11:1062–1065

Norton N, Moskvina V, Morris DW, Bray NJ, Zammit S, Williams NM, Williams HJ, Preece AC, Dwyer S, Wilkinson JC et al (2006). Evidence that interaction between neuregulin 1 and its receptor erbB4 increases susceptibility to schizophrenia. Am J Med Genet B Neuropsychiatr Genet 141B:96–101

Numakawa T, Yagasaki Y, Ishimoto T, Okada T, Suzuki T, Iwata N, Ozaki N, Taguchi T, Tatsumi M, Kamijima K et al (2004). Evidence of novel neuronal functions of dysbindin, a susceptibility gene for schizophrenia. Hum Mol Genet 13:2699–2708

Owen F, Cross AJ, Crow TJ, Longden A, Poulter M, Riley GJ (1978). Increased dopamine-receptor sensitivity in schizophrenia. Lancet 2:223–226

Ozeki Y, Tomoda T, Kleiderlein J, Kamiya A, Bord L, Fujii K, Okawa M, Yamada N, Hatten ME, Snyder SH et al (2003). Disrupted-in-schizophrenia-1 (DISC-1): mutant truncation prevents binding to NudE-like (NUDEL) and inhibits neurite outgrowth. Proc Natl Acad Sci U S A 100:289–294

Perkins DO, Jeffries CD, Jarskog LF, Thomson JM, Woods K, Newman MA, Parker JS, Jin J, Hammond SM (2007). MicroRNA expression in the prefrontal cortex of individuals with schizophrenia and schizoaffective disorder. Genome Biol 8:R27

Reynolds GP, Czudek C, Andrews HB (1990). Deficit and hemispheric asymmetry of GABA uptake sites in the hippocampus in schizophrenia. Biol Psychiatry 27:1038–1044

Riley B (2004). Linkage studies of schizophrenia. Neurotox Res 6:17–34

Rosenthal D, Wender PH, Kety SS, Welner J, Schulsinger F (1971). The adopted-away offspring of schizophrenics. Am J Psychiatry 128:307–311

Ross CA, Margolis RL, Reading SA, Pletnikov M, Coyle JT (2006). Neurobiology of schizophrenia. Neuron 52:139–153

Saha S, Chant D, Welham J, McGrath J (2005). A systematic review of the prevalence of schizophrenia. PLoS Med 2:e141

Sanders AR, Duan J, Levinson DF, Shi J, He D, Hou C, Burrell GJ, Rice JP, Nertney DA, Olincy A et al (2008). No significant association of 14 candidate genes with schizophrenia in a large European ancestry sample: implications for psychiatric genetics. Am J Psychiatry 165:497–506

Sartorius N, Jablensky A, Korten A, Ernberg G, Anker M, Cooper JE, Day R (1986). Early manifestations and first-contact incidence of schizophrenia in different cultures. A preliminary report on the initial evaluation phase of the WHO Collaborative Study on determinants of outcome of severe mental disorders. Psychol Med 16:909–928

Schurov IL, Handford EJ, Brandon NJ, Whiting PJ (2004). Expression of disrupted in schizophrenia 1 (DISC1) protein in the adult and developing mouse brain indicates its role in neurodevelopment. Mol Psychiatry 9:1100–1110

Schwab SG, Albus M, Hallmayer J, Honig S, Borrmann M, Lichtermann D, Ebstein RP, Ackenheil M, Lerer B, Risch N et al (1995). Evaluation of a susceptibility gene for schizophrenia on chromosome 6p by multipoint affected sib-pair linkage analysis. Nat Genet 11:325–327

Schwab SG, Knapp M, Mondabon S, Hallmayer J, Borrmann-Hassenbach M, Albus M, Lerer B, Rietschel M, Trixler M, Maier W et al (2003). Support for association of schizophrenia with genetic variation in the 6p22.3 gene, dysbindin, in sib-pair families with linkage and in an additional sample of triad families. Am J Hum Genet 72:185–190

Schwarz E, Bahn S (2008). Biomarker discovery in psychiatric disorders. Electrophoresis. 29(30):2884–2890

Seeman P, Chau-Wong M, Tedesco J, Wong K (1975). Brain receptors for antipsychotic drugs and dopamine: direct binding assays. Proc Natl Acad Sci U S A 72:4376–4380

Seeman P, Lee T (1975). Antipsychotic drugs: direct correlation between clinical potency and presynaptic action on dopamine neurons. Science 188:1217–1219

Shifman S, Bronstein M, Sternfeld M, Pisante-Shalom A, Lev-Lehman E, Weizman A, Reznik I, Spivak B, Grisaru N, Karp L et al (2002). A highly significant association between a COMT haplotype and schizophrenia. Am J Hum Genet 71:1296–1302

Shifman S, Johannesson M, Bronstein M, Chen SX, Collier DA, Craddock NJ, Kendler KS, Li T, O'Donovan M, O'Neill FA et al (2008). Genome-wide association identifies a common variant in the reelin gene that increases the risk of schizophrenia only in women. PLoS Genet 4:e28

Silberberg G, Darvasi A, Pinkas-Kramarski R, Navon R (2006). The involvement of ErbB4 with schizophrenia: association and expression studies. Am J Med Genet B Neuropsychiatr Genet 141B:142–148

Spielman RS, Ewens WJ (1996). The TDT and other family-based tests for linkage disequilibrium and association. Am J Hum Genet 59:983–989

Stefansson H, Sarginson J, Kong A, Yates P, Steinthorsdottir V, Gudfinnsson E, Gunnarsdottir S, Walker N, Petursson H, Crombie C et al (2003). Association of neuregulin 1 with schizophrenia confirmed in a Scottish population. Am J Hum Genet 72:83–87

Stefansson H, Sigurdsson E, Steinthorsdottir V, Bjornsdottir S, Sigmundsson T, Ghosh S, Brynjolfsson J, Gunnarsdottir S, Ivarsson O, Chou TT et al (2002). Neuregulin 1 and susceptibility to schizophrenia. Am J Hum Genet 71:877–892

Straub RE, Jiang Y, MacLean CJ, Ma Y, Webb BT, Myakishev MV, Harris-Kerr C, Wormley B, Sadek H, Kadambi B et al (2002). Genetic variation in the 6p22.3 gene DTNBP1, the human ortholog of the mouse dysbindin gene, is associated with schizophrenia. Am J Hum Genet 71:337–348

Straub RE, MacLean CJ, O'Neill FA, Burke J, Murphy B, Duke F, Shinkwin R, Webb BT, Zhang J, Walsh D et al (1995). A potential vulnerability locus for schizophrenia on chromosome 6p24–22: evidence for genetic heterogeneity. Nat Genet 11:287–293

Sullivan PF (2005). The genetics of schizophrenia. PLoS Med 2:e212

Sullivan PF, Lin D, Tzeng JY, van den Oord E, Perkins D, Stroup TS, Wagner M, Lee S, Wright FA, Zou F et al (2008). Genomewide association for schizophrenia in the CATIE study: results of stage 1. Mol Psychiatry 13:570–584

Talbot K, Eidem WL, Tinsley CL, Benson MA, Thompson EW, Smith RJ, Hahn CG, Siegel SJ, Trojanowski JQ, Gur RE et al (2004). Dysbindin-1 is reduced in intrinsic, glutamatergic terminals of the hippocampal formation in schizophrenia. J Clin Invest 113:1353–1363

The Wellcome Trust Case Control Consortium (2007). Genome-wide association study of 14,000 cases of seven common diseases and 3,000 shared controls. Nature 447:661–678

Tkachev D, Mimmack ML, Ryan MM, Wayland M, Freeman T, Jones PB, Starkey M, Webster MJ, Yolken RH, Bahn S (2003). Oligodendrocyte dysfunction in schizophrenia and bipolar disorder. Lancet 362:798–805

Tochigi M, Iwamoto K, Bundo M, Komori A, Sasaki T, Kato N, Kato T (2008). Methylation status of the reelin promoter region in the brain of schizophrenic patients. Biol Psychiatry 63:530–533

Tunbridge EM, Harrison PJ, Weinberger DR (2006). Catechol-O-methyltransferase, cognition, and psychosis: Val158Met and beyond. Biol Psychiatry 60:141–151

Van Den Bogaert A, Schumacher J, Schulze TG, Otte AC, Ohlraun S, Kovalenko S, Becker T, Freudenberg J, Jonsson EG, Mattila-Evenden M et al (2003). The DTNBP1 (dysbindin) gene contributes to schizophrenia, depending on family history of the disease. Am J Hum Genet 73:1438–1443

Vawter MP, Crook JM, Hyde TM, Kleinman JE, Weinberger DR, Becker KG, Freed WJ (2002). Microarray analysis of gene expression in the prefrontal cortex in schizophrenia: a preliminary study. Schizophr Res 58:11–20

Walsh T, McClellan JM, McCarthy SE, Addington AM, Pierce SB, Cooper GM, Nord AS, Kusenda M, Malhotra D, Bhandari A et al (2008). Rare structural variants disrupt multiple genes in neurodevelopmental pathways in schizophrenia. Science 320:539–543

Wang Q, Liu L, Pei L, Ju W, Ahmadian G, Lu J, Wang Y, Liu F, Wang YT (2003). Control of synaptic strength, a novel function of Akt. Neuron 38:915–928

Wei ML (2006). Hermansky-Pudlak syndrome: a disease of protein trafficking and organelle function. Pigment Cell Res 19:19–42

Weickert CS, Straub RE, McClintock BW, Matsumoto M, Hashimoto R, Hyde TM, Herman MM, Weinberger DR, Kleinman JE (2004). Human dysbindin (DTNBP1) gene expression in normal brain and in schizophrenic prefrontal cortex and midbrain. Arch Gen Psychiatry 61:544–555

Weinberger DR (1995a). From neuropathology to neurodevelopment. Lancet 346:552–557

Weinberger DR (1995b). Neurodevelopmental perspectives on schizophrenia. In: Blood FE, Kupfer DJ (eds) Psychopharmacology: the fourth generation of progress. Raven Press, Ltd, New York p 1171–1183

Weinberger DR (1996). On the plausibility of "the neurodevelopmental hypothesis" of schizophrenia. Neuropsychopharmacology 14:1S–11S

Wender PH, Rosenthal D, Kety SS, Schulsinger F, Welner J (1974). Crossfostering. A research strategy for clarifying the role of genetic and experiential factors in the etiology of schizophrenia. Arch Gen Psychiatry 30:121–128

Williams NM, Glaser B, Norton N, Williams H, Pierce T, Moskvina V, Monks S, Del Favero J, Goossens D, Rujescu D et al (2008). Strong evidence that GNB1L is associated with schizophrenia. Hum Mol Genet 17:555–566

Williams NM, Norton N, Williams H, Ekholm B, Hamshere ML, Lindblom Y, Chowdari KV, Cardno AG, Zammit S, Jones LA et al (2003). A systematic genomewide linkage study in 353 sib pairs with schizophrenia. Am J Hum Genet 73:1355–1367

Xu B, Roos JL, Levy S, van Rensburg EJ, Gogos JA, Karayiorgou M (2008). Strong association of de novo copy number mutations with sporadic schizophrenia. Nat Genet. 40:880–885

Autism Spectrum Disorders

Sabine M Klauck

Contents

1 Introduction.. 81
2 Molecular Genetic Screening.. 84
    2.1 Genome-Wide Screens and Fine-Mapping Approaches................ 84
    2.2 Chromosomal Aberrations... 88
    2.3 Candidate Genes 89
3 Future Directions 91
References.. 92

Abstract Autism spectrum disorders are highly heritable with a complex neuro-developmental phenotype characterized by distinct impairments of cognitive function in the field of social interaction and speech development together with restrictive and repetitive behavior. Manifold approaches have been undertaken worldwide to identify susceptibility loci or genes to understand the underlying molecular, cellular and pathomorphological processes. Despite the identification of promising candidate genes no clear conclusions can be made today about genetic loci involved in these disorders. This chapter will focus on relevant results from the last decade of research with emphasis on whole genome screens and association studies.

1 Introduction

Autism is only one entity within the heterogeneous group of neurodevelopmental childhood syndromes defined as autism spectrum disorders (ASDs). The term *autism* was first used in the middle of the last century by Leo Kanner (1943) and Hans Asperger (1944) describing a very specific psychopathology recognized in children. Since then, especially in the last 20 years, a much more comprehensive

---

S.M. Klauck
German Cancer Research Center (DKFZ), Division of Molecular Genome Analysis,
Im Neuenheimer Feld 580, 69120, Heidelberg, Germany
s.klauck@dkfz.de

D.B. Wildenauer (ed.), *Molecular Biology of Neuropsychiatric Disorders*,
Nucleic Acids and Molecular Biology, © Springer-Verlag Berlin Heidelberg 2009

view of the autistic symptomatology has been established. Today, the neurodevelopmental childhood disorders are defined and summarized in the *Diagnostic and Statistical Manual of Mental Disorders* (DSM-IV) (American Psychiatric Association 1994) and *International Classification of Mental and Behavioral Disorders* (ICD-10) (World Health Organization 1992) under the generic name pervasive developmental disorder (PDD) or autism spectrum disorder (ASD): (1) Autism or autistic disorder, (2) Asperger syndrome (AS), (3) Rett syndrome (RTT), (4) Childhood disintegrative disorder (CDD), and (5) Pervasive developmental disorder-not otherwise specified (PDD-NOS) or atypical autism. The three prominent characteristic areas of malfunction of ASD are: (1) impairments in social interaction, (2) impairments in verbal and non-verbal communication, and (3) restricted repetitive and stereotyped patterns of behavior, interests and activities (Fig. 1). ASD symptoms are recognized typically within the first 3 years of age with a lifelong persistence. Diagnostic tools for ASD have been internationally standardized for reliable diagnoses in view of the worldwide genetic studies. In addition to the diagnostic criteria of DSM-IV or ICD-10, this has been accomplished by the release and actual development of the Autism Diagnostic Interview-Revised (ADI-R) (Lord et al. 1994), a parents' or caregivers' questionnaire, and the Autism Diagnostic Observation Schedule-Generic (ADOS-G) (Lord et al. 2000), a direct testing tool of the patients' current behavioral pattern. The prevalence for autistic disorder representing the narrow phenotype is 0.1–0.3% and 0.3–0.6%

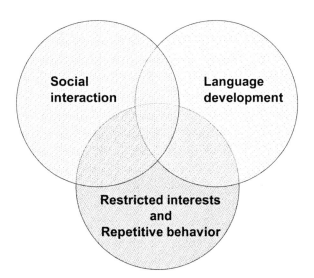

Fig. 1 Symptom areas of autism spectrum disorders according to ICD-10 (World Health Organization 1992). For a defined diagnosis of ASD, impairments in these areas should be evident before the age of 3 years

for the broader ASDs (Fombonne 2005). There has been some concern about increased rates of ASD over time, but due to refinement of diagnostic methodology and ascertainment strategies during the last 20 years comparison of reported prevalence rates are difficult to interpret. ASD is a neurodevelopmental disorder caused by mainly genetic factors, based on the observation of much higher concordance rates, considering cognitive deficits and social abnormalities for a narrow or broad definition of the phenotype, of 60–92% in monozygotic twins in contrast to 0–10% in dizygotic twins in a larger British twin study (Bailey et al. 1995). The heritability from this study is estimated to be more than 90%, but the influence of environmental factors for the specific affected individual towards the ASD phenotype may still be considerable. Several diagnosable medical conditions show symptomatoloy of autism spectrum disorders as well (e.g., fragile X syndrome, tuberous sclerosis complex, neurofibromatosis). These cases only account for less than 10% of patients with an autistic phenotype, while for the majority of "idiopathic" ASD the underlying genetic causes are unknown. It is a well-accepted hypothesis that several susceptibility genes are interacting together with a complex mode of inheritance leading to the typical phenotype(s) of the autism spectrum disorder. There may be at least 3 to 4 genes involved (Pickles et al. 1995) but up to 100 genes have also been discussed (Pritchard 2001). Approximately four times more males than females are affected (Smalley 1997) pointing towards a possible involvement of the sex chromosomes and imprinting effects in the etiology of the disorder, but no specific genes have so far been conclusively implicated. Rett syndrome showing mostly female cases has a separate status within the group of ASD, because the genetic cause of RTT has been uncovered by the identification of mutations in the *MECP2* gene located in Xq28 (Amir et al. 1999) in about 80% of cases. Recently, some atypical cases of RTT showed mutations in the *CDKL5* gene (reviewed in Weaving et al. 2005). Hence, a major gene defect is disease-causing in the majority of RTT patients in contrast to the as yet unidentified susceptibility genes in autistic disorder, Asperger syndrome, and CDD.

There are three main approaches to locate ASD susceptibility genes: whole genome screens searching for linkage or association in families with affected sibling pairs, gene association studies including selective candidate gene and pathway-based analyses, and cytogenetic studies revealing chromosomal abnormalities in mostly rare cases to pinpoint possible genetic loci with relevance for a broader spectrum of ASD patients. With the advent of technical advances to assess copy number variation (CNV) through microarray chip technology or single nucleotide polymorphism (SNP) genotyping for genome-wide association (GWA) in the human genome, this approach has most recently contributed to the identification of a large number of novel candidate loci for ASD. Furthermore, animal modeling of autism could give valuable hints toward the understanding of the molecular, cellular and pathomorphological processes involved.

This chapter will summarize the most relevant findings of the last years with emphasis on those studies dealing mainly with the more narrow phenotype of autistic disorder within the group of autism spectrum disorders.

2 Molecular Genetic Screening

2.1 Genome-Wide Screens and Fine-Mapping Approaches

The main goal of conducting whole-genome screens is to define regions with putative susceptibility genes for ASD for further fine mapping by association studies and consecutive detailed candidate gene screening. The study design is based on usage of families with multiple affected members, typically affected sibling pairs but also affected relative pairs. A whole set of 300–400 polymorphic markers evenly distributed over all chromosomes is genotyped within the families. Increased allele sharing "identity by descent (IBD)" with disease status of the affected family members is expressed after statistical analysis as a maximum multipoint logarithm of the odds (lod) score (MLS) or a non-parametric linkage (NPL) score value indicative of a putative susceptibility region. Fine-mapping by family-based association testing for alleles in linkage disequilibrium (LD) with the susceptibility variant is now standard in most of the recent studies to circumvent population stratification bias. Association methods have the advantage of being more powerful than linkage methods at a certain locus by detecting genes of weaker effect (Risch and Merikangas 1996).

Since the first published whole-genome screen by the International Molecular Genetic Study of Autism Consortium (IMGSAC) (1998) another 15 genome-wide screens for autism spectrum disorder have been conducted including families residing in Europe and/or North America (Barrett et al 1999; Philippe et al. 1999; Risch et al. 1999; Buxbaum et al. 2001; Liu et al. 2001; Shao et al. 2002a; Auranen et al. 2002; Ylisaukko-oja et al. 2004; Cantor et al. 2005; McCauley et al. 2005; Schellenberg et al. 2006; Ylisaukko-oja et al. 2006; Lauritsen et al. 2006; Ma et al. 2007; The Autism Genome Project Consortium 2007). Most of the screens used independent samples, but some have also been overlapping depending on the availability of central collections to multiple investigators. Furthermore, follow-up whole-genome screens have been performed by four investigation centers after successively increasing the sample size and adding more markers in the regions of interest from the first screens (International Molecular Genetic Study of Autism Consortium 2001a; Alarcón et al. 2002, 2005; Yonan et al. 2003; Rehnström et al. 2007). There have been suggestive linkage findings with an MLS > 1 for 19 of the autosomal chromosomes and the X chromosome suggesting the presence of genetic heterogeneity between studies. Highest LOD scores from each of the genome-wide screens are listed in Table 1. Only three studies reached a genome-wide significance level above the threshold of MLS 3.6 at marker D2S2188 on 2q31.1 with MLS 4.8 (International Molecular Genetic Study of Autism Consortium 2001a), at marker D3S3037 on 3q26.32 with $Z_{max\,dom}$ of 4.31 (Auranen et al. 2002), and at marker D3S2432 on 3p23–22.3 with NPL_{all} = 3.83 (Rehnström et al 2006).

The genome-wide scan using 17 Finnish families with a strict diagnosis of Asperger syndrome observed positive linkage findings in a two-stage approach on chromosomal regions 1q21-q22, 3p14–24 and 13q31-q33 (Ylisaukko-oja et al. 2004). In a follow-up study using 12 new extended Asperger syndrome families, the linkage findings to 3p21-q25 have been replicated (Rehnström et al. 2006). The

Autism Spectrum Disorders

Table 1 Genome-wide linkage studies in autism spectrum disorders listing the highest LOD scores using the complete set of families

Research group	Number of affected sibling pairs	Chromosomal region	Marker	Highest LOD scores	Reference
IMGSAC (subset of Lamb et al. (2001))	87	7q32.1	D7S530-D7S684	2.53	IMGSAC (1998)
CLSA	75	13q22.1	D13S800	3.0	Barrett et al. (1999)
PARIS	51	6q21	D6S283	2.23	Philippe et al. (1999)
Stanford	90	1p13.2	D1S1675	2.15	Risch et al. (1999)
AGRE (subset of Alarcón et al. (2002))	95	2q31.3	D2S364	2.39	Buxbaum et al. (2001)
AGRE	110	Xq26.1	DXS1047	2.67	Liu et al. (2001)
IMGSAC (subset of Lamb et al. (2001)	152	2q31.1	D2S2188	3.74	IMGSAC (2001a)
CAT	99	Xq21.33	DXS6789	2.54	Shao et al. (2002a)
Finland	38	3q26.32	D3S3037	4.31	Auranen et al. (2002)
AGRE (subset of Alarcón et al. (2005))	152	7q35	D7S1824	2.98	Alarcón et al. (2002)
AGRE	345	17q11.2	D17S1800	2.83	Yonan et al. (2003)
Finland (subset of Rehnström et al. (2007))	17 (Asperger syndrome)	3p23–p22.3	D3S2432	3.32	Ylisaukko-oja et al. (2004)
AGRE	291	3q22.1	D3S3045-D3S1763	3.10	Alarcón et al. (2005)
AGRE	91	17q21.2	D17S1299	1.9	Cantor et al. (2005)
CLSA + AGRE	158	17q11.2	D17S1294	2.85	McCauley et al. (2005)
IMGSAC	219	2q31.1-q31.3	D2S2314-D2S2310	2.54	Lamb et al. (2005)
CPEA	222	10p15.1	D10S591	0.0014 (p-value)	Schellenberg et al. (2006)
AGRE + Finland	314	4q21-q31	D4S1591	2.53	Ylisaukko-oja et al. (2006)

(continued)

Table 1 (continued)

Research group	Number of affected sibling pairs	Chromosomal region	Marker	Highest LOD scores	Reference
Faroe Islands	12 (cases)	3p25.3	D3S3594	0.00007 (p-value)	Lauritsen et al. (2006)
Duke + Vanderbilt	26 (extended families)	12q14.2	Rs1445442	3.02	Ma et al. (2007)
AGP	1,181	11p13	Rs2421826	3.57	The Autism Genome Project Consortium (2007)
Finland	29 (Asperger syndrome)	3p23-p22.3	D3S2432	3.83	Rehnström et al. (2006)

Abbeviations: AGRE, Autism Genetic Resource Exchange; CAT, Collaborative Autism Team; CLSA; Collaborative Linkage Study of Autism; CPEA, Collaborative Programs of Excellence in Autism; IMGSAC, International Molecular Genetic Study of Autism Consortium; PARIS, Paris Autism Research International Sib Pair Study

first two loci are replicative findings of previously published autism susceptibility regions. The loci on 1q and 13q overlap with reported schizophrenia susceptibility loci. These results underline the possibility that certain broader genome regions contain susceptibility genes for several distinct neuropsychiatric disorders; some of them may be shared justifying additional analyses.

Several other regions of interest have emerged in more than one study, including regions on chromosomes 1p, 5q, 7q, 15q, 16p, 17q, 19p, and Xq. The other loci are either unique or may represent false positives. The variable results between studies are certainly also a matter of sample heterogeneity by inclusion of patients with a variety of classification schemes. Despite using the standardized diagnostic interviews, studies differ by incorporating only sibling pairs with the diagnosis autistic disorder or also Asperger syndrome and PDD-NOS. In addition, population differences may play a role in the background of evolutionary development of marker profiles in ethnically distinct surroundings.

Two of the most interesting regions with frequent and strong evidence of linkage between studies are located on chromosomes 2q and 7q (Table 1), but remain still broad with 25 cM (centiMorgan) and 60 cM, respectively. Fine-mapping linkage screening of the 2q and 7q regions, especially under consideration of the endophenotype for delayed onset of speech, resulted in further support through suggestive linkage findings in several studies (Alarcón et al. 2005; Ashley-Koch et al. 1999; International Molecular Genetic Study of Autism Consortium 2001b; Bradford et al. 2001; Shao et al. 2002b). Attempts to locate specific candidate genes within these chromosomal regions resulted in positive association findings of the mitochondrial aspartate/glutamate carrier *SLC25A12* (Ramoz et al 2004) and *CMYA3* genes (Faham et al. 2005), respectively, in 2q24-q33, but negative findings are reported from an independent autism family sample (Blasi et al. 2006a). Ambiguous results are also known from analyses of the reelin (*RELN*) gene in the chromosome 7q22 linkage region, being a factor influencing neuronal migration in brain development. Positive association for a 5'UTR triplet repeat polymorphism (Persico et al. 2001; Zhang et al. 2002) and different SNPs (Skaar et al. 2005 Serajee et al. 2006) could not otherwise be replicated (Krebs et al. 2002; Bonora et al. 2003; Devlin et al. 2004; Li et al. 2004). The low frequency of four different missense changes identified in a sample of 315 ASD families (Bonora et al. 2003) cannot account for the relatively strong linkage findings in 7q. Postmortem brain studies still suggest a role of the Reelin protein in the brain causing structural and cognitive deficits in autistic disorder (Fatemi et al. 2005). Interestingly, the gene *engrailed 2 (EN2)* in 7q36 with impact on cerebellar development showed positive association of intronic single nucleotide polymorphisms (SNPs) in two independent family data sets (Benayed et al. 2005). Additional proof of relevance for the ASD phenotype remains to be shown for all of these candidate genes in independent autism samples.

Sex and parent of origin linkage modeling suggested possible sex-limited effects of the susceptibility loci on 7q, 15q and 16p, but no such effect for loci on chromosomes 2 and 9 (Lamb et al. 2005). In the same study, parent-of-origin effects could be shown by maternal IBD sharing on chromosome 7q and 9p and paternal IBD sharing on 7q. A different study identified a male-specific linkage peak of MLS 4.3 at 17q11 (Stone et al. 2004). The gender specific analyses and results raise the possibility of sex-specific genetic factors potentially influenced by hormone levels at the sex-limited loci and a role of imprinted gene(s) at the parent-of-origin loci, a finding which needs further evaluation in extended family samples.

Using trait subsets of the autism patient samples with specific phenotype characteristics attempt to decrease sample heterogeneity. Several linkage findings by means of endophenotypes are currently known: (1) Families with obsessive-compulsive behavior showed suggestive linkage in 1q24.2 (Buxbaum et al. 2004), and support for previously identified linkage loci on 6q14.3 (Philippe et al. 1999) and 19p13 (International Molecular Genetic Study of Autism Consortium 1998, Philippe et al. 1999; Liu et al. 2001). (2) The factor "developmental milestones" led to evidence for the susceptibility loci 17q11.2 and 19p13 (McCauley et al. 2005) replicating in part other 17q11 regional findings (Stone et al. 2004). (3) A sample of 34 affected sib pairs with a history of developmental regression was analyzed and the regions 21q21.1 and 7q36.1 (Molloy et al. 2005) have been found, the latter supporting a previous linkage finding at 165 cM on 7q from one of the whole genome-screens (Liu et al. 2001). (4) Application of the statistical method of ordered-subset analysis (OSA) using the factor "insistence of sameness" from the repetitive-stereotyped patterns domain of the ADI-R as covariate (Shao et al. 2003) identified positive linkage in the 15q11-q13 region, a chromosomal region known to be the most common site of chromosomal abnormalities in this disorder (Vorstman et al. 2006). (5) Screening for savant skills only led to contradictory positive and negative linkage findings for 15q11-q13, respectively (Nurmi et al. 2003; Ma et al. 2005). Altogether, besides the promising outcome of the approaches by integrating categorical phenotypic measures, all studies have been lacking larger numbers of cases, and the validity of the results remains to be elucidated in further replication studies.

Recently, a large linkage scan has been performed by the Autism Genome Project (AGP) Consortium (http://autismgenome.org) on 1,181 families with at least two affected individuals using Affymetrix 10K SNP arrays while also analyzing copy number variation in these families (The Autism Genome Project Consortium 2007). This study has been established as a meta-analysis making use of previous collected families of different autism consortia throughout the United States and Europe. Linkage and CNV analyses implicate chromosome 11p12-p13 and neurexins, respectively, and other promising candidate loci.

2.2 Chromosomal Aberrations

Chromosomal abnormalities detected by cytogenetic assays are of major aid in locating relevant genes for any monogenic or polygenic disease. A number of such visible breakpoints, translocations, duplications, and deletions have been reported for predominantly individual cases of ASD spreading over all chromosomes as was recently extensively reviewed (Vorstman et al. 2006; Castermans et al. 2004). At this point, no direct correlation could be made towards a genomic region inheriting a susceptibility gene for autism. However, integration of data from linkage analyses, GWA studies to pinpoint copy number variation, and reports of chromosomal abnormalities are useful to narrow down genomic regions of interest for fine mapping of susceptibility genes.

The chromosomal region 15q11-q13 has gained much attention due to frequent reports of duplications of mainly maternal origin (Vorstman et al. 2006). The region of interest hosts a cluster of γ-amino butyric acid ($GABA_A$) receptor subunit genes (*GABRB3*, *GABRA5*, and *GABRG3*). Any malfunction of these genes may have implications for the inhibition of excitatory neural pathways as well as during early brain development and therefore be pathological for autism. A couple of linkage and association studies reported limited evidence for involvement of the $GABA_A$ receptors, where the most common positive linkage finding was within the *GABRB3* gene (Cook et al. 1998; Martin et al. 2000; Buxbaum et al. 2002; McCauley et al. 2004; Curran et al. 2005).

One other region of interest has been the subtelomeric region of 2q37 with a higher frequency of deletions in comparison to other chromosomal areas (Ghaziuddin and Burmeister 1999; Wolff et al. 2002; Lukusa et al. 2004). Detailed analysis of a larger patient sample for variants in the *CENTG2* gene revealed a limited number of autism specific non-synonymous variants but no evidence for linkage disequilibrium (Wassink et al. 2005). Therefore, the involvement of other genes in the 2q37 region should be considered.

Technology developments such as matrix-based comparative genomic hybridization (m-CGH) (Solinas-Toldo et al. 1997) and representational oligonucleotide microarray analysis (Lucito et al. 2003) allow the pinpointing of autism-related genomic loci by screening of larger patient samples for much smaller duplication or deletion regions. Applying array-CGH, a set of 29 patients with syndromic

ASD have been studied at 1 Mb resolution (Jacquemont et al. 2006). Six deletions and two duplications have been found with altered segments ranging in size from 1.4 to 16 Mb, but lacking recurrent abnormality. A large scale study of genome CNVs has been performed on a sample of 264 families with idiopathic autism which revealed a strong association of de novo CNVs, mostly deletions, with the disorder (Sebat et al. 2007). The frequency of these spontaneous mutations was 10% in the sample of sporadic cases and 3% in cases from multiplex families, whereas the frequency of CNVs in unaffected individuals was 1%. In agreement with Jacquemont et al. (2006), none of the genomic variants has been observed more than twice in the sample. This reflects the assumption that the autism phenotype results from a large variety of genetic defects. A second study with a large sample of 427 unrelated ASD cases showed 277 unbalanced CNVs in 44% of ASD families (Marshall et al. 2008). De novo CNVs have been found in 7 and 2% of idiopathic families having one child, or two or more ASD siblings, respectively. Several of the CNVs have been found in previously reported ASD loci, such as *NLGN4* and 22q11.2, 15q11-q13, *SHANK3* and *NRXN1*, supporting the role of these postsynaptic density genes. In addition, a CNV at 16p11.2 has been discovered at 1% frequency overlapping with a known mental retardation site (Ballif et al. 2007). The 16p11.2 microdeletion was evident in 0.6% of 712 cases analyzed by Kumar et al. (2008). This novel microdeletion and a reciprocal microduplication has been reported from a GWA study including more than 2,000 patients with ASD (Weiss et al. 2008). It remains to be disentangled whether the identified genomic imbalances are truly involved with the disease or represent large-scale copy number polymorphisms in the human genome (Sebat et al. 2004; Iafrate et al. 2004).

2.3 Candidate Genes

Different criteria make a gene or genetic locus eligible for association studies or further screening for variants or mutations. The gene or gene product (1) is thought to be of relevance for behavior in humans, (2) belongs to a neurodevelopmental pathway in the brain by expression in fetal brain tissues, (3) has been implicated through studies of animal models, (4) has been identified through a chromosomal abnormality, and (5) has been located positionally by linkage studies.

Over the last decade, manifold candidate studies have been conducted following up on potential susceptibility loci in ASD. The majority of them did not reveal a clear picture of either positive association at a certain gene or genomic locus or identification of disease-relevant variations or mutations by comparison of individual studies. This might be the result of allelic heterogeneity, sample heterogeneity, small samples sizes or ethnically distinct backgrounds. More than 100 functional or positional candidate genes have been tested directly, but lack conclusive evidence of involvement in ASD. It is impossible to review all studies here, but examples of four genes or gene families should demonstrate the worldwide efforts and remaining difficulty in discovering their role in ASD.

Genes involved in the physiological pathway of serotonin are strong candidates for autism as serotonin serves as a neurotransmitter in the brain responsible for a couple of cognitive functions. Hyperserotonemia has been reported in autism leaving the serotonin transporter (*SLC6A4 or 5-HTT*) gene responsible for serotonin re-uptake at the presynaptic membrane of neurons as primary target of extensive investigations. The results of the numerous studies are inconsistent, reporting either association of different alleles of the functional promoter polymorphism 5-HTTLPR (short or long allele) involved in gene expression or no evidence for association (reviewed in Devlin et al. 2005). The finding that the recently identified single nucleotide polymorphism (SNP) rs25531 with its A variant in the long allele of 5-HTTLPR sequence is responsible for the high *5-HTT* mRNA level may explain the ambiguous results of previous association studies with the need for more detailed genotyping in the future (Wendland et al. 2006). Other SNPs in *5-HTT* show strong transmission disequilibrium with autism (Kim et al. 2002). Positive linkage findings from genome screens at the 17q11.2-q12 locus (International Molecular Genetic Study of Autism Consortium 2001a, Yonan et al. 2003; McCauley et al. 2005) also spot this gene locus. Extensive variant screening identified several rare coding (e.g., Gly56Ala) and non-coding variants in *5-HTT* with a strong correlation to the endophenotype of rigid-compulsive behavior (Sutcliffe et al. 2005). The report of another rare functional coding variant Ile425Leu leading to higher mRNA expression (Kilic et al. 2003) in two families with complex neuropsychiatric phenotypes (obsessive-compulsive disorder, Asperger syndrome, social phobia, anorexia nervosa, tic disorder, alcohol and other substance abuse/dependence) (Ozaki et al. 2003) suggested a potential role for ASD. A replication study involving 210 individuals with autism or Asperger syndrome did not find this variant in the sample excluding it from relevance for ASD (Wendland et al. 2008).

Hints from linkage studies (Shao et al. 2002a; Auranen et al. 2002) and reports from chromosomal deletions in three autistic females (Thomas et al. 1999) led to the screening of the genes neuroligin 3 (*NLGN3*) at Xp22.3 and 4 (*NLGN4*) at Xq13. The neuroligins are cell-adhesion proteins with important function in synaptogenesis during brain development and in connection of pre- and postsynaptic membranes. A frameshift mutation in *NLGN4* and a missense mutation in *NLGN3* in two separate families have been found (Jamain et al. 2003) leading to functional inactivation of neuroligins (Chih et al. 2004). Another 2-base-pair deletion within *NLGN4* was found in a large family segregating with X-linked mental retardation including three males with ASD (Laumonnier et al. 2004). Mutations in these neuroligin genes seem to be rather rare events. Extensive screening of other large patient samples only revealed four other missense mutations with questionable function in *NLGN4* (Yan et al. 2005), but otherwise negative results have been reported (Vincent et al. 2004; Talebizadeh et al. 2004; Gauthier et al. 2005; Ylisaukko-oja et al. 2005; Blasi et al. 2006b; Wermter et al. 2008).

Neuroligins interact with ß-neurexins during synaptogenesis. The AGP consortium identified a hemizygous deletion of coding exons for *NRXN1* in a pair of affected siblings (The Autism Genome Project Consortium 2007). Furthermore, two putative missense structural variants of neurexin 1ß have been found in 4 of a total of 203 patients (Feng et al. 2006), and subtle changes of *NRXN1* through a balanced translocation disrupting the gene in two patients with ASD are suggested by Kim et al. (2008) to contribute to ASD susceptibility. In addition, the Contactin Associated Protein-Like 2 gene (*CNTNAP2*), a member of the Neurexin family, has been associated with language-related autism QTL in AGRE (Autism Genetic Resource Exchange) trios and an independent sample of 72 multiplex families (Alarcón et al. 2008; Arking et al. 2008), respectively. A resequencing approach in 635 patients revealed a total of 27 nonsynonymous changes, from which 13 were rare and unique to patients and 8 of these are potential deleterious by bioinformatic approaches (Bakkaloglu et al. 2008).

The fourth gene is *SHANK3* on chromosome 22q13 encoding a protein of the postsynaptic density (PSD) complex of excitatory synapses. It is a binding partner of neuroligins and regulates the structural organization of dendritic spines. Extensive screening by FISH (fluorescence in situ hybridization) and sequencing detected three families with ASD and alterations of 22q13 or *SHANK 3* (Durand et al. 2007). These comprised a de novo deletion of 142 kb of terminal 22q13, insertion of a guanine nucleotide in exon 21 causing a frameshift and a terminal deletion of 22q13 in a girl with autism and severe language delay, and a 22qter partial trisomy in her brother with Asperger syndrome. Consecutively, one de novo mutation and two gene deletions have been found in an independent sample of 400 ASD cases (Moessner et al. 2007). These data support the idea that haploinsufficiency of *SHANK3* can cause a monogenic form of autism. Moreover, variants in the ribosomal protein gene RPL10 have recently be described in two sib-pair families (Klauck et al. 2006). Functional studies of the variants in yeast suggest a change in translational function with respect to the regulation of the translation process. This may have an impact on synaptogenesis when a high level of protein synthesis is required for dendritic spine development. It remains to be shown whether other genes with function in synaptogenesis and development of dendritic spines and which act together with the neuroligins are involved in ASD.

3   Future Directions

Through the last decade a lot of information has been gained towards the identification of susceptibility genes for autism spectrum disorders. The diagnostic criteria for ASD have been refined to facilitate detailed analyses, including endophenotypes, together with knowledge from systematic molecular genetic screening approaches such as whole genome screens, copy number variant screening, association studies and candidate gene screenings. Despite much progress, the final definition of susceptibility genes underlying ASD is still a challenge for the future. The ultimate goal is to define a series of genetic variants to be responsible for a specific symptomatology within the whole spectrum of disabilities in ASD. To accomplish this, technologies such as whole genome association studies making use of high throughput genotyping methods are promising to support the identification of disease genes for complex disorders, keeping in mind problems with multiple testing, study design, definition of intermediate phenotypes, and interaction between polymorphisms (Carlson et al. 2004). The International HapMap Project determines the common patterns of DNA sequence variants in the human genome, the degree of association between them in terms of strong linkage disequilibrium, known as haplotype blocks, and gains insights into structural variation and recombination (The International HapMap Consortium 2005). This information is needed to integrate disease relevant variants with knowledge of common population variants for the autism projects as well.

Much more emphasis should be put into gene function and pathway analyses to understand the development of brain structures and their function in cognitive processing. From this knowledge, it may be possible to develop therapeutic targets for drug treatments, but also, very importantly, screening diagnostics that would allow for early intervention with behavioral therapies of individuals inheriting risk factors for autism spectrum disorders.

References

Alarcón M, Cantor RM, Liu J, Gilliam TC, the Autism Genetic Resource Exchange Consortium, Geschwind DH (2002) Evidence for a language quantitative trait locus on chromosome 7q in multiplex autism families. Am J Hum Genet 70:60–71

Alarcón M, Yonan AL, Gilliam TC, Cantor RM, Geschwind DH (2005) Quantitative genome scan and ordered-subsets analysis of autism endophenotypes support language QTLs. Mol Psychiatry 10:747–757

Alarcón M, Abrahams BS, Stone JL, Duvall JA, Perederiy JV, Bomar JM, Sebat J, Wigler M, Martin CL, Ledbetter DH, Nelson SF, Cantor RM, Geschwind DH (2008) Linkage, association, and gene-expression analyses identify *CNTNAP2* as an autism-susceptibility gene. Am J Hum Genet 82:150–159

American Psychiatric Association (1994) Diagnostic and statistical manual of mental disorders, 4th edn. American Psychiatric Association, Washington, DC

Amir RE, Van den Veyver IB, Wan M, Tran CQ, Francke U, Zoghbi HY (1999) Rett syndrome is caused by mutations in X-linked *MECP2*, encoding methyl-CpG-binding protein 2. Nat Genet 23:185–188

Arking DE, Cutler DJ, Brune CW, Teslovich TM, West K, Ikeda M, Rea A, Guy M, Lin S, Cook EH Jr, Chakravarti A (2008) A common genetic variant in the neurexin superfamily member *CNTNAP2* increases familial risk of autism. Am J Hum Genet 82:160–164

Ashley-Koch A, Wolpert CM, Menold MM, Zaeem L, Basu S, Donnelly SL, Ravan SA, Powell CM, Qumsiyeh MB, Aylsworth AS, Vance JM, Gilbert JR, Wright HH, Abramson RK, DeLong GR, Cuccaro ML, Pericak-Vance MA (1999) Genetic studies of autistic disorder and chromosome 7. Genomics 61:227–236

Asperger H (1944) Die Autistischen Psychopathen im Kindesalter. Arch Psychiatr Nervenkr 117:76–136

Auranen M, Vanhala R, Varilo, Ayers K, Kempas E, Yliksaukko-oja T, Sinsheimer JS, Peltonen L, Järvelä I (2002) A genomewide screen for autism-spectrum disorders: evidence for a major susceptibility locus on chromosome 3q25-27. Am J Hum Genet 71:777–790

Bailey A, Le Couteur A, Gottesman I, Bolton P, Simonoff E, Yuzda E, Rutter M (1995) Autism as a strongly genetic disorder: evidence from a British twin study. Psychol Med 25:63–78

Bakkaloglu B, O'Roak BJ, Louvi A, Gupta AR, Abelson JF, Morgan TM, Chawarska K, Klin A, Ercan-Sencicek AG, Stillman A, Tanriover G, Abrahams BS, Duvall JA, Robbins EM, Geschwind DH, Biederer T, Gunel M, Lifton RP, State MW (2008) Molecular cytogenetic analysis and resequencing of *Contactin Associated Protein-Like 2* in autism spectrum disorders. Am J Hum Genet 82:165–173

Ballif B, Hornor SA, Jenkins E, Madan-Khetarpal S, Surti U, Jackson KE, Asamoah A, Brock PI, Gowans GC, Conway RL, Graham JM Jr, Medne L, Zackai EH, Shaikh TH, Geoghegan J, Selzer RR, Eis PS, Bejjani BA, Shaffer LG (2007) Discovery of a previously unrecognized microdeletion syndrome of 16p11.2-p12.2. Nat Genet 39:1071–1073

Barrett S, Beck JC, Bernier R, Bisson E, Braun TA, Casavant TL, Childress D, Folstein SE, Garcia M, Gardiner MB, Gilman S, Haines JL, Hopkins K, Landa R, Meyer NH, Mullane JA, Nishimura DY, Palmer P, Piven J, Purdy J, Santangelo SL, Searby C, Sheffield V, Singleton J, Slager S, Struchen T, Svenson S, Vieland V, Wang K, Winklosky B (1999) An autosomal genomic screen for autism. Collaborative Linkage Study of Autism. Am J Med Genet 88:609–615

Benayed R, Gharani N, Rossman I, Mancuso V, Lazar G, Kamdar S, Bruse SE, Tischfield S, Smith BJ, Zimmerman RA, DiCicco-Bloom E, Brzustowicz LM, Millonig JH (2005) Support for the homeobox transcription factor gene *ENGRAILED 2* as an autism spectrum disorder susceptibility locus. Am J Hum Genet 77:851–868

Blasi F, Bacchelli E, Carone S, Toma C, Monaco AP, Bailey AJ, Maestrini E, International Molecular Genetic Study of Autism Consortium (IMGSAC) (2006a) *SLC25A12* and *CMYA3* gene variants are not associated with autism in the IMGSAC multiplex sample. Eur J Hum Genet 14:123–126

Blasi F, Bacchelli E, Pesaresi G, Carone S, Bailey AJ, Maestrini E, International Molecular Genetic Study of Autism Consortium (IMGSAC) (2006b) Absence of coding mutations in the X-linked genes *Neuroligin 3* and *Neuroligin 4* in individuals with autism from the IMGSAC collection. Am J Med Genet Part B (Neuropsychiatr Genet) 141B:220–221.

Bonora E, Beyer KS, Lamb JA, Parr JR, Klauck SM, Benner A, Paolucci M, Abbott A, Ragoussis I, Poustka A, Bailey AJ, Monaco AP, International Molecular Genetic Study of Autism Consortium (IMGSAC) (2003) Analysis of reelin as a candidate gene for autism. Mol Psychiatry 8:885–892

Bradford Y, Haines J, Hutcheson H, Gardiner M, Braun T, Sheffield V, Cassavant T, Huang W, Wang K, Vieland V, Folstein S, Santangelo S, Piven J (2001) Incorporating language phenotypes strengthens evidence of linkage to autism. Am J Med Genet (Neuropsychiatr Genet) 105:539–547

Buxbaum JD, Silverman JM, Smith CJ, Kilifarski M, Reichert J, Hollander E, Lawlor BA, Fitzgerald M, Greenberg DA, Davis KL (2001) Evidence for a susceptibility gene for autism on chromosome 2 and for genetic heterogeneity. Am J Hum Genet 68:1514–1520

Buxbaum JD, Silverman JM, Smith CJ, Greenberg DA, Kilifarski M, Reichert J, Cook EH Jr, Fang Y, Song C-Y, Vitale R (2002) Association between a *GABRB3* polymorphism and autism. Mol Psychiatry 7:311–316

Buxbaum JD, Silverman J, Keddache M, Smith CJ, Hollander E, Ramoz N, Reichert JG (2004) Linkage analysis for autism in a subset of families with obsessive-compulsive behaviors: evidence for an autism susceptibility gene on chromosome 1 and further support for susceptibility genes on chromosome 6 and 19. Mol Psychiatry 9:144–150

Cantor RM, Kono N, Duvall JA, Alvarez-Retuerto A, Stone JL, Alarcón M, Nelson SF, Geschwind DH (2005) Replication of autism linkage: fine-mapping peak at 17q21. Am J Hum Genet 76:1050–1056

Carlson CS, Eberle MA, Kruglyak L, Nickerson DA (2004) Mapping complex disease loci in whole-genome association studies. Nature 429:446– 452

Castermans D, Wilquet V, Steyaert J, Van de Ven W, Fryns J-P, Devriendt K (2004) Chromosomal anomalies in individuals with autism. Autism 8:141–161

Chih B, Afridi SK, Clark L, Scheiffele P (2004) Disorder-associated mutations lead to functional inactivation of neuroligins. Hum Mol Genet 13:1471–1477

Cook EH Jr, Courchesne RY, Cox NJ, Lord C, Gonen D, Guter SJ, Lincoln A, Nix K, Haas R, Leventhal BL, Courchesne E (1998) Linkage-disequilibrium mapping of autistic disorder, with 15q11–13 markers. Am J Hum Genet 62:1077–1083

Curran S, Roberts S, Thomas S, Veltman M, Browne J, Medda E, Pickles A, Sham P, Bolton PF (2005) An association analysis of microsatellite markers across the Prader-Willi/Angelman critical region on chromsome 15 (q11–13) and autism spectrum disorder. Am J Med Genet Part B (Neuropsychiatr Genet) 137B:25–28

Devlin B, Bennett P, Dawson G, Figlewicz DA, Grigorenko EL, McMahon W, Minshew N, Pauls D, Smith M, Spence MA, Rodier PM, Stodgell C, Schellenberg GD, The CPEA Genetics Network (2004) Alleles of a reelin CGG repeat do not convey liability to autism in a sample from the CPEA network. Am J Med Genet Part B (Neuropsychiatr Genet) 126B:46–50

Devlin B, Cook Jr EH, Coon H, Dawson G, Grigorenko EL, McMahon W, Minshew N, Pauls D, Smith M, Spence MA, Rodier PM, Stodgell C, Schellenberg GD, The CPEA Genetics Network (2005) Autism and the serotonin transporter: the long and short of it. Mol Psychiatry 10:1110–1116

Durand CM, Betancur C, Boeckers TM, Bockmann J, Chaste P, Fauchereau F, Nygren G, Rastam M, Gillberg IC, Anckarsäter H, Sponheim E, Goubran-Botros H, Delorme R, Chabane N, Mouren-Simeoni M-C, de Mas P, Bieth E, Rogé B, Héron D, Burglen L, Gillberg C, Leboyer M, Bourgeron T (2007) Mutations in the gene encoding the synaptic scaffolding protein SHANK3 are associated with autism spectrum disorders. Nat Genet 39:25–27

Faham M, Zheng J, Moorhead M, Fakhrai-Rad H, Namsaraev E, Wong K, Wang Z, Chow SG, Lee L, Syenaga K, Reichert J, Boudreau A, Eberle J, Bruckner C, Jain M, Karlin-Neumann G, Jones HB, Willis TD, Buxbaum JD, Davis RW (2005) Multiplexed variation scanning for 1,000 amplicons in hundreds of patients using mismatch repair detection (MRD) on tag arrays. Proc Natl Acad Sci USA 102:14717–14722

Fatemi SH, Snow AV, Stary JM, Araghi-Niknam M, Reutiman TJ, Lee S, Brooks AI, Pearce DA (2005) Reelin signalling is impaired in autism. Biol Psychiatry 57:777–787

Feng J, Schroer R, Yan J, Song W, Yang C, Bockholt A, Cook EH Jr, Skinner C, Schwartz CE, Sommer SS (2006) High frequency of neurexin 1ß signal peptide structural variants in patients with autism. Neurosci Lett 409:10–13

Fombonne E (2005) Epidemiology of autistic disorder and other pervasive developmental disorders. J Clin Psychiatry 66 (suppl 10):3–8

Gauthier J, Bonnel A, St-Onge J, Karemera L, Laurent S, Mottron L, Fombonne E, Joober R, Rouleau GA (2005) *NLGN3/NLGN4* gene mutations are not responsible for autism in the Quebec population. Am J Med Genet Part B (Neuropsychiatr Genet) 132B:74–75

Ghaziuddin M, Burmeister M (1999) Deletion of chromosome 2q37 and autism: a distinct subtype? J Autism Dev Disord 29:259–263

Iafrate AJ, Feuk L, Rivera M, Listewnik ML, Donahoe PK, Qi Y, Scherer SW, Lee C (2004) Detection of large-scale variation in the human genome. Nat Genet 36:949–951

International Molecular Genetic Study of Autism Consortium (1998) A full genome screen for autism with evidence for linkage to a region on chromosome 7q. Hum Mol Genet 7:571–578

International Molecular Genetic Study of Autism Consortium (IMGSAC) (2001a) A genomewide screen for autism: strong evidence for linkage to chromosomes 2q, 7q, and 16p. Am J Hum Genet 69:570–581

International Molecular Genetic Study of Autism Consortium (IMGSAC) (2001b) Further characterization of the autism susceptibility locus AUTS1 on chromosome 7q. Hum Mol Genet 10:973–982

Jacquemont M-L, Sanlaville D, Redon R, Raoul O, Cormier-Daire V, Lyonnet S, Amiel J, Le Merrer M, Heron D, de Blois M-C, Prieur M, Vekemans M, Carter NP, Munnich A, Colleaux L, Philippe A (2006) Array-based comparative genomic hybridisation identifies high frequency of cryptic chromosomal rearrangements in patients with syndromic autism spectrum disorders. J Med Genet 43:843–849

Jamain S, Quach H, Betancour C, Råstam M, Colineaux C, Gillberg IC, Soderstrom H, Giros B, Leboyer M, Gillberg C, Bourgeron T, Paris Autism Research International Sibpair Study (2003) Mutations of the X-linked genes encoding neuroligins NLGN3 and NLGN4 are associated with autism. Nat Genet 34:27–29

Kanner L (1943) Autistic disturbances of affective contact. Nerv Child 12:217–250

Kilic F, Murphy DL, Rudnick G (2003) A human serotonin transporter mutation causes constitutive activation of transport activity. Mol Pharmacol 64:440-446

Kim SJ, Cox N, Courchesne R, Lord C, Corsello C, Akshoomoff N, Guter S, Leventhal BL, Courchesne E, Cook EHJR (2002) Transmission disequilibrium mapping at the serotonin transporter gene (*SLC6A4*) region in autistic disorder. Mol Psychiatry 7:278–288

Kim HG, Kishikawa S, Higgins AW, Seong IS, Donovan DJ, Shen Y, Lally E, Weiss LA, Najm J, Kutschke K, Descartes M, Holt L, Braddock S, Troxell R, Kaplan L, Volkmar F, Klin A, Tsatsanis K, Harris DJ, Noens I, Pauls DL, Daly MJ, MacDonald ME, Morton CC, Quade BJ, Gusella JF (2008) Disruption of neurexin 1 associated with autism spectrum disorder. Am J Hum Genet 82:199–207

Klauck SM, Felder B, Kolb-Kokocinski A, Schuster C, Chiocchetti A, Schupp I, Wellenreuther R, Schmötzer G, Poustka F, Breitenbach-Koller L, Poustka A (2006) Mutations in the ribosomal protein gene *RPL10* suggest a novel modulating disease mechanism for autism. Mol Psychiatry 11:1073–1084

Krebs MO, Betancur C, Leroy S, Bourdel MC, Gillberg C, Leboyer M, Paris Autism Research International Sibpair (PARIS) Study (2002) Absence of association between a polymorphic GGC repeat in the 5′ untranslated region of the reelin gene and autism. Mol Psychiatry 7:801–804

Kumar RA, KaraMohamed S, Sudi J, Conrad DF, Brune C, Badner JA, Gilliam TC, Nowak NJ, Cook EH Jr, Dobyns WB, Christian SL (2008) Recurrent 16p11.2 microdeletion in autism. Hum Mol Genet 17:628–638

Lamb JA, Barnby G, Bonora E, Sykes N, Bacchelli E, Blasi F, Maestrini E, Broxholme J, Tzenova J, Weeks D, Bailey AJ, Monaco AP, International Molecular Genetic Study of Autism Consortium (IMGSAC) (2005) Analysis of IMGSAC autism susceptibility loci: evidence for sex limited and parent of origin specific effects. J Med Genet 42:132–137

Laumonnier F, Bonnet-Brilhault F, Gomot M, Blanc R, David A, Moizard M-P, Raynaud M, Ronce N, Lemonnier E, Calvas P, Laudier B, Chelly J, Fryns J-P, Ropers H-H, Hamel BCJ, Andres C, Barthélémy C, Moraine C, Briault S (2004) X-linked mental retardation and autism are associated with a mutation in the *NLGN4* gene, a member of the neuroligin family. Am J Hum Genet 74:552–557

Lauritsen MB, Als TD, Dahl HA, Flint TJ, Wang AG, Vang M, Kruse TA, Ewald H, Mors O (2006) A genome-wide search for alleles and haplotypes associated with autism and related pervasive developmental disorders on the Faroe Islands. Mol Psychiatry 11:37–46

Li J, Nguyen L, Gleason C, Lotspeich L, Spiker D, Risch N, Myers RM (2004) Lack of evidence for an association between WNT2 and RELN polymorphisms and autism. Am J Med Genet Part B (Neuropsychiatr Genet) 126B:51–57

Liu J, Nyholt DR, Magnussen P, Parano E, Pavone P, Geschwind D, Lord C, Iversen P, Hoh J, Ott J, Gilliam TC, Autism Genetic Resoruce Exchange Consortium (2001) A genomewide screen for autism susceptibility loci. Am J Hum Genet 69:327–340

Lord C, Rutter M, Le Couteur A (1994) Autism diagnostic interview-revised: a revised version of a diagnostic interview for caregivers of individuals with possible pervasive developmental disorders. J Autism Dev Disord 24:659–685

Lord C, Risi S, Lambrecht L, Cook EH Jr, Leventhal BL, DiLavore BL, Pickles A, Rutter M (2000) The autism diagnostic observation schedule-generic: a standard measure of social and communication deficits associated with the spectrum of autism. J Autism Dev Disord 30:205–223

Lucito R, Healy J, Alexander J, Reiner A, Esposito D, Chi M, Rodgers L, Brady A, Sebat J, Troge J, West JA, Rostan S, Nguyen KCQ, Powers S, Ye KQ, Olshen A, Venkatraman E, Norton L, Wigler M (2003) Representational oligonucleotide microarray analysis: a high-resolution method to detect genome copy number variation. Genome Res 13:2291–2305

Lukusa T, Vermeesch JR, Holvoet M, Fryns JP, Devriendt K (2004) Deletion 2q37.3 and autism: molecular cytogenetic mapping of the candidate region for autistic disorder. Genet Counsel 15:293–301

Ma DQ, Jaworski J, Menold MM, Donnelly S, Abramson RK, Wright HH, Delong GR, Gilbert JR, Pericak-Vance MA, Cuccaro ML (2005) Ordered-subset analysis of savant skills in autism for 15q11-q13. Am J Med Genet Part B (Neuropsychiatr Genet) 135B:38–41

Ma DQ, Cuccaro ML, Jaworski JM, Haynes CS, Stephan DA, Parod J, Abramson RK, Wright HH, Gilbert JR, Haines JL, Pericak-Vance MA (2007) Dissecting the locus heterogeneity of autism: significant linkage to chromosome 12q14. Mol Psychiatry 12:376–384

Marshall CR, Noor A, Vincent JB, Lionel AC, Feuk L, Skaug J, Shago M, Moessner R, Pinto D, Ren Y, Thiruvahindrapduram B, Fiebig A, Schreiber S, Friedman J, Ketelaars CEJ, Vos YJ, Ficicioglu C, Kirkpatrick S, Nicolson R, Sloman L, Summers A, Gibbons CA, Teebi A, Chitayat D, Weksberg R, Thompson A, Vardy C, Crosbie V, Luscombe S, Baatjes R, Zwaigenbaum L, Roberts W, Fernandez B, Szatmari P, Scherer SW (2008) Structural variation of chromosomes in autism spectrum disorder. Am J Hum Genet 82:477–488

Martin ER, Menold MM, Wolpert CM, Bass MP, Donnelly SL, Ravan SA, Zimmerman A, Gilbert JR, Vance JM, Maddox LO, Wright HH, Abramson RK, DeLong GR, Cuccaro ML, Pericak-Vance MA (2000) Analysis of linkage disequilibrium in γ-aminobutyric acid receptor subunit genes in autistic disorder. Am J Med Genet (Neuropsychiatr Genet) 96:43–48

McCauley JL, Olson LM, Delahanty R, Amin T, Nurmi EL, Organ EL, Jacobs MM, Folstein SE, Haines JL, Sutcliffe JS (2004) A linkage disequilibrium map of the 1-Mb 15q12 GABAA receptor subunit cluster and association to autism. Am J Med Genet Part B (Neuropsychiatr Genet) 131B:51–59

McCauley JL, Li C, Jiang L, Olson Lm Crockett G, Gainer K, Folstein SE, Haines JL, Sutcliffer JS (2005) Genome-wide and ordered-subset linkage analyses provide support for autism loci on 17q and 19p with evidence of phenotypic and interlocus genetic correlates. BMC Med Genet 6:1

Moessner R, Marshall CR, Sutcliffe JS, Skaug J, Pinto D, Vincent J, Zwaigenbaum L, Fernandez B, Roberts W, Szatmari P, Scherer SW (2007) Contribution of *SHANK3* mutations to autism spectrum disorder. Am J Hum Genet 81:1289–1297

Molloy CA, Keddache M, Martin LJ (2005) Evidence for linkage on 21q and 7q in a subset of autism characterized by developmental regression. Mol Psychiatry 10:741–746

Nurmi EL, Dowd M, Tadevosyan-Leyfer O, Haines JL, Folstein SE, Sutcliffe JS (2003) Exploratory subsetting of autism families based on savant skills improves evidence of genetic linkage to 15q11-q13. J Am Acad Child Adolesc Psychiatry 42:856–863

Ozaki N, Goldman D, Kaye WH, Plotnicov K, Greenberg BD, Lappalainen J, Rudnick G, Murphy DL (2003) Serotonin transporter missense mutation associated with a complex neuropsychiatric phenotype. Mol Psychiatry 8:933–936

Persico AM, D'Agruma L, Maiorano N, Totaro A, Militerni R, Bravaccio C, Wassink TH for the CLSA, Schneider C, Melmed R, Trillo S, Montecchi F, Palermo M, Pascucci T, Puglisi-Allegra S, Reichelt K-L, Conciatori M, Marino R, Quattrocchi CC, Baldi A, Zelante L, Gasparini P, Keller F (2001) Reelin gene alleles and haplotypes as a factor predisposing to autistic disorder. Mol Psychiatry 6:150–159

Philippe A, Martinez M, Guilloud-Bataille M, Gillberg C, Råstam M, Sponheim E, Coleman M, Zappella M, Aschauer H, Van Maldergen L, Penet C, Feingold J, Brice A, Leboyer M, and the Paris Autism Research International Sibpair Study (1999) Genome-wide scan for autism susceptibility genes. Hum Mol Genet 8:805–812

Pickles A, Bolton P, Macdonald H, Bailey A, Le Couteur A, Sim C-H, Rutter M (1995) Latent-class analysis of recurrence risks for complex phenotypes with selection and measurement error: a twin and family history study of autism. Am J Hum Genet 57:717–726

Pritchard JK (2001) Are rare variants responsible for susceptibility to complex diseases? Am J Hum Genet 69:124–137

Ramoz N, Reichert JG, Smith CJ, Silverman JM, Bespalova IN, Davis KL, Buxbaum JD (2004) Linkage and association of the mitochondrial aspartate/glutamate carrier *SLC25A12* gene with autism. Am J Psychiatry 161:662–669

Rehnström K, Ylisaukko-oja T, Nieminen-von Wendt T, Sarenius S, Källman T, Kempas E, von Wendt L, Peltonen L, Järvelä I (2006) Independent replication and initial fine mapping of 3p21-24 in Asperger syndrome. J Med Genet 43(2):e6

Risch N, Merikangas K (1996) The future of genetic studies of complex human disease. Science 273:1516–1517

Risch N, Spiker D, Lotspeich L, Nouri N, Hinds D, Hallmayer J, Kalaydjieva L, McCague P, Dimiceli S, Pitts T, Nguyen L, Yang J, Harper C, Thorpe D, Vermeer S, Young H, Hebert J, Lin A, Ferguson J, Chiotti C, Wiese-Slater S, Rogers T, Salmon B, Nicholas P, Petersen PB, Pingree C, McMahon W, Wong DL, Cavalli-Sforza LL, Kraemer HC, Myers RM (1999) A genomic screen of autism: evidence for a multilocus etiology. Am J Hum Genet 65:493–507

Schellenberg GD, Dawson G, Sung YJ, Estes A, Munson J, Rosenthal E, Rothstein J, Flodman P, Smith M, Coon H, Leong L, Yu C-E, Stodgell C, Rodier PM, Spence MA, Minshew N, McMahon WM, Wijsman EM (2006) Evidence for multiple loci from a genome scan of autism kindreds. Mol Psychiatry 11:1049–1060

Sebat J, Lakshmi B, Troge J, Alexander J, Young J, Lundin P, Månér S, Massa H, Walker M, Chi M, Navin N, Lucito R, Healy J, Hicks J, Ye K, Reiner A, Gilliam TC, Trask B, Patterson N, Zetterberg A, Wigler M (2004) Large-scale copy number polymorphism in the human genome. Science 305:525–528

Sebat J, Lakshmi B, Malhotra D, Troge J, Lese-Martin C, Walsh T, Yamrom B, Yoon S, Krasnitz A, Kendall J, Leotta A, Pai D, Zhang R, Lee Y-H, Hicks J, Spence SJ, Lee AT, Puura K, Lehtimäki T, Ledbetter D, Gregersen PK, Bregman J, Sutcliffe JS, Jobanputra V, Chung W, Warburton D, King M-C, Skuse D, Geschwind DH, Gilliam TC, Ye K, Wighler M (2007) Strong association of de novo copy number mutations with autism. Science 316:445–449

Serajee FJ, Zhong H, Mahbubul Huq AHM (2006) Association of reelin gene polymorphisms with autism. Genomics 87:75–83

Shao Y, Wolpert CM, Raiford KL, Menold MM, Donnelly SL, Ravan SA, Bass MP, McClain C, von Wendt L, Vance JM, Abramson RH, Wright HH, Ashley-Koch A, Gilbert J, DeLong RG, Cuccaro ML, Pericak-Vance MA (2002a) Genomic screen and follow-up analysis for autistic disorder. Am J Med Genet (Neuropsychiatr Genet) 114:99–105

Shao Y, Raiford KL, Wolpert CM, Cope HA, Ravan SA, Ashley-Koch AA, Abramson RK, Wright HH, DeLong RG, Gilbert JR, Cuccaro M, Pericak-Vance MA (2002b) Phenotypic homogeneity provides increased support for linkage on chromosome 2 in autistic disorder. Am J Hum Genet 70:1058–1061

Shao Y, Cuccaro ML, Hauser ER, Raiford KL, Menold MM, Wolpert CM, Ravan SA, Elston L, Decena K, Donnelly SL, Abramson RK, Wright HH, DeLong GR, Gilbert JR, Pericak-Vance MA (2003) Fine mapping of autistic disorder to chromosome 15q11-q13 by use of phenotypic subtypes. Am J Hum Genet 72:539–548

Skaar DA, Shao Y, Haines JL, Stenger JE, Jaworski J, Martin ER, DeLong GR, Moore JH, McCauley JL, Sutcliffe JS, Ashley-Koch AE, Cuccaro ML, Folstein SE, Gilbert JR, Pericak-Vance MA (2005) Analysis of the *RELN* gene as genetic risk factor for autism. Mol Psychiatry 10:563–571

Smalley SL (1997) Genetic influences in childhood-onset psychiatric disorders: autism and attention-deficit/hyperactivity disorder. Am J Hum Genet 60:1276–1282

Solinas-Toldo S, Lampel S, Stilgenbauer S, Nickolenko J, Benner A, Döhner H, Cremer T, Lichter P (1997) Matrix-based comparative genomic hybridisation: biochips to screen for genomic imbalances. Genes Chromosomes Cancer 20: 399–407

Stone JL, Merriman B, Cantor R, Yonan AL, Gilliam TC, Geschwind DH, Nelson SF (2004) Evidence for sex-specific risk alleles in autism spectrum disorder. Am J Hum Genet 75:1117–1123

Sutcliffe JS, Delahanty RJ, Prasad HC, McCauley JL, Han Q, Jiang L, Li C, Folstein SE, Blakely RD (2005) Allelic heterogeneity at the serotonin transporter locus (SLC6A4) confers susceptibility to autism and rigid-compulsive behaviors. Am J Hum Genet 77:265–279

Talebizadeh Z, Bittel DC, Veatch O, Butler M, Takahashi TN, Miles JH (2004) Do known mutations in neuroligin genes (*NLGN3* and *NLGN4*) cause autism? J Autism Dev Disord 34:735–736

The Autism Genome Project Consortium (2007) Mapping autism risk loci using genetic linkage and chromosomal rearrangements. Nat Genet 39:319–328

The International HapMap Consortium (2005) A haplotype map of the human genome. Nature 437:1299–1320

Thomas NS, Sharp AJ, Browne CE, Skuse D, Hardie C, Dennis NR (1999) Xp deletions associated with autism in three females. Hum Genet 104:43–48

Vincent JB, Kolozsvari D, Roberts WS, Bolton PF, Gurling HMD, Scherer SW (2004) Mutation screening of X-chromosomal neuroligin genes: no mutations in 196 autism probands. Am J Med Genet Part B (Neuropsychiatr Genet) 129B:82–84

Vorstman JA, Staal WG, van Daalen van Engeland H, Hochstenbach PFR, Franke L (2006) Identification of novel autism candidate regions through analysis of reported cytogenetic abnormalities associated with autism. Mol Psychiatry 11:18–28

Wassink TH, Piven J, Vieland VJ, Jenkins L, Frantz R, Bartlett CW, Goedken R, Childress D, Spence MA, Smith M, Sheffield VC (2005) Evaluation of the chromosome 2q37.3 gene *CENTG2* as an autism susceptibility gene. Am J Med Genet Part B (Neuropsychiatr Genet) 136B:36–44

Weaving LS, Ellaway CJ, Gécz J, Christodoulou J (2005) Rett syndrome: clinical review and genetic update. J Med Genet 42:1–7

Weiss LA, Shen Y, Korn JM, Arking DE, Miller DT, Fossdal R, Saemundsen E, Stefansson H, Ferreira MAR, Green T, Platt OS, Ruderfer DM, Walsh CA, Altshuler D, Chakravarti A, Tanzi RE, Stefansson K, Santangelo SL, Gusella JF, Sklar P, Wu B-L, Daly MJ, for the Autism Consortium (2008) Association between microdeletion and microduplication at 16p11.2 and autism. N Engl J Med 358:667–675

Wendland JR, Martin BJ, Kruse MR, Lesch K-P, Murphy D (2006) Simultaneous genotyping of four functional loci of human *SLC6A4*, with reappraisal of *5-HTTLPR* and rs25531. Mol Psychiatry 11:224–226

Wendland JR, DeGuzman TB, McMahon F, Rudnick G, Detera-Wadleigh SD, Murphy DL (2008) SERT Ileu425Val in autism, Asperger syndrome and obsessive-compulsive disorder. Psychiatric Genet 18:31–39

Wermter A-K, Kamp-Becker I, Strauch K, Schulte-Körner G, Remschmidt H (2008) No evidence for involvement of genetic variants in the X-linked neuroligin genes *NLGN3* and *NLGN4X* in probands with autism spectrum disorder on high functioning level. Am J Med Genet Part B (Neuropsychiatr Genet) 147B:535–537

Wolff DJ, Clifton K, Karr C, Charles J (2002) Pilot assessment of the subtelomeric regions of children with autism: detection of a 2q deletion. Genet Med 4:10–14

World Health Organization (1992) International classification of mental and behavioral disorders. Clinical descriptions and diagnostic guidelines, 10th edn. World Health Organization, Geneva

Yan J, Oliveira G, Coutinho A, Yang C, Feng J, Katz C, Sram J, Bockholt A, Jones IR, Craddock N, Cook EH Jr, Vicente A, Sommer SS (2005) Analysis of the *neuroligin 3* and *4* genes in autism and other neuropsychiatric patients. Mol Psychiatry 10:329–335

Yonan AL, Alarcon M, Cheng R, Magnusson PKE, Spence SJ, Palmer AA, Grunn A, Juo S-HH, Terwilliger JD, Liu J, Cantor RM, Geschwind DH, Gilliam TC (2003) A genomewide screen of 345 families for autism-susceptibility loci. Am J Hum Genet 73:886–897

Ylisaukko-oja T, Nieminen-von Wendt T, Kempas E, Sarenius S, Varilo T, von Wendt L, Peltonen L, Järvelä I (2004) Genome-wide scan for loci of Asperger syndrome. Mol Psychiatry 9:161–168

Ylisaukko-oja T, Rehnström K, Auranen M, Vanhala R, Alen R, Kempas E, Ellonen P, Turunen JA, Makkonen I, Riikonen R, Nieminen-von Wendt T, von Wendt L, Peltonen L, Järvelä I (2005) Analysis of four neuroligin genes as candidates for autism. Eur J Hum Genet 13:1285–1292

Ylisaukko-oja T, Alarcón M, Cantor RM, Auranen M, Vanhala R, Kempas E, von Wendt L, Järvelä I, Geschwind DH, Peltonen L (2006) Search for autism loci by combined analysis of Autism Genetic Resource Exchange and Finnish families. Ann Neurol 59:145–155

Zhang H, Liu X, Zhang C, Mundo E, Macciardi F, Grayson DR, Guidotti AR, Holden JJA (2002) Reelin gene alleles and susceptibility to autism spectrum disorders. Mol Psychiatry 7:1012–1017

Molecular Genetics of ADHD

Virginia L. Misener and Cathy L. Barr

Contents

1	Complexities of the ADHD Phenotype	100
	1.1 Developmental Shifts	101
	1.2 Gender Effects	102
2	Genetic Basis of ADHD	104
3	Environmental Risk Factors	105
4	Molecular Approaches to Gene Identification	106
	4.1 Candidate Gene Studies in ADHD	107
	4.2 Molecular Genetic Studies Using Population-Based Samples	130
	4.3 Genome Scans for ADHD	131
5	Relationship of the Molecular Genetic Findings in ADHD to Reading Disabilities	133
6	Interpretation of the Molecular Genetic Studies	135
7	After the Linkage or Association Finding	135
8	Implications of the Molecular Genetic Findings for Diagnosis and Treatment	138
References		139

Abstract The search for genes influencing the development of attention-deficit hyperactivity disorder (ADHD) has only recently begun, and the preliminary results are very promising. Thus far, over 20 genes have been reported as associated with ADHD by at least one study. Several genome scans have provided significant and suggestive evidence for linkage; however, overlap between regions across independent genome scans is sparse at this point. The quick progress in the genetic findings was initially surprising given the complexity of the phenotype and the relatively small sample sizes used in the initial studies. The studies thus far support the perspective that dysregulation of the neurotransmitter systems, particularly the dopamine and serotonin systems, are involved in ADHD. The majority of studies have focused on the dopaminergic system because of the strong neurobiological and pharmacological support for this system in ADHD. Positive association findings for

---

C.L. Barr (✉)
Room MP14-302, Genetics and Development Division, The Toronto Western Hospital,
399 Bathurst St., Toronto, ON, M5T 2S8, Canada
CBarr@uhnres.utoronto.ca

D.B. Wildenauer (ed.), *Molecular Biology of Neuropsychiatric Disorders*,
Nucleic Acids and Molecular Biology, © Springer-Verlag Berlin Heidelberg 2009

the genes for the dopamine transporter and dopamine receptors D4 and D5 have been reported and replicated in a number of samples, as well as supported by meta-analyses. Not all studies have replicated association findings across genes, however, and nonreplication may result from low power due to small sample sizes, clinical heterogeneity between the samples, or differences in ethnic composition influencing the amount of linkage disequilibrium between the markers studied and the DNA change(s) contributing to the phenotype. Alternatively, false positive findings can occur and, thus, results from genetic studies should be interpreted cautiously. Thus far, most studies have relied on only one or a few markers per gene, with only a few studies having comprehensively examined the genetic information across the gene. Assuming that the polymorphisms tested are not the functional change contributing to the genetic susceptibility, the linkage disequilibrium between the markers tested and the causal variant will not likely be complete, resulting in an underestimate of the contribution of a gene and in false negative findings. Given that the genes studied so far are estimated to contribute only weakly or moderately to the risk for the development of ADHD, additional genes are predicted to be involved, and molecular genetic studies are continuing.

1 Complexities of the ADHD Phenotype

Attention-deficit hyperactivity disorder (ADHD) is the most prevalent psychiatric disorder of childhood, with prevalence estimated at 9% in boys and 3% in girls (Szatmari et al. 1989). The hallmark of ADHD is developmentally inappropriate and impairing levels of inattention, overactivity and impulsivity that begin in the pre-school years. Currently, ADHD is defined by the American Psychiatric Association: Diagnostic and Statistical Manual of Mental Disorders, Fourth Edition (DSM-IV) (American Psychiatric Association 1994) as having three subtypes; inattentive, hyperactive/impulsive, and combined. These subtype designations are established according to a checklist of nine possible symptoms within each of two domains. To meet the criteria for the inattentive subtype requires the presence of at least six of nine possible inattentive symptoms. Similarly, the hyperactive/impulsive subtype requires at least six of nine hyperactive/impulsive symptoms. The combined type requires meeting both of the subtype criteria. In addition, for any diagnosis of ADHD, the DSM-IV requires the presence of some symptoms before the age of 7, the persistence of symptoms for at least 6 months, and evidence for impairment due to these symptoms in two or more settings. The current criteria further stipulate that the symptoms do not occur exclusively during the course of a pervasive developmental disorder, schizophrenia, or other psychotic disorder, and are not better accounted for by another mental disorder such as mood disorder, anxiety disorder, or a personality disorder.

A number of characteristics of ADHD are problematic for genetic studies. Foremost is that the presence of symptoms is based solely on informant report (e.g., parent, self, teacher, clinician). Thus, an informant's bias, based on expectations of age- and situation-appropriate behavior, can influence the reporting of symptoms.

Additional aspects of the phenotype with potential to influence genetic studies include developmental changes in symptoms from early childhood to adulthood, gender differences in prevalence of ADHD, and uncertainties regarding genetic relationships with comorbid disorders.

1.1 Developmental Shifts

Developmental studies of ADHD have repeatedly documented a change in behavioral symptoms over time, with the majority of these studies focused on school-age children and adolescents. Cross-sectional and longitudinal studies indicate that hyperactivity/impulsivity appears earlier and is more prominent in pre-school and school-age children, but often declines as a person matures into adulthood. Inattention problems appear later, at around the time of school entry, and tend to persist beyond adolescence (Hart et al. 1995; Biederman et al. 2000). In adults, symptoms can diminish significantly, with many individuals no longer meeting criteria for the disorder, though rarely with complete loss of all symptoms (Barkley et al. 2002a). This characteristic of ADHD can be problematic for genetic studies where diagnosis of adults is necessary, for example, for studies in extended families or segregation analyses.

The mechanism by which these developmental shifts occur is not known. From a genetic standpoint, it could be that the underlying genetic influences change over the course of development. For example, it is possible that new expression of previously unexpressed genes could contribute to changes in symptoms or symptom severity. The genotype could also manifest itself differently during development, having a different impact on symptoms over time, in combination with other genes or the environment. Support for the emergence of new genetic effects contributing to symptom changes in adolescence comes from a longitudinal study of twins, assessed initially at age 8–9 and assessed again at age 13–14 (Larsson et al. 2004). That study found that the stability of ADHD symptoms was mainly due to the same genetic effects operating at both time points, but with new genetic and nonshared environmental effects in early adolescence contributing to the change in symptoms between childhood and early adolescence. These data indicate that some genes play a role over this entire age range (8–14), while others may emerge during specific developmental periods. Similar longitudinal twin studies covering the other critical time period occurring from adolescence to adulthood have not been published at this point, although family studies indicate that persistence of the disorder into adolescence and adulthood may define a more familial type of ADHD (Faraone et al. 2000).

A number of neurotransmitter systems relevant to ADHD are known to change during development, particularly the dopaminergic and serotonergic systems (Biegon and Greuner 1992; Pick et al. 1999). For example, postmortem studies indicate peak expression of the dopamine transporter protein at 9–10 years of age, decreasing at 50–63 years of age (Meng et al. 1999). These age-related changes could be contributing to the changes in ADHD symptoms over time, and thus, developmental shifts in gene expression should be considered in the design of molecular studies. For example, two Gα subunit proteins are involved in the coupling of dopamine D1 receptors to adenylyl cyclase, $G\alpha_s$ and $G\alpha_{olf}$. In adult rat brains, $G\alpha_{olf}$ is highly expressed in the striatum (nigrostriatial pathway, involved in motor function and in cognitive processes important to ADHD such as motor response inhibition), whereas $G\alpha_s$ is barely detectable in the striatum but highly expressed in the cortex (mesocortical pathway, involved in cognitive functions). In contrast, during development, the striatum expresses $G\alpha_s$ but not $G\alpha_{olf}$ before the first postnatal week in rats, followed by a progressive switch to $G\alpha_{olf}$ during the first 3 postnatal weeks (Zhuang et al. 2000). In our clinical sample of school-age ADHD children (ages 6–16), we have evidence for a significant association of the $G\alpha_{olf}$ gene (*GNAL*) (Laurin et al. 2008). However, based on the expression pattern of this gene in animal models, and assuming similarity in expression in humans, we would not predict $G\alpha_{olf}$ to be a contributing factor in the development of early hyperactivity symptoms in very young children because it is not expressed at this time. $G\alpha_s$ may instead be a candidate.

Environmental risks may also have a different impact across development. For instance, some prenatal and perinatal risk factors may have less impact over time, while effects of family environment may have increasing effects during early childhood, but less during late adolescence. Alternatively, some prenatal risks, such as low birth weight, have shown increasing effects over time (sleeper effects) (McGrath et al. 2000). Animal models of hypoxia/ischemia at birth have found long-term abnormalities in dopamine levels, dopamine turnover, and expression of dopamine receptors, some of which normalize over time. For example, in one model of perinatal hypoxia in rats, there are early reductions in D1 and D2 dopamine receptor mRNA; however, by early adulthood, normal levels are recovered (Gross et al. 2000). Animal models of prenatal and perinatal exposure to nicotine also show alterations in multiple neurotransmitter systems, some of which normalize over time (Ernst et al. 2001). Therefore, environmental risks may have time-limited effects, but as yet this has not been investigated in relation to ADHD symptoms. Given the developmental changes in inattention and hyperactivity/impulsivity symptoms across the lifespan, changes in genetic and environmental influences underlying the symptoms over time are a distinct possibility. Age-related changes should be considered both in the design of experiments, as well as the interpretation of results derived from different age groups.

1.2 Gender Effects

More males than females are affected by ADHD, with sex ratios reported in the range of 3:1–9:1 (Levy et al. 2005). Although ADHD symptoms are similar in boys and girls, there are differences in phenotypic expression of the disorder. Girls are more likely to have the predominantly inattentive type and less likely to have a learning disability in reading or math, comorbid depression, conduct disorder or oppositional defiant disorder (Biederman et al. 2002). The reasons for these gender differences are unclear, and a number of factors may come into play, including referral bias for clinical studies and differential reporting by parents and/or teachers (i.e., a boy may be more likely to be reported as having ADHD than a girl). However, these factors cannot entirely explain the observed gender differences, and epidemiological studies suggest that true gender effects are present in the population. For the most part, twin studies have found no evidence for genetic or environmental differences between boys and girls (ACE parameters are equal between the two), indicating that the risk genes will be the same (Gjone et al. 1996; Thapar et al. 2000; Price et al. 2001; Rietveld et al. 2004; Hudziak et al. 2005; Kuntsi et al. 2005; McLoughlin et al. 2007). One study of DZ twins suggests that a polygenic multiple threshold model (an individual is affected with ADHD if he/she exceeds a certain threshold of genetic liability) best explains the sex differences in ADHD prevalence (Rhee et al. 2001). That is, females are less likely to be affected because they require a greater threshold to express the disorder. Supporting this model, higher rates of affected relatives have been observed for female probands than for male probands in family studies (Rhee et al. 2001).

Gender differences are found in a number of autosomal inherited disorders. It is particularly interesting to note that disorders thought to be dopamine related, e.g., Tourette Syndrome, Parkinson's disorder and schizophrenia, are more prevalent in males than females, and a protective role of estrogen has been proposed. Estrogens have effects on dopaminergic (McEwen and Alves 1999; Dluzen 2000; Zhou et al. 2002), noradrenergic (McEwen and Alves 1999; Herbison et al. 2000), and serotonergic transmission (McEwen and Alves 1999; Zhou et al. 2002), and there is evidence that estrogen inhibits the dopamine transporter (Dluzen 2000). This may provide a protective effect in females by increasing synaptic dopamine. Estrogen has both regional and temporal effects on striatal dopamine receptor expression. In particular, the D1, D5, and D2 receptors have been shown to be upregulated by estrogen (Dluzen 2000). The mechanism for this is poorly understood, however for some genes, such as the dopamine D1 receptor gene (*DRD1*), increased transcription is due to an estrogen response element in the promoter (Lee and Mouradian 1999). Based on this biological information, predictions about gender effects can be modeled for molecular studies when risk alleles are identified. For example, one could speculate that if a *DRD1* risk allele contributes to decreased expression of the receptor, then females would be more likely to require two copies (rather than a single copy) of the risk allele for development of the ADHD phenotype, since the upregulation of *DRD1* expression by estrogen might compensate, to some extent, for the down-regulatory effect of the risk allele. Further, one would predict a reduction in symptoms at/after puberty in females. Although these effects can be modeled in molecular analyses, current studies are hampered by low numbers of girls in clinical samples. To obtain power for gender-specific analyses, much larger sample sizes or specific recruitment of girls is required.

2 Genetic Basis of ADHD

Numerous family studies have been published for ADHD, particularly during the 1990s. Those studies, in total, provide strong evidence for the familiality of ADHD, and segregation analyses have been consistent with a major locus effect with low penetrance and polygenic transmission (Faraone et al. 1992). A particular focus of these studies was the genetic relationship of ADHD with comorbid disorders (Biederman et al. 1991a, b, 1992; Wozniak et al. 1995; Faraone and Biederman 1997; Faraone et al. 1991, 1993, 1997, 1998). For example, family studies of conduct disorder (CD) and ADHD indicate that these disorders fall along a continuum of increasing levels of familial aetiological factors and, correspondingly, severity of illness (Faraone et al. 1991). Family risk analysis indicates that mood disorders and ADHD share common familial etiological factors (Biederman et al. 1992), while anxiety and ADHD segregate independently (Biederman et al. 1991a).

Twin studies are more powerful, in that they allow for the disentanglement of environmental factors and provide for more precise determination of the genetic contribution to the phenotype. Further, multivariate analyses can provide estimates of the genetic correlation (r_g), an estimate of the percentage of genes that are common to the phenotypes. These analyses allow for the inheritance to be appropriately modeled for the molecular genetic studies, thereby increasing power. We can conclude from twin studies conducted thus far that both ADHD, defined categorically, and ADHD symptoms are highly heritable (70–90%) (Thapar et al. 1995, 2000; Gjone et al. 1996; Eaves et al. 1997; Levy et al. 1997; Sherman et al. 1997). Heritability is high for quantitative analysis of hyperactivity scores (Thapar et al. 1995), item analysis of individual symptoms (Levy et al. 2001), and latent class analysis of DSM-IV symptoms (Todd et al. 2001c; Rasmussen et al. 2002). Because the heritability is similar for the extreme end (ADHD as a diagnosis), as for the entire distribution of ADHD symptoms, ADHD is thought to represent the extreme of the normal distribution of symptoms, and that genes contributing to ADHD as a clinical diagnosis will contribute to ADHD symptoms as quantitative trait loci in the population.

Discrepancies among informants of ADHD behavior have been extensively documented, but both parent- and teacher-rated ADHD symptoms show a high degree of heritability (Thapar et al. 2000; Martin et al. 2002). Interestingly, several studies have provided evidence for a common genetic factor underlying parent and teacher ratings of symptoms, together with additional, informant-specific genetic influences (Thapar et al. 2000; Martin et al. 2002). Multivariate analyses indicate that there are common as well as unique genes contributing to the symptom dimensions of inattention and hyperactivity/impulsivity ($r_g = 0.57$–0.62) (Levy et al. 1997; Sherman et al. 1997; McLoughlin et al. 2007), supporting the utility of analyzing the symptom dimensions of inattention and hyperactivity/impulsivity separately in genetic studies.

Multivariate genetic analyses have also been used to investigate comorbid disorders. Such analyses indicate that genetic factors for symptoms of ADHD and reading ability will overlap, particularly in the case of inattention symptoms ($r_g \sim 0.36$–0.70) (Willcutt et al. 2007). For CD and ADHD, the genetic correlation indicates very high genetic overlap of 50–100% (Nadder et al. 1998; Thapar et al. 2001). Analysis of CD and oppositional defiant disorder (ODD) indicate that they represent the same genetic liability (Nadder et al. 2002).

3 Environmental Risk Factors

As summarized above, twin studies indicate that environmental factors are also involved in ADHD, but that these account for far less of the variance than do genetic factors. However, additive genetic and environmental estimates can be inflated or obscured by gene × environment (G × E) interactions (Rutter and Silberg 2002). Recently, several high impact publications sparked renewed attention to the role of G × E interactions in psychiatric disorders (Caspi et al. 2002, 2003), including ADHD (Kahn et al. 2003; Thapar et al. 2005; Brookes et al. 2006a; Laucht et al. 2007; Neuman et al. 2007). The contribution of specific, measured, environmental risk factors to disorder susceptibility is of extreme interest to geneticists, as the correct incorporation of a specific environmental risk into the model can increase power, as well as provide valuable information on the development of risk. More critically, the understanding of environmental risk factors holds promise for prevention. For example, maternal smoking during pregnancy has been indicated as a risk factor for development of ADHD and ADHD symptoms in the offspring in a number of studies (Mick et al. 2002; Batstra et al. 2003; Kotimaa et al. 2003; Linnet et al. 2003; Thapar et al. 2003; Barman et al. 2004; Braun et al. 2006).

Before specific environmental risks are incorporated into the molecular analyses, it must first be determined if there is an interaction or correlation effect, so that the environmental risks can be modeled appropriately (Moffitt et al. 2005). G × E interaction indicates that there is a genetic sensitivity, or susceptibility to the environment. Thus far, twin analyses testing for specific G × E interactions in ADHD have been limited, and interaction effects have not been found to date (Knopik et al. 2005). Therefore, it is unknown if environmental risk factors work together with genetic variation to create risk. Some risks could be completely environmental, resulting in phenocopies (Sprich-Buckminster et al. 1993), or they could act additively, increasing risk to genetically predisposed individuals (Knopik et al. 2005). In support of an additive effect, the twin study of Knopik and colleagues (2005) examined the relationship between genetic transmission and parental alcohol dependency, smoking, and low birth weight in female twin pairs. Mother and father alcohol dependency, maternal heavy alcohol use during pregnancy, and low birth weight predicted ADHD in the offspring. Low birth weight did not act as a mediator in that sample. Interestingly, risk was not significantly associated with maternal smoking once other risk factors were accounted for, and no significant evidence for G × E interaction was identified.

A number of environmental risk factors have been found to be associated with ADHD, but no one specific factor has been unequivocally identified (Milberger et al. 1997b; Ernst et al. 2001; Linnet et al. 2003; Braun et al. 2006). For example, complications in pregnancy, delivery and infancy (PDICs), and chronic exposure during pregnancy (smoking, alcohol, illicit drugs) have been reported to increase risk (Milberger et al. 1997b; Linnet et al. 2003), but this has not been found in all studies (Sprich-Buckminster et al. 1993; Weissman et al. 1999; Linnet et al. 2003). In one study, the relationship with PDICs was strongest for ADHD with comorbidities and for nonfamilial cases, suggesting that these risks may produce phenocopies (Sprich-Buckminster et al. 1993).

Low birth weight has also been indicated as a risk factor for ADHD (Szatmari et al. 1990; Breslau et al. 1996; Botting et al. 1997; Nadeau et al. 2001; Bhutta et al. 2002; Foulder-Hughes and Cooke 2003; Hultman et al. 2007), although some studies did not reveal such an association (e.g., Chandola et al. 1992; Sommerfelt et al. 1996). A recent study of monozygotic twins discordant for ADHD symptoms indicated that lower birth weight twins had significantly higher ADHD symptoms than heavier twins (Hultman et al. 2007). The effects were similar in monozygotic compared to dizygotic twins, indicating no evidence for a strong genetic contribution. Furthermore, there was no indication of a genetic correlation, indicating that these effects were not due to shared genetic influences on intrauterine growth and ADHD (Hultman et al. 2007). Psychosocial adversity (low SES, family conflict, maternal psychopathology) has also been associated with ADHD, with risk increasing as the number of risk factors increased (Biederman et al. 1995, 2002).

We note that these reported environmental risks are not specific to ADHD, and have been indicated as risks for a number of neurodevelopmental disorders and cognitive deficits (McGrath et al. 2000; Ernst et al. 2001). The current assumption is that specificity arises from impact of the environmental factor in genetically predisposed individuals.

In addition to there being limited information on the contribution of specific environmental risks to ADHD, the capacity to effectively incorporate these factors into molecular analyses is also limited. For example, to investigate interaction effects involving genetic variation and maternal smoking during pregnancy, a substantial number of these families would need to be recruited to the study. Estimates in the mid-1990s indicated that 19–27% of mothers in the US smoked during pregnancy (DiFranza and Lew 1995). Further, there is some indication of gender differences in the risk to offspring, potentially reducing power (Weissman et al. 1999). With increased education to mothers on the risks of smoking and substance abuse during pregnancy, these types of risk factors are, fortunately, decreasing. Thus, to study these risks, family collection must be extremely large, generally requiring multiple recruitment sites, or must focus on high-risk populations.

4 Molecular Approaches to Gene Identification

Based on the high heritability estimates, a number of groups worldwide have moved to molecular studies to identify genes that contribute to susceptibility to develop ADHD. Generally, the search for genes contributing to a disorder takes two approaches. One is a directed test of certain genes, based on specific biological hypotheses. This is the candidate gene approach. To test genes for linkage or association, DNA markers are used. These are variants in the DNA code, each with a known position on the genetic or physical map. The critical feature of DNA markers is that they are polymorphic, i.e., each is characterized by different DNA sequence variants (alleles) in the population. The nonrandom inheritance of these DNA variants in families (linkage) or populations (association) suggests a relationship of the gene studied to the disorder. In the second approach, rather than testing specific candidate genes, chromosomal regions that are likely to contain susceptibility genes are identified. This is achieved by examining the inheritance of DNA markers in families, and using a statistical approach to determine if chromosomal regions are shared by affected individuals more often than expected by chance. This often takes the form of a genome scan, whereby DNA markers spaced at regular intervals across the chromosomes are tested, such that the inheritance of each chromosomal region can be determined. This leads to the identification of regions containing susceptibility genes, which in turn can be identified by fine mapping of the target region and/or by selecting, for further analysis, plausible candidate genes in the region. Both the candidate gene and genome scan approaches have been used successfully in genetic studies of ADHD. A number of susceptibility genes, as well as chromosomal regions that point to the location of additional genes, have been identified.

4.1 Candidate Gene Studies in ADHD

Most genetic studies of ADHD, thus far, have used the candidate gene approach, in which candidates are tested based on biological plausibility, usually derived from animal models, pharmacological treatments used in ADHD, neuropsychological findings, and/or brain systems involved in attention and hyperactive behavior. In total, there is much evidence to suggest that genetic factors contributing to changes in synaptic levels of neurotransmitters, especially dopamine, norepinephrine, and serotonin, and/or to changes in neurotransmitter signalling responses, are likely to be involved in the biological underpinnings of ADHD. As a result, genes controlling the dopaminergic, noradrenergic, and serotonergic systems have been a major focus in genetic studies. In particular, genes encoding the receptors for these neurotransmitters, those encoding proteins involved in neurotransmitter release and re-uptake, and those encoding enzymes involved in neurotransmitter biosynthesis and degradation have been the most widely studied. Among these, genes involved in the dopamine system have received the most attention. Thus far, the genetic results support a long-held hypothesis that subtle abnormalities in these systems could account for the symptoms of ADHD (Levy and Swanson 2001). Candidate gene studies have used both case-control and family-based designs. In the case-control approach, allele and/or genotype frequencies of a marker are compared between ADHD cases and screened controls. In the family-based approach, alleles not passed to the offspring serve as the ethnically matched control for the case. Unless a causal variant is assayed directly, both types of association study rely on linkage disequilibrium (LD) between the assayed variant and causal variant contributing to ADHD susceptibility. LD refers to the situation that arises when a particular marker allele lies very close to the susceptibility allele, such that the alleles are inherited together over many generations. As a result, the presence of the causal variant will be detectable by the same marker allele in multiple affected individuals, coming from multiple, apparently unrelated, families. In case-control analyses, allelic association is identified by a significantly increased or decreased frequency of a particular marker allele in cases compared to controls, and in family-based analyses, by preferential transmission of the marker allele from parents to children with the trait. In both instances, genetic association represents deviations from the random occurrence of the alleles with respect to the studied phenotype. A statistic commonly used in family-based tests of association is the transmission disequilibrium test (TDT) (Spielman and Ewens 1996), in which the frequency with which an allele is transmitted from heterozygous parents to affected offspring is compared to that of an alternative allele. Under the null hypothesis of no association, each allele has an equal chance of being transmitted. Deviation from equal transmissions suggests that the DNA variant that distinguishes the allele either is, or is in LD with, the DNA variant contributing to the measured phenotype or disorder. Often, multiple markers across a gene are used in the analysis, as this may increase power to detect association. The combination of alleles on a chromosome received by an individual from one parent is termed the haplotype, for haploid genotype. Haplotypes are often used in genetic studies, as this can provide more information, in many instances, than a single marker alone.

4.1.1 The Dopamine Transporter

The first molecular genetic study of ADHD, published in 1995 (Cook et al. 1995), examined the potential role of the dopamine transporter gene, *DAT1* (official nomenclature is *SLC6A3*; solute carrier family 6, neurotransmitter transporter, dopamine, member 3). *DAT1* was considered to be a strong candidate for involvement in ADHD, as the dopamine transporter regulates synaptic dopamine levels through re-uptake, and is the primary target for methylphenidate. In their study of *DAT1*, Cook and co-workers reported a significant association of ADHD with the 10-repeat (480 bp PCR fragment) allele of the variable number of tandem repeats (VNTR) polymorphism located in the 3'-untranslated region (3'-UTR) of the gene (Cook et al. 1995). This finding was later replicated in a number of independent studies (Gill et al. 1997; Waldman et al. 1998; Daly et al. 1999; Curran et al. 2001; Chen et al. 2003; Brookes et al. 2006a; Lim et al. 2006), though several nonreplications have also been reported (Palmer et al. 1999; Holmes et al. 2000; Curran et al. 2001; Roman et al. 2001; Todd et al. 2001a; Muglia et al. 2002a; Smith et al. 2003; Qian et al. 2004; Kim et al. 2005b; Langley et al. 2005; Simsek et al. 2005; Cheuk et al. 2006b)

Molecular Genetics of ADHD

(Barr et al. 2001c; Kustanovich et al. 2004; Bakker et al. 2005; Bobb et al. 2005; Simsek et al. 2006; Das and Mukhopadhyay 2007; Genro et al. 2007).

A number of meta-analyses and analyses of pooled odds ratios across ADHD studies have been conducted for *DAT1*, based on the VNTR alleles (Maher et al. 2002; Faraone et al. 2005; Purper-Ouakil et al. 2005; Li et al. 2006a; Yang et al. 2007). These studies have indicated the contribution of this gene to be from none to modest, with odds ratios in the range of 1.04–1.27. However, these analyses, based on transmission of the 3'-UTR VNTR alleles, will underestimate the effect of this gene if the 3'-UTR VNTR polymorphism is not, itself, the causal variant.

The 3'-UTR VNTR has been the focus of the majority of molecular genetic studies of the *DAT1* gene in ADHD because it was the first polymorphism identified for this gene and some *in vitro* expression assays indicated that the VNTR alleles influence transcription or protein levels. However, not all studies have confirmed this. For example, one study found expression to be significantly higher when the 3'-UTR containing the 10-repeat allele was transfected into the African Green Monkey kidney cell line (COS7), compared to constructs harboring either the 7- or 9-repeat alleles (Fuke et al. 2001), while another study found the opposite result, with greater expression observed when constructs containing the 9-repeat allele were transfected into human embryonic kidney (HEK-293) cells (Miller and Madras 2002). Further, the study by Miller and colleagues identified a DNA variant (T/C) located 134 bp downstream of the VNTR, recognized by the restriction enzyme *Dra*I, which was reported to influence levels of protein expression in a promoter-dependent manner. Two additional studies, using either a mouse dopaminergic cell line (Greenwood and Kelsoe 2003) or human neuroblastoma and human embryonic kidney cell lines (Mill et al. 2005a), found no differential effects on transcription by the 10- or 9-repeat alleles. The differing results from the *in vitro* studies may be a function of differences in assay design, such as promoter specificities of the reporter constructs, cell types used for transfection and expression of the constructs, and/or sequences around the VNTR included in the construct. For example, one study found that inclusion of the ~800 bp sequence extending downstream from the stop codon to just 5' of the VNTR, but not including the VNTR, can itself influence *DAT1* transcription, mRNA stability, or translation, as evidenced by lower transporter protein density in an *in vitro* assay (VanNess et al. 2005). Thus, inclusion of this sequence in a VNTR construct may also contribute independently to the amount of protein produced.

In our ADHD sample, we did not find significant evidence for association with the VNTR 10-repeat (480 bp) allele; however, we did find significant evidence for association with one of the haplotypes containing this allele (Barr et al. 2001c). Our finding of stronger evidence for association of a haplotype containing the 10-repeat allele, as compared to the allele alone, argues that this allele is unlikely to be the sole contributor to the *DAT1* association with ADHD. In our 2005 study, we continued to explore the relationship of this gene to ADHD (Feng et al. 2005b), genotyping DNA variants around the VNTR region to investigate whether other DNA variation in the *DAT1* 3'-UTR could be contributing to ADHD. One of the variants analyzed was the *Dra*I DNA change (T/C) that had previously been reported to influence *DAT1* expression levels (Miller and Madras 2002). We also screened the VNTR region in ADHD probands, by direct resequencing, to determine whether there was sequence variation within the repeat units that could account for the association. We did not find any variation in the sequence for either the 10- or 9-repeat alleles in the probands screened, nor did we observe the reported *Dra*I (T/C) variation (Feng et al. 2005b). Thus, we can rule out the *Dra*I sequence change and sequence variation within the repeats as contributors.

Our most significant finding for the *DAT1* gene, in our sample of ADHD families, is with a polymorphism, rs27072 (Feng et al. 2005b), located 442 bp 5′ of the VNTR, in a region reported to influence gene regulation, as discussed above (VanNess et al. 2005). We have since replicated the association of rs27072 in a population-based twin sample from Quebec, finding significant evidence for association to inattention and hyperactive/impulsive symptoms with the rs27072 polymorphism, but not with the VNTR (Ouellet-Morin et al. 2007).

Recently, some other studies that have begun to examine additional polymorphisms within the *DAT1* gene have also yielded more significant results than those for the VNTR alone (Hawi et al. 2003; Galili-Weisstub et al. 2005; Brookes et al. 2006a; Asherson et al. 2007). The largest of these studied two samples, one from the UK and the other from Taiwan, and found, in each sample, significant evidence for association with both the 3′-UTR VNTR and an intron 8 VNTR, as well as a specific haplotype of the two (Brookes et al. 2006a). The authors recently confirmed this specific haplotype association in an independent study sample, with an overall odds ratio of 1.4 across all three samples (Asherson et al. 2007).

The implication of these studies is that the 3′-UTR VNTR of *DAT1* is not the functional variant involved in ADHD, or if it is contributing to function, it is not the sole contributor within this gene. The predominant use of a single *DAT1* marker for ADHD studies has been a major limitation in the interpretation of association findings for this gene. A number of imaging (Heinz et al. 2000; Jacobsen et al. 2000; Martinez et al. 2001) and pharmacogenetic (Winsberg and Comings 1999; Roman et al. 2002b; Kirley et al. 2003) studies, and recently G × E interaction studies (Kahn et al. 2003; Brookes et al. 2006a), based on the 3′-UTR VNTR polymorphism, have been published, but with conflicting findings. Therefore, the conclusions of these papers regarding the role of *DAT1* in these phenotypes should not be considered definitive until additional polymorphisms are more comprehensively studied. Based on the recommendations of the *DAT1* Workshop at the International Society of Psychiatric Genetics Meeting 2005, (Chaired by Dr. Barr), continued study of this gene should concentrate on polymorphisms that tag the major haplotypes for *DAT1*, as well as the reported functional polymorphisms (several promoter polymorphisms and several nonsynonymous amino acid changes) (Greenwood and Kelsoe 2003; Kelada et al. 2005). The importance of expanding the *DAT1* analyses to include polymorphisms other than those in the 3′ region, has been taken into account in some recent studies (Langley et al. 2005; Brookes et al. 2006b; Ohadi et al. 2006; Friedel et al. 2007; Genro et al. 2007). Among these, evidence for ADHD association with two promoter

Molecular Genetics of ADHD

polymorphisms, -839C/T and -67A/T, has been reported (Ohadi et al. 2006; Genro et al. 2007), and in a gene-wide study of 30 SNPs, significant evidence for association of polymorphisms located in the fourth intron of the gene was obtained (Friedel et al. 2007). Further detailed investigations of the *DAT1* gene in relation to ADHD are clearly warranted.

4.1.2 Dopamine Receptor D4

Another gene showing strong evidence for involvement in ADHD is the gene for dopamine receptor D4 (*DRD4*). Association between ADHD and the 7-repeat allele of a 48 bp VNTR located in the third exon of the *DRD4* gene was first reported in 1996, using a case-control approach (LaHoste et al. 1996), and has since been replicated in a number of studies (Rowe et al. 1998; Smalley et al. 1998; Swanson et al. 1998; Faraone et al. 1999; Holmes et al. 2000; Sunohara et al. 2000; Tahir et al. 2000; Mill et al. 2001; Roman et al. 2001). While several studies failed to replicate association with the 7-repeat allele (Castellanos et al. 1998; Eisenberg et al. 2000; Hawi et al. 2000a; Kotler et al. 2000; Kirley et al. 2002; Smith et al. 2003; Bakker et al. 2005), others have identified association with other alleles of this polymorphism in different ethnic groups (Kotler et al. 2000; Manor et al. 2002a; Leung et al. 2005). In some cases, the ability to replicate may have been hindered by small sample size. For example, while an initial study of 41 cases and 56 controls by Castellanos et al (1998) did not find evidence for association with the exon 3 VNTR, evidence for association with the 7-repeat allele has since been obtained in a follow-up study, by the same group, using an expanded study sample of 166 cases and 282 controls (Gornick et al. 2007). Several meta-analyses published for this gene support the association between the *DRD4* 7-repeat allele and ADHD (Faraone et al. 2001; Maher et al. 2002; Li et al. 2006a), with pooled analyses for studies of this polymorphism, up to 2005, indicating a significant association for case-control (OR = 1.45 [95% CI 1.27–1.65]) and family-based (OR = 1.16 [95% CI 1.03–1.30]) association studies (Faraone et al. 2005).

The 48 bp VNTR is located in the third intracellular loop of the receptor, and has been shown to influence dopamine receptor affinity (Asghari et al. 1994) and intracellular signalling (Asghari et al. 1995). These studies indicate that the 7-repeat allele is slightly less sensitive to dopamine than the 4-repeat allele, leading to the suggestion that the 48 bp VNTR polymorphism is the functional change contributing to the ADHD phenotype, i.e. susceptibility being attributable to a hypodopaminergic state caused by the 7-repeat. Of note, there is no direct relationship between the allele length *per se* and changes in pharmacology or functional activity (Jovanovic et al. 1999). For example, whereas the longer (7-repeat) allele is slightly *less* sensitive to dopamine than the shorter (4-repeat) allele, the longest (10-repeat) allele is slightly *more* sensitive compared to the shortest (2-repeat) allele (Jovanovic et al. 1999). Thus, some studies of ADHD that have grouped "long" (e.g. 6–8 repeats) versus "short" (e.g. 2–5 repeats) alleles (Eisenberg et al. 2000; Kotler et al. 2000) would have pooled alleles that are not functionally similar, likely confounding the association results. Moreover, such studies have not all used the same grouping scheme to pool the alleles (Eisenberg et al. 2000; Kotler et al. 2000; Qian et al. 2004; Cheuk et al. 2006a).

Further confounding association studies is the finding that there are dramatic differences in allele frequencies of the 48 bp VNTR across populations, as well as sequence variation within the repeat units (Lichter et al. 1993). For example, the frequency of the 7-repeat allele is found to range from as high as 48.3% in Amerindians to as low as 1.9% in East and South Asians (Chang et al. 1996), with differences in the distribution of the repeat units across populations as well (Lichter et al. 1993). Thus far, the possible contribution of the sequence variation within the repeat units to receptor function has not been tested, and association by repeat length alone may obscure functional differences within the repeats.

The existence of different allele frequencies in different populations will influence the power to detect association, and has the potential to confound case-control association studies if the ethnic composition of cases and controls is not exactly matched. Thus, ethnicity of the sample is an important consideration for the study of the *DRD4* gene. It has been suggested, by some, that the 2-repeat allele, which is rare in Caucasians but found at much higher frequency in many Asian populations, may be a functionally similar derivative of the 7-repeat, and might therefore be expected to be associated with ADHD in Asian study samples (Brookes et al. 2005b; Leung et al. 2005). However, only one study, based on a small number of Han Chinese ADHD subjects ($n = 32$), has found evidence for this (Leung et al. 2005). Other studies of ADHD from Chinese, Taiwanese, and Korean populations (all with considerably larger sample sizes) have not found evidence for association with the 2-repeat allele (Qian et al. 2004; Brookes et al. 2005b; Kim et al. 2005b; Cheuk et al. 2006a). Thus, the balance of evidence, to date, does not support a relationship between the 2-repeat allele and ADHD in these populations.

In addition to the 48 bp VNTR in exon 3, several other *DRD4* mutations/polymorphisms exist, and are either known or likely to cause a change in function. For example, there are several coding region variants in the first exon of the gene, including a 13 bp (Nothen et al. 1994) and a 21 bp (Cichon et al. 1995) deletion, and a 12 bp repeat (Catalano et al. 1993). There are also a number of variants in the promoter region, including a 120 bp repeat (Seaman et al. 1999; D'Souza et al. 2004), a 27 bp deletion (Szantai et al. 2005), and several SNPs, such as −616C/G, −521C/T and −376C/T (Okuyama et al. 1999; Szantai et al. 2005). The −521C/T SNP and the 120 bp repeat variant have each been found to influence expression levels in *in vitro* transcription assays (Okuyama et al. 1999; D'Souza et al. 2004).

Several studies have investigated the inheritance of *DRD4* polymorphisms other than the 48 bp VNTR in ADHD (Barr et al. 2000b, 2001a; McCracken et al. 2000; Payton et al. 2001a; Todd et al. 2001b; Mill et al. 2003; Lowe et al. 2004b; Bakker et al. 2005; Kereszturi et al. 2007; Yang et al. 2008), some of which have provided further evidence for association with polymorphisms and haplotypes. McCracken and colleagues (McCracken et al. 2000; Kustanovich et al. 2004) investigated the 120 bp repeat polymorphism located in the promoter region of the gene, and found a significant association with the 2-repeat allele, an allele

Molecular Genetics of ADHD

that has been correlated with lower transcriptional activity in *in vitro* transfection assays (D'Souza et al. 2004). However, the association with this allele was not replicated in other clinical (Barr et al. 2001a; Mill et al. 2003; Lowe et al. 2004b; Brookes et al. 2005b; Kereszturi et al. 2007) or population-based (Todd et al. 2001b) samples, and in one such study, association with the opposite allele was found (Kereszturi et al. 2007). Other studies have found evidence for association with the C allele of −616C/G (Lowe et al. 2004b) and the T allele of −521C/T (Yang et al. 2008), but these findings have also not been replicated (Barr et al. 2001a; Payton et al. 2001a; Mill et al. 2003; Lowe et al. 2004b; Kereszturi et al. 2007). We (Barr et al. 2000b) and others (Mill et al. 2003; Bakker et al. 2005) have also analyzed a (G)n mononucleotide repeat polymorphism located in intron 1 (Petronis et al. 1994). None of these studies found evidence for association with the (G)n repeat individually and, in a 2-marker haplotype analysis of the (G)n repeat and 48 bp VNTR, by Bakker and colleagues (2005), evidence for association was also not identified (Bakker et al. 2005). By contrast, our haplotype analysis of 3 markers showed significant evidence for biased transmission of one particular haplotype containing the 7-repeat allele of the 48 bp VNTR, the 140 bp allele of the intronic (G)n repeat, and the 2-repeat allele of the 12 bp repeat polymorphism in exon 1 (Barr et al. 2000b). Also significant was the observation of nontransmission of a haplotype containing the 4-repeat allele of the 48 bp VNTR, the 142 bp allele of the (G)n repeat, and the 2-repeat allele of the 12 bp repeat, indicating the possibility of a protective allele on this haplotype (Barr et al. 2000b). Similarly, in a separate study of the 48 bp VNTR together with three promoter polymorphisms, the 120 bp repeat and two SNPs, −616C/G (recognized by the restriction enzyme *Ava*II) and −521C/T (recognized by *Fsp*I), we also found significant evidence for biased transmission of a particular 7-repeat-containing haplotype (2-repeat of 120 bp, −616C, −521T and 7-repeat of the 48 bp VNTR), as well as a trend for nontransmission of a particular 4-repeat-containing haplotype (2-repeat of 120 bp, −616G, −521T and 4-repeat of the 48 bp VNTR) (Barr et al. 2001a). Further support for the existence of a protective allele has also been provided by a recent meta-analysis indicating a protective effect of the 4-repeat allele, with an OR of 0.90 (95% CI 0.84–0.97) (Li et al. 2006a). We note, however, that in a 5-marker haplotype analysis, Mill and colleagues (2003) found evidence for overtransmission of a particular haplotype comprised of the 2-repeat of 120 bp, −616C, −521C, the 142 bp allele of the (G)n repeat, and the 4-repeat of the exon 3 VNTR. Thus, it would appear that neither risk conferred by the 7-repeat nor protection conferred by the 4-repeat is likely to be a clear cut phenomonen. Rather, it seems more likely that such effects will depend on the particular combination of variants making up the haplotype.

The importance of considering "risk" alleles in the context of the haplotype in which they occur has recently been illustrated in the study of the adrenergic β2 receptor gene and asthma (Drysdale et al. 2000). Several nonsynonymous coding SNPs have been associated with asthma, and these variants alter various aspects of receptor function, including binding, coupling to stimulatory G protein, and trafficking. By examining these coding variants together with promoter polymorphisms in the context of the haplotypes on which they commonly occur, Drysdale and colleagues first demonstrated that, in asthmatics, specific haplotype pairs were associated with mean bronchodilator response to the β agonist albuterol. This response differed by 2-fold for the haplotypes, but there was no relationship to individual SNPs. Second, they demonstrated that the haplotypes were correlated with the expression of mRNA and protein, measured using *in vitro* expression assays, and that this was consistent with the direction and magnitude of the in vivo response to albuterol. The results of these *in vitro* studies differed, however, from the results obtained for the SNPs when studied in isolation. Thus, analogous to this study of haplotypes of the adrenergic β2 receptor gene in relation to asthma, elucidation of the combinatorial effects of multiple functional variants within the context of naturally occurring haplotypes is an avenue of research that needs to be explored in functional studies of the *DRD4* gene in relation to ADHD. It is conceivable, for example, that a combination of multiple "low-expressor" promoter alleles together on the same haplotype as the 7-repeat allele of the exon 3 VNTR could result in lower expression of a protein subsensitive to dopamine, thereby conferring risk for ADHD. Investigations into the functional nature of *DRD4* haplotypes have not yet been reported.

4.1.3 Dopamine Receptor D5

Collective evidence from a number of studies also supports the involvement of the dopamine receptor D5 gene, *DRD5*, in ADHD. The first of these studies (Daly et al. 1999) found significant evidence for an association between ADHD and the 148 bp allele of a microsatellite repeat polymorphism located 18.5 kilobases (kb) 5′ of the *DRD5* gene (Sherrington et al. 1993), and this was replicated in a later study (Manor et al. 2004). We and others detected trends for biased transmission of this allele, though these trends did not reach statistical significance (Barr et al. 2000c; Tahir et al. 2000; Payton et al. 2001a; Kustanovich et al. 2004). While other studies found no evidence of association of this allele (Mill et al. 2004a; Bakker et al. 2005; Bobb et al. 2005), a joint analysis of 14 independent samples, using both previously published studies as well as unpublished data, has strongly supported the association between ADHD and the 148 bp microsatellite allele of *DRD5* ($P = 0.00005$, OR = 1.24) (Lowe et al. 2004a). Several other meta-analyses and pooled analyses also support involvement of this gene in ADHD (Maher et al. 2002; Faraone et al. 2005; Li et al. 2006a), with the most recent study indicating a significant relationship with the 148 bp allele of the (CA) repeat [$p = 8 \times 10^{-8}$; OR = 1.34 (95% CI 1.21–1.50)], and the suggestion of a protective effect of the 136 bp allele [OR = 0.57 (95% CI 0.34–0.96)] (Li et al. 2006a). No evidence of biased transmission of the 146 bp allele was found.

Recent studies have sought to extend these findings by testing additional markers in the *DRD5* gene (Hawi et al. 2003; Mill et al. 2004a). Analysis of two additional microsatellite markers upstream of the gene and a marker located at position 1,481 in the 3′-UTR identified significant evidence for ADHD association to alleles at

Molecular Genetics of ADHD   115

two of the markers, and to haplotypes of these markers that included the previously identified risk allele (148 bp of the CA repeat) (Hawi et al. 2003). In a separate study, analysis of a marker located in the promoter region of the gene, and therefore closer to the gene than the original CA repeat marker, did not provide evidence for association; however, a trend with an allele of a marker located 131 kb 3′ of the gene has been observed (Mill et al. 2004a).

While further studies of this gene are clearly needed, analysis of additional polymorphisms within the gene and screening for DNA variants by sequence analysis are hampered by the existence of two pseudogenes with high (94%) homology to *DRD5* (Nguyen et al. 1991). Primers must be carefully designed and the experiments appropriately controlled in order to avoid confounding of experiments by these pseudogenes.

4.1.4 Dopamine Receptors D1, D2, and D3

Among the other three receptors for dopamine (dopamine receptors D1, D2, and D3), all of which have been tested as candidates for involvement in ADHD, replicated evidence for association has been obtained for *DRD1*, but not for *DRD2* or *DRD3*.

The dopamine receptor D1 (designated D_{1A} in rodents) is the founding member of the D1 subfamily of dopamine receptors – receptors that mediate adenylyl cyclase activation and phosphoinositide hydrolysis via coupling to heterotrimeric G proteins, Gs and Gq, respectively (Wang et al. 1995; Jin et al. 2001). D1 receptors are prevalent in brain regions implicated in ADHD, including the prefrontal cortex (PFC) and striatum (Jackson and Westlind-Danielsson 1994; Missale et al. 1998). Another line of evidence implicating *DRD1* in the etiology of ADHD is the hyperactive phenotype of D_{1A}-knockout mice (Xu et al. 1994a, b; Clifford et al. 1998). We have found significant evidence for association of *DRD1* in our sample of ADHD families (Misener et al. 2004). We found suggestive evidence for association in single marker analyses, however the most significant finding was for a 4-marker haplotype, D1P.5(G)/rs35916350, D1P.6/rs265981(T), D1.1/rs4532(G), and D1.7/rs686(C). We further found that the association was only with inattention symptoms, but not with hyperactive/impulsive symptoms (Misener et al. 2004). We recently confirmed this finding in an independent sample of families collected through probands with reading difficulties, also finding significant evidence for association of this haplotype with inattention symptoms in that sample (Luca et al. 2007). Further support for the involvement of *DRD1* in ADHD comes from a recent case-control study, in which significantly higher frequencies of the D1P.6/ rs265981(T) and D1.1/rs4532(G) alleles were found among individuals with ADHD, as compared to controls (Bobb et al. 2005). While two other family-based studies of ADHD have not replicated the association with *DRD1*, they may have been hampered by small sample sizes and/or the analysis of individual markers instead of haplotypes (Kirley et al. 2002; Bobb et al. 2005).

Involvement of the *DRD2* gene in ADHD was initially suggested by a case-control study in which association of the A1 allele of the *Taq*1A SNP was observed (Comings et al. 1991). However, association of the *Taq*1A SNP has not been replicated in any family-based study to date (Rowe et al. 1999; Waldman et al. 1999; Kirley et al. 2002; Kustanovich et al. 2004; Misener et al. submitted), and analyses of additional *DRD2* polymorphisms (Kirley et al. 2002; Qian et al. 2007; Misener et al. submitted) have also failed to support a role for this gene in ADHD.

The *Taq*1A polymorphism, located 3′ of the *DRD2* gene, was originally thought to be intergenic, but is now known to be located within the coding sequence of a neighbouring gene, ankyrin repeat and kinase domain containing 1 (ANKK1), altering the amino acid sequence of the encoded ANKK1 protein (Neville et al. 2004). A number of neuropsychiatric disorders have been reported to be associated with the *Taq*1A marker (Comings et al. 1991; Blum et al. 1995), and thus there has been substantial controversy surrounding some of these findings, in light of the ANKK1 discovery. Furthermore, unless very carefully controlled for, the interpretation of case-control studies is likely complicated by the variable frequency of the A1 allele across populations, which has been documented to range from as low as 0.09 in Yemenite Jews to as high as 0.75 in Muskoke Amerindians (Barr and Kidd 1993).

Studies of the *DRD3* gene in ADHD have to date focused primarily on the non-synonymous coding polymorphism, Ser9Gly. Among four studies that have analyzed this polymorphism, none have found evidence for association with ADHD (Barr et al. 2000d; Payton et al. 2001a; Kirley et al. 2002; Muglia et al. 2002b). In addition, our analysis of a second polymorphism, located in intron 5, also failed to show evidence for association with the disorder (Barr et al. 2000d).

4.1.5 Noradrenergic System

There is a considerable body of evidence supporting noradrenergic mechanisms in ADHD. The noradrenergic (norepinephrine) system is known to be involved in the improvement of visual attention, modulation of vigilance (sustained attention), initiation of adaptive responses, and learning and memory (Arnsten et al. 1996; Pliszka et al. 1996; Biederman and Spencer 1999). Stimulants can potentiate the actions of norepinephrine, in addition to dopamine, in the synapse, and in the treatment of ADHD, maximal efficacy has been found for drugs that affect both of these neurotransmitters simultaneously (Solanto 1998). In addition, the improvement of ADHD symptoms with tricyclic antidepressants has been attributed to the actions of these drugs in the reuptake of norepinephrine (Biederman and Spencer 1999), further supporting noradrenergic mechanisms underlying ADHD. Interestingly, Atomoxetine (Strattera), a selective norepinephrine reuptake inhibitor, has also been shown to be effective in the treatment of ADHD (Spencer et al. 1998). Thus, analogous to the rationale for studying *DAT1*, the norepinephrine transporter gene, *NET1* (official nomenclature: solute carrier family 6 member 2, *SLC6A2*), is a strong candidate for involvement in ADHD, as this transporter is the primary target for Atomoxetine.

While the majority of studies, thus far, have not found significant evidence for *NET1* association with ADHD (Barr et al. 2002; McEvoy et al. 2002; De Luca et al. 2004a; Xu et al. 2005a; Brookes et al. 2006b), two recent studies have (Bobb et al. 2005; Kim et al. 2006). One study identified association with 2 intronic SNPs, rs998424 and rs3785157 (Bobb et al. 2005), and a second study identified an association with an A/T polymorphism located 3081 bp upstream of the transcription initiation site (−3081A/T) (Kim et al. 2006). The authors of the latter study further showed that the −3081T allele was correlated with reduced transcription in an *in vitro* assay, and creates a putative E2-box consensus binding site. Functional studies using super shift gel assays and cotransfection analyses indicated that the T allele creates a binding site for the repressors Slug and Scratch, thus providing a possible functional explanation for the association. While most of the *NET1* studies, to date, have been limited to only a few polymorphisms, some investigators have recently undertaken more detailed analyses of comprehensive sets of SNPs spanning the entire gene. In one such study, a trend for the rs3785157 SNP was observed (Xu et al. 2005a), and in another, trends for two other intronic SNPs, rs3785143 and rs11568324 (but not rs3785157), were observed (Brookes et al. 2006b). These findings did not reach statistical significance, however, owing to adjustments for multiple tests given the large numbers of SNPs analyzed. Further comprehensive studies of the *NET1* gene in ADHD, including those with a particular emphasis on the −3081A/T promoter polymorphism described above, are clearly needed.

Genes that encode the adrenergic receptors have also been studied in relation to ADHD. The first of these studies was based on a sample of 274 individuals diagnosed with Tourette Syndrome (TS), 144 of whom also met DSM-IV criteria for ADHD, and 62 controls. A combined analysis of the *ADRA2A* and *ADRA2C* genes showed significant correlation between these genes and scores for ADHD, history of learning disabilities and poor grade-school performance (Comings et al. 1999). However, when the results were analyzed for individual genes, and were corrected for multiple tests, the significance was modest (Comings et al. 1999).

Since that first report, a number of groups have investigated adrenergic receptor genes as candidates for involvement in ADHD. The *ADRA2A* gene has been the most extensively studied, with mixed results, to date. A Brazilian study of the *ADRA2A* gene did not find significant evidence for association between ADHD diagnosis and the −1291 C/G promoter polymorphism, rs1800544 (recognized by restriction enzyme *Msp*I). However, based on a further analysis of inattention and combined symptom scores, which showed higher mean scores for the group homozygous for the minor allele (G allele), the authors suggested that this gene may have a small effect on susceptibility, or may modulate the severity of the disorder (Roman et al. 2003). Two further studies by that group also provided evidence for association. In their second study of this polymorphism, using an independent sample, they again identified evidence for association of the G allele with inattention symptoms (Roman et al. 2006). In their third study, a non-clinically referred sample was used. This sample was collected by selecting, from public schools, children with elevated inattention symptoms (at least four inattention symptoms and at most three hyperactive/impulsive symptoms), followed by a clinical interview to further select those children meeting DSM-IV criteria for the inattentive subtype. That study also found evidence for association of the $-1291/Msp$I polymorphism (homozygosity of the G allele) using case-control analysis, but no evidence for biased allele transmission using family-based methods (Schmitz et al. 2006).

Other polymorphisms of the *ADRA2A* gene have also been investigated in relation to ADHD. Parks and colleagues conducted a study of three polymorphisms, the $-1291/Msp$I promoter SNP (rs1800544) and SNPs in the 5'-UTR (rs1800545; recognized by *Hha*I) and 3'-UTR (rs553668; recognized by *Dra*I), using a small, primarily Caucasian ADHD sample from the U. S. In single marker analyses, only the *Dra*I T allele showed evidence for association with ADHD, although a trend for overtransmission of the G allele of $-1291/Msp$I was observed, and haplotype analyses found evidence for association of a haplotype containing both of these alleles. Quantitative TDT analyses of symptom scores yielded evidence for association of the $-1291/Msp$I G and *Dra*I T alleles with both inattentive and hyperactive-impulsive symptoms (Park et al. 2005). Using a much larger study sample of Han Chinese families, Wang and co-workers also investigated the $-1291/Msp$I and *Dra*I SNPs. These investigators found no evidence for association of either of these markers individually, however a trend for association of one haplotype comprised of the $-1291/Msp$I C allele and the *Dra*I C allele was reported for the combined subtype of ADHD (Wang et al. 2006b), i.e., the opposite alleles to those identified in the Brazilian and U.S. studies described above. Positive findings for the opposite alleles have also been reported recently in a second (independent) U.S. study. In that analysis of these same markers, Deupree and co-workers found evidence for overtransmission of the *Dra*I C allele, individually, as well as overtransmission of the haplotype comprised of $-1291/Msp$I C, *Hha*I G and *Dra*I C (Deupree et al. 2006). In addition, using radioligand binding assays in platelets to investigate functional properties of the α-2A receptor, this group found evidence for reduced binding affinity of the receptor in children with ADHD, relative to unaffected children, but similar receptor densities between the two groups. In a further genetic analysis, they also observed a relationship between these binding parameters and the $-1291/Msp$I and *Dra*I polymorphisms (Deupree et al. 2006). Thus, the finding of lower affinity of the receptor in children with ADHD indicates the possibility of an associated coding region change influencing binding.

In our own published studies of the adrenergic receptor genes, *ADRA2A*, *ADRA1C* and *ADRA2C* (Barr et al. 2001b; Xu et al. 2001), we did not find evidence for association with ADHD. However, a limitation of those studies was that only one polymorphism was used for each gene, as these were the only known polymorphisms at that time. Our most recent work analyzed 32 markers across the genes for all of the adrenergic receptors, and no evidence for association was identified (Feng et al., unpublished). The only other study to examine the *ADRA2C* gene found no evidence for association in a sample of ADHD adults (De Luca et al. 2004b).

Elegant work using animal models and pharmacological analyses has shown that the adrenergic system, particularly the α2 adrenergic receptor, is critical for short-term maintenance of visuo-spatial information mediated by the prefrontal

Molecular Genetics of ADHD

cortex (Arnsten et al. 1996). Therefore, the noradrenergic system may contribute, more specifically, to the cognitive deficits in ADHD, for example, deficits in working memory (Arnsten and Li 2005; Lijffijt et al. 2005; Luman et al. 2005; Martinussen et al. 2005; Willcutt et al. 2005). It is conceivable then, that for genetic studies of the adrenergic system, working memory may be a more informative phenotype than the categorical diagnosis of ADHD. While our preliminary results do not show evidence for a relationship between the adrenergic genes and verbal working memory, at this point, further study is required.

Also relevant to the proposed involvement of adrenergic genes in working memory is a recent study investigating the *ADRA2A* gene in relation to ADHD and reading disabilities (RD) (Stevenson et al. 2005). Working memory deficits are common in ADHD and in RD, and it is thought that this may be a reflection of shared genetic influences between the two disorders. Interestingly, previous studies of the noradrenergic metabolite, 3-methoxy-4-hydroxyphenylglycol in the plasma of children with ADHD had found elevated levels in the ADHD children with comorbid RD (Halperin et al. 1997). On the basis of these findings, as well as the role of the adrenergic system in working memory, Stevenson and colleagues investigated the *ADRA2A* gene in a sample of children with ADHD, some of whom were comorbid for RD. Examination of the −1291/MspI polymorphism identified association of the G allele with the subsample having ADHD + RD, but not with the entire ADHD sample (Stevenson et al. 2005). Further comprehensive studies of the adrenergic system genes in relation to ADHD, ADHD + RD comorbidity and associated working memory deficits are clearly warranted, given the strong hypothesis for the involvement of this system in these behavioral and cognitive phenotypes.

4.1.6 Serotonin System

In addition to the catecholamines, there is also support for the serotonin system in ADHD. There is extensive interaction between the dopamine and serotonin neurotransmitter systems, and disruption of one affects the other. The evidence for serotonergic involvement in ADHD is gleaned from extensive documentation of the role of serotonin in impulse control and evidence from animal models, particularly knockout models, of the receptors (Quist and Kennedy 2001). Support for serotonergic involvement in hyperactivity is found in evidence from the *DAT1* knockout mouse, whereby the calming effects of stimulants in this animal model of hyperactivity (analogous to the effects of stimulants in ADHD) is mediated through the serotonergic system (Gainetdinov et al. 1999). Studies, so far, of genes in this system are promising. Several groups have analyzed a functional 44 bp insertion/deletion polymorphism located within a repeat element in the promoter region of the serotonin transporter gene, *HTT* (official nomenclature: solute carrier family 6 member 4, *SLC6A4*), in which the short (44 bp deleted) allele is reported to reduce transcription and production of the transporter (Lesch et al. 1996). Studies using case-control and haplotype relative risk-based methods have reported significant underrepresentation of the short/short genotype and/or overrepresentation of the long/long genotype in ADHD probands relative to controls (Manor et al. 2001; Seeger et al. 2001; Zoroglu et al. 2002; Beitchman et al. 2003), though we note that in one of these studies, significant results were obtained only for the DSM-IV combined subtype of ADHD (Manor et al. 2001). Other studies found either no association or only trends for the 44 bp insertion/deletion polymorphism (Kent et al. 2002; Langley et al. 2003; Kim et al. 2005a; Xu et al. 2005b; Zhao et al. 2005; Banerjee et al. 2006; Guimaraes et al. 2007; Heiser et al. 2007), and one study found association with the opposite allele (short allele) in Han Chinese (Li et al. 2007). Of note is that the allele frequencies of this promoter polymorphism are quite different across populations, with the short allele frequencies estimated at approximately 0.40 in North American Caucasians and 0.70 in Chinese (Gelernter et al. 1999).

Additional studies have also begun to investigate *HTT* markers located elsewhere in the gene, including a VNTR polymorphism located in intron 2 and a G/T SNP, rs3813034, located in the 3′-UTR. In one study in which all three markers were investigated by single-marker TDT analysis, only the 3′-UTR SNP (T allele) showed evidence for association with ADHD (Kent et al. 2002). However, this finding of ADHD association with the 3′-UTR SNP has not been replicated in other studies (Xu et al. 2005b; Wigg et al. 2006; Heiser et al. 2007). Kent and co-workers also investigated the transmission of 2-marker haplotypes, and found evidence for overtransmission of two of these, the haplotype defined by the VNTR 10-repeat allele and the 3′-UTR T allele, and the haplotype defined by the promoter long allele and the 3′-UTR T allele (Kent et al. 2002). The intron 2 VNTR polymorphism, with alleles of 9, 10, 11 and 12 repeats of a 17 bp repeat unit, has been reported to function as a cell type-specific and inducible enhancer using *in vitro* reporter assays, with higher expression exhibited by the 12-repeat allele construct (Fiskerstrand et al. 1999). Further, a comparison of the 10- and 12-repeat alleles found differential gene expression in the rostral hindbrain in E10.5 mouse embryos, with constructs containing the 12-repeat allele demonstrating higher transcription than those containing the 10-repeat allele (MacKenzie and Quinn 1999). Thus, this polymorphism is of significant interest for genetic studies. Thus far, however, there has been little evidence to support an association between ADHD and the intron 2 VNTR when tested as a single marker. While one study reported association of the 10-repeat allele (12/12 genotype less frequent in cases) (Zoroglu et al. 2002), another reported association of the 12-repeat allele (Banerjee et al. 2006), and several others have not found any evidence for association (Kent et al. 2002; Beitchman et al. 2003; Langley et al. 2003; Xu et al. 2005b; Heiser et al. 2007; Li et al. 2007). Other ADHD studies that have investigated haplotypes involving this marker have produced mixed results. Evidence for overtransmission of a 2-marker haplotype consisting of the promoter long allele and the intron 2 10-repeat allele has been reported in a Korean study (Kim et al. 2005a), and while some support for this was provided by a trend for the same haplotype in the study of Kent et al. (2002), the investigation of these markers in a Han Chinese study found a trend for overtransmission of a different haplotype, the haplotype containing the promoter long allele and 12-repeat allele of the VNTR (Li et al. 2007). Finally, in an investigation of 3-marker haplotypes of the promoter, intron 2 and 3′-UTR polymorphisms in both U.K. and Taiwanese ADHD samples, no evidence for biased transmission of any of the major haplotypes was observed (Xu et al. 2005b).

A recent development in the story of the 44 bp insertion/deletion promoter polymorphism of *HTT* is the identification of a single nucleotide change within the repeat that changes the function. An A to G base pair substitution, rs25531, located within one of the imperfect repeat sequences was demonstrated to be functionally significant (Hu et al. 2006). That study found that the long allele with the G substitution functions like the short allele, i.e., with reduced transcription and decreased transporter expression. Because this additional polymorphism was unknown until recently, it was not taken into consideration in previous studies of the 44 bp insertion/deletion, and this may have confounded the interpretation of the relationship between the *HTT* gene and ADHD as analyses of the long vs. short alleles would have combined functionally dissimilar alleles. In our own study of 209 ADHD families, we have since investigated both the long/short promoter polymorphism and the rs25531 A/G variant within it, as well as the 3'-UTR SNP, rs3813034 (Kent et al. 2002), and a rare, nonsynonymous coding SNP (Ozaki et al. 2003) We did not find evidence for association of these polymorphisms, or haplotypes of these polymorphisms, with ADHD in our Toronto sample, and the rare coding SNP was not observed (Wigg et al. 2006). Thus, neither the 44 bp insertion/deletion nor the A/G variant within it appears to be major factors in our sample.

The serotonin receptor genes have also been investigated as candidates for involvement in ADHD. Evidence for the serotonin receptor 1B gene, *HTR1B*, as a susceptibility gene in ADHD was first reported by us (Quist et al. 1999, 2003), whereby we found preferential transmission of the G allele of the rs6296 (G861C, *Hinc*II) polymorphism of *HTR1B* in ADHD families, particularly for paternal transmissions. This was supported by a study of the same polymorphism in a large sample of 273 families (Hawi et al. 2002). Another independent study suggested that variation in the *HTR1B* gene may primarily affect the inattentive subtype of ADHD, with paternal overtransmission of the 861G allele to offspring with ADHD being largely attributable to inattentive cases (Smoller et al. 2006). A pooled analysis of all three of these studies indicated an OR of 1.35 (95% CI 1.13–1.62) for the G861C polymorphism (Smoller et al. 2006). Further investigation of the *HTR1B* gene was also conducted by Smoller and colleagues, through the analysis of 5 additional SNPs, −261T/G (rs11568817), −161A/T (rs130058), 129C/T (rs6298), 371T/G (rs130060) and 1180A/G (rs6297) that, together with the C861G (rs6296) SNP, were found to form a haplotype block spanning the *HTR1B* gene. Among these, the −261T/G and −161A/T SNPs have been characterized as functional polymorphisms, based on allele-specific differences in transcription factor binding and transcriptional activity observed in *in vitro* assays (Duan et al. 2003). In their investigation of the polymorphisms individually, Smoller et al. (2006) found evidence for association with the two synonymous coding variants, 129C/T and 861G/C, and with the 3'-UTR variant, 1180A/G. In haplotype analyses, overtransmission of a particular 6-marker haplotype containing the 861G allele and the −261T and −161A ("high expressor") allele combination was identified. In addition, biased transmission of another SNP, rs2000292, located downstream of the gene, was observed. All of the evidence for association in this study was restricted to the inattentive subtype of ADHD, and was not found in analyses of the entire ADHD sample (Smoller et al. 2006). In other studies, absence of association between the *HTR1B* gene and ADHD, based on analyses of the −161A/T (rs130058), 129C/T (rs6298) and/or 861G/C (rs6296) polymorphisms, has been reported (Bobb et al. 2005; Li et al. 2005; Heiser et al. 2007). However, none of those studies examined maternal and paternal transmissions separately, and only one examined ADHD subtypes. In that study of a Han Chinese ADHD sample, some evidence for over-transmission of the 861G allele was observed for the inattentive subtype, but this did not reach statistical significance (Li et al. 2005). In our own recent study, we expanded our original investigation of the relationship between the *HTR1B* gene and ADHD, using a larger sample of families and genotyping the 861G/C SNP as well as 5 additional polymorphisms, several of which (rs130058, rs6298, rs6297 and rs2000292) had been included in some of the earlier studies. No evidence for association of the *HTR1B* gene with ADHD diagnosis as a categorical trait or with the inattentive and hyperactive-impulsive symptoms as quantitative traits was obtained in the analyses of either single markers or haplotypes, even when maternal and paternal transmissions were analysed separately (Ickowicz et al. 2007). Thus, our original finding is not supported in the expanded sample.

Several groups, including our own, have also investigated the serotonin receptor 2A gene, *HTR2A*, as a candidate for involvement in ADHD. The serotonin 2A receptor is of particular interest, owing to pharmacological analyses that have indicated a role for this receptor in the modulation of extracellular dopamine levels. In those analyses, the serotonin 2A receptor was found to be important in facilitating stimulus (amphetamine)-induced dopamine release, but not basal release (Porras et al. 2002b). Evidence also indicates that dopamine-mediated hyperlocomotor behavior in mice, induced by agents such as cocaine and amphetamine, requires activation of the serotonin 2A receptor (O'Neill et al. 1999).

In our initial analysis of the *HTR2A* gene, based on our first 115 ADHD families, we found biased transmission of the Tyr allele of the His452Tyr polymorphism, indicating an association with the disorder (Quist et al. 2000). Although the initial association finding was quite weak ($p = 0.03$), we have recently updated our genotyping of the *HTR2A* His452Tyr polymorphism, in our now larger sample ($n = 213$), and the results are still significant ($p = 0.03$, unpublished). Interestingly, there is some indication that the His to Tyr change causes a functional change in the receptor. Studies in platelets have found the Tyr variant of the receptor to be associated with a blunted Ca^{2+} mobilization response, i.e., a lower peak amplitude and longer time course, relative to the His variant, suggesting possible receptor desensitization associated with the Tyr form of the protein (Ozaki et al. 1997). An additional study of this polymorphism, using NIH3T3 cells transfected with either the His or Tyr form of the receptor, found a reduced ability of the Tyr variant to activate phospholipases C and D, indicating that signaling through $G_{q/11}$ and $G_{12/13}$ is impaired (Hazelwood and Sanders-Bush 2004). Another polymorphism, T102C, which is not in LD with His452Tyr, provided no evidence for association with ADHD in our study sample (Quist et al. 2000).

Molecular Genetics of ADHD

There have been a number of follow-up studies of our initial report on *HTR2A* (Hawi et al. 2002; Zoroglu et al. 2003; Bobb et al. 2005; Li et al. 2006b; Guimaraes et al. 2007; Heiser et al. 2007), several of which have also investigated the His452Tyr polymorphism. The study by Hawi and colleagues analyzed this polymorphism in two samples, British and Irish, finding no association overall (Hawi et al. 2002). However, when the Irish sample was examined alone, there was some evidence for association with the 452His allele, the opposite allele to that identified in our sample. The recent study of Guimaraes et al. (2007) also reported an association with the His allele in a Brazilian ADHD sample; however, the association was only for males in that sample (Guimaraes et al. 2007). In two other studies of the His452Tyr polymorphism, no evidence for association was identified (Bobb et al. 2005; Heiser et al. 2007). A number of studies have also investigated other polymorphisms of the *HTR2A* gene, including the T102C SNP, mentioned above, as well as a promoter region SNP, −1438G/A. Neither of these SNPs have shown evidence for association with ADHD in any of the studies (Quist et al. 2000; Zoroglu et al. 2003; Bobb et al. 2005; Li et al. 2006d; Guimaraes et al. 2007; Heiser et al. 2007). Interestingly, although the study by Li et al. (2006a–g) found no evidence for association of either the T102C SNP or the −1438G/A SNP with childhood diagnosis of ADHD, a longitudinal study by the same investigators found that a higher frequency of the −1438A allele in adolescents was correlated with reduction of symptoms from childhood (Li et al. 2006b).

The association in males in the Guimaraes et al. (2007) study is interesting, given evidence for estrogen regulation of the *HTR2A* gene (Fink and Sumner 1996; Cyr et al. 1998). Also of relevance, in light of the findings of Li et al. (2006b), is the observation of age-dependent changes in receptor binding levels that have demonstrated lower receptor expression in newborns, highest expression in 6-year-olds, and a sharp decline in adolescents between the ages of 13 and 17 (Biegon and Greuner 1992). This suggests that the *HTR2A* gene may be a factor in changes in symptoms over adolescence, and thus, careful consideration should be given to the age group of subjects studied. Further complicating the study of this gene is evidence for differential allelic expression, obtained using lymphocytes (Pastinen et al. 2004), that may result from monoallelic expression of this gene in some tissues or cell types (Bunzel et al. 1998; Bray et al. 2004). Thus, *HTR2A* involvement in genetic risk for ADHD may be complicated, potentially influenced by lower levels of transcription resulting from monoallelic expression in some tissues, age-related changes in expression, as well as a protective effect in females due to increased transcription of the gene in response to estrogen.

Studies of other serotonergic receptor genes in ADHD have been very limited, to date, despite some strong biological evidence that would implicate such genes as possible contributors (Bonhomme et al. 1995; Porras et al. 2002a, b). Most of the studies, so far, have been conducted by Li and colleagues, using their large sample of Han Chinese families. While one study of the *HTR4* gene found no evidence for association with any of 3 single markers, analyses of haplotypes indicated an association (Li et al. 2006c). Studies of a limited number of polymorphisms in the *HTR5A* and *HTR6* genes did not find evidence for association (Li et al. 2006d). In their study of the *HTR1D* gene, two SNPs were investigated, with one (1236A/G, located in the 3'-UTR), showing evidence for association with ADHD (Li et al. 2006g). Li and colleagues have also investigated the *HTR2C* gene, focusing on two promoter region SNPs, −759C/T and −697G/C. Both single marker and haplotype analyses provided evidence for association of the −759C and −697G alleles with ADHD in their study sample (Li et al. 2006e). In one other investigation of the *HTR2C* gene, analysis of a nonsynonymous coding SNP, Cys23Ser (rs6318), in a US study sample, found no evidence for association with ADHD (Bobb et al. 2005).

4.1.7 Genes Regulating Neurotransmitter Synthesis and Degradation

Genes that encode enzymes involved in neurotransmitter synthesis and degradation have also been tested as candidates for involvement in ADHD. These include tyrosine hydroxylase (*TH*), which converts tyrosine to L-DOPA (the rate-limiting step in dopamine biosynthesis), dopa decarboxylase (*DDC*), which converts L-DOPA to dopamine, and dopamine β-hydroxylase (*DBH*), which, in noradrenergic neurons, converts dopamine to norepinephrine. Also included are tryptophan hydroxylase (*TPH*), which catalyzes the first of two steps in the conversion of tryptophan to serotonin (the second step being carried out by dopa decarboxylase), as well as catechol-*O*-methyltransferase (*COMT*) and monoamine oxidase (*MAO*), which are involved in the degradation of neurotransmitters. Among these, *TH*, *COMT*, *DBH* and *MAO* have received the most attention.

Findings from three independent family-based studies of the *TH* gene in ADHD have been largely negative (Barr et al. 2000b; Payton et al. 2001a; Kirley et al. 2002). In one study, an analysis of parent of origin effects found some evidence for biased paternal transmissions, though this was based on a relatively small number of transmissions (Kirley et al. 2002).

Studies of ADHD and *COMT*, thus far, have focused on a single functional polymorphism, Val158Met. Despite an initial positive finding in a family-based study of *COMT* (Eisenberg et al. 1999), evidence for association did not hold up once the sample size was increased (Manor et al. 2000), and in several subsequent studies neither we (Barr et al. 1999a) nor others (Hawi et al. 2000b; Tahir et al. 2000; Payton et al. 2001a; Kirley et al. 2002; Taerk et al. 2004; Bellgrove et al. 2005; Turic et al. 2005) detected evidence for association with *COMT*. The study of *COMT* in a Chinese sample found no evidence for association in the entire sample, but did report association in certain subsets, including males in the sample, males without comorbid conditions, and the inattentive subsample; however, the sample sizes of the latter two groups were quite small (Qian et al. 2003). Meta-analyses of 11 published studies found no evidence for association of this gene with ADHD (OR = 0.99 [95% CI 0.88–1.12]) (Cheuk and Wong 2006).

In focusing on only a single polymorphism, ADHD studies of *COMT* have ignored the complexity of the gene, and this may confound the interpretation of results. Specifically, additional genetic variation contributing to differential expres-

Molecular Genetics of ADHD

sion of this gene has recently been identified (Bray et al. 2003; Craddock et al. 2006). The merging of functionally distinct haplotypes in the analysis of a single polymorphism would likely obscure association, if indeed it does exist.

Plasma levels of dopamine β-hydroxylase (DβH) are, to a large extent, regulated by the *DBH* gene itself, and several polymorphisms that correlate with plasma DβH levels have been identified (Cubells and Zabetian 2004). These include the *TaqIA* (rs2519152), 1603C/T (rs6271) and −1021C/T (rs1611115) SNPs, as well as insertion/deletion and microsatellite polymorphisms located in the 5′ region (Wei et al. 1997; Cubells et al. 2000; Cubells and Zabetian 2004; Tang et al. 2006). However, incomplete correlation between DβH plasma levels and these polymorphisms indicates the possibility of a DNA change in LD with these variants as the potential functional change. The polymorphism at −1021 exhibits the strongest correlation with plasma levels, with the T allele being associated with lower expression. However, the addition of a nonsynonymous polymorphism, R535C, to this analysis increased the correlation by 0.02 ($p = 0.0024$) (Cubells and Zabetian 2004).

Results for the *DBH* gene and ADHD have been mixed. Positive results have been reported for the A2 allele (T) of the *TaqIA* polymorphism located in intron 5 of the *DBH* gene (Comings et al. 1999; Daly et al. 1999; Roman et al. 2002a). We did not find significant evidence for association of the *DBH* intron 5 polymorphism with ADHD; however, there was a trend for the *Taq*1 A2 allele (Wigg et al. 2002). Two other markers were negative, however, including the insertion/deletion and microsatellite markers, located 4.7 kb 5′ to the start site, that are correlated with DβH plasma levels. One additional study found evidence for association to the *TaqIA* polymorphism (Smith et al. 2003), but to the opposite (A1, or C) allele than previously reported. Other negative studies, including studies of the −1021 polymorphism have also been reported (Payton et al. 2001a; Inkster et al. 2004; Bhaduri and Mukhopadhyay 2006), and additional markers genotyped on the sample of Daly and colleagues did not reveal any other associated markers (Hawi et al. 2003).

Tests for involvement of the *MAO* genes, *MAOA* and *MAOB* (located on the X chromosome), originated by studying a nearby microsatellite marker, DXS7. One group reported evidence of association with the 157 bp allele of this dinucleotide repeat polymorphism (Jiang et al. 2000) and another reported no association (Lowe et al. 2001), in samples of Chinese and Irish ADHD families, respectively. Lowe and colleagues subsequently reanalyzed the results of Jiang et al. after they observed that two paternal alleles were included in the analyses when, for X linked markers, only a single allele exists in males. Although still significant, there was a marked reduction in the significance of the finding (Lowe et al. 2001). Jiang and co-workers followed up their analysis of DXS7 by testing two additional dinucleotide repeat polymorphisms, a (CA) repeat in the *MAOA* gene and a (GT) repeat in the *MAOB* gene (Jiang et al. 2001). Consistent with their DXS7 results, these investigators found significant evidence for association to the *MAOA* marker (global $p < 0.05$), but results for the *MAOB* marker were not significant. Analysis of the *MAOA* gene in a sample of ADHD families from the UK showed trends for association to the 122 bp allele, but not to the 114 bp allele, of the *MAOA* (CA) repeat polymorphism (Payton et al. 2001a).

Two other *MAOA* polymorphisms, an exon 8 SNP (941T/G), associated with altered enzymatic activity of the MAOA protein (Hotamisligil and Breakefield 1991), and a 5′ region 30 bp VNTR, associated with altered transcriptional activity of the *MAOA* gene (Sabol et al. 1998; Deckert et al. 1999; Denney et al. 1999), have also been investigated. The 5′ VNTR is characterized by at least 5 alleles (2, 3, 3.5, 4 and 5 repeats) of an imperfect 30 bp repeat located 1.2 kb upstream of the *MAOA* gene. Four of the 5 alleles were tested for transcriptional activity using *in vitro* reporter assays, with the 3.5- and 4-repeat alleles found to have 2–10 times more transcriptional activity than the 3- or 5-repeat alleles (Sabol et al. 1998). However, a later study reported higher activity correlating with the 5-repeat allele, as well as the 3.5- and 4-repeat alleles, indicating that the "long" alleles (i.e. 3.5, 4 and 5) were correlated with higher activity (Deckert et al. 1999).

Alleles of the *MAOA* 30 bp VNTR have been tested for association with ADHD, with differing results. Both the case-control and family-based analyses in the Israeli study showed a significant association between ADHD and the 4- and 5-repeat alleles, which were combined and termed "long" (Manor et al. 2002b). A study in an Indian population also examined the 30 bp VNTR polymorphism and found association to the 3-repeat allele in males (designated as 3.5-repeat allele by the authors), associated with low activity (Das et al. 2006). Further, much like the *DRD4* exon 3 repeats and the *HTT* long/short promoter polymorphism, sequence variants were identified within this *MAOA* polymorphism (Das et al. 2006), which might also contribute to differences in function. In the UK study, while no associations were found for ADHD alone, case-control analysis of a male subset with ADHD and comorbid conduct disorder revealed a significant association with the "low activity" (3-repeat) allele (Lawson et al. 2003). A similar trend was observed in the family sample, but the sample size in this case was too small to be conclusive. This finding is interesting in view of previous reports of this marker being associated with aggressive/impulsive behavior (Manuck et al. 2000).

The *MAOA* exon 8 polymorphism, 941T/G (recognized by restriction enzyme *Fnu*4HI), is a synonymous polymorphism associated with low (941T) or high (941G) activity (Hotamisligil and Breakefield 1991). The first study of this polymorphism in ADHD found no evidence of association (Lawson et al. 2003). A study in Irish families identified association with the 941G allele and with haplotypes containing this allele, but not with alleles of the 30 bp promoter VNTR associated in the Manor study, nor with the (CA)n repeat in intron 2 or a T/C SNP in intron 13 (Domschke et al. 2005). An association with the 941G allele was also observed in a sample of Taiwanese families, but no association with alleles of the 30 bp VNTR was observed (Xu et al. 2007). Haplotypes of the 941G allele (high activity) and the 3-repeat (low activity) allele were also associated.

The association of ADHD with the *MAOA* "high" activity alleles (exon 8 941G and promoter VNTR 4- and 5-repeat alleles), and thus increased degradation of monoamines, is consistent with multiple lines of evidence supporting a hypodopaminergic state as contributing to ADHD (Levy and Swanson 2001), and suggests a plausible explanation for the risk associated with this gene. However, the findings of association of different alleles associated with low activity in the Indian sample and in the UK ADHD+CD comorbid sample, and of haplotypes with both

Molecular Genetics of ADHD

high and low activity-correlated alleles are hard to reconcile at this point. Further, if the risk is simply related to low or high activity, then association should appear across polymorphisms correlated with activity. Interestingly, a recent study examined the relationship between genotypes of the VNTR alleles and brain MAOA activity in adult males, measured using PET, and found no evidence for a relationship (Fowler et al. 2007). Thus, the association with alleles at the 5′ VNTR marker cannot be explained by a simple relationship of alleles to transcription levels. Clearly there is more to understand concerning this gene.

ADHD association studies of the *MAOB* gene have been negative, thus far, although a comprehensive study of the gene has yet to be published. Studies, so far, have examined the (CA)n repeat in intron 2 (rs3838196) in the Chinese sample discussed above (Jiang et al. 2001), and this same polymorphism as well as a SNP (rs1799836), in an Irish sample (Domschke et al. 2005).

With regards to dopa decarboxylase, only a single group has studied the *DDC* gene to date (Hawi et al. 2001; Kirley et al. 2002). In the family-based study of *DDC* in ADHD (Hawi et al. 2001), significant associations with two markers as well as haplotypes were observed.

Only a few studies of ADHD have investigated the tryptophan hydroxylase (*TPH*) genes, encoding the enzymes responsible for the rate-limiting step in the synthesis of serotonin from tryptophan. A complication in the genetic study of *TPH* in psychiatric disorders has recently emerged. Studies of knockout mice indicated that the known *TPH* gene on 11p15.2 (now called *TPH1*) was not highly expressed in brain (Walther et al. 2003). This led to the cloning of a second gene, *TPH2* on 12q21.1, which codes for the form of the enzyme highly expressed in brain (Walther et al. 2003). Initial indications that *TPH1* was not expressed in brain at all, led to a call for reconsideration of all previous genetic findings (see Shaltiel et al. 2005). However, further studies indicated that both genes are expressed in human brain, with differences in regional expression (Zill et al. 2007).

Both the *TPH1* and *TPH2* genes have been studied for their relationship to ADHD. No evidence of association was reported for *TPH1* and ADHD in one study (Tang et al. 2001), and an additional study found association only with the combined subtype of ADHD and a rare haplotype (Li et al. 2006f). The findings for *TPH2* have been primarily positive, with several studies reporting association (Sheehan et al. 2005; Walitza et al. 2005; Brookes et al. 2006b). While one other study found no association, we note that the sample size for that study was quite small (*n* = 108 families, 63 of these full trios) (Sheehan et al. 2007). Thus, although preliminary, the *TPH2* gene appears to be a promising candidate.

4.1.8 Genes Regulating Neurotransmitter Release

Investigations into genes regulating neurotransmitter release were prompted by the mouse mutant strain *coloboma*. This strain is hemizygous for a 2 cM deletion that includes the gene for the synaptic vesicle docking fusion protein, synaptosomal-associated protein of 25 kDa (*SNAP25*), and displays spontaneous hyperactivity that is responsive to dextroamphetamine but not to methylphenidate (Hess et al. 1996). Of note is that the 2 cM deletion that encompasses the *Snap25* gene in the *coloboma* mouse also deletes several other genes. The insertion of a *Snap25* transgene into the *coloboma* mouse line rescues the hyperactive phenotype but not the other phenotypes (head bobbing, ocular dysmorphology) identified in this mouse line, indicating that the hyperactivity is directly attributable to the *Snap25* gene (Hess et al. 1996). *Snap25* has also been implicated in axonal growth and synaptic plasticity, and may have a role in learning and memory (Osen-Sand et al. 1993). Further, the *coloboma* mouse strain exhibits developmental delays in motor skills, although the mutant mice eventually catch up with their normal littermates (Heyser et al. 1995). This provides an interesting parallel to ADHD children where problems in developmental coordination are often noted, particularly in fine motor control. Approximately 50% of ADHD children meet criteria for developmental coordination disorder (DCD). A recent twin study identified shared genetic factors for DCD and ADHD, particularly for the inattentive subtype (Martin et al. 2006).

The phenotype of the *coloboma* mouse strain originally appeared to indicate a straightforward relationship, with the deletion of a single copy of the *Snap25* gene resulting in a hyperlocomoter phenotype (deletion of both copies of the gene results in nonviable offspring). However, knockout of a single copy of this gene on a different mouse strain background (C57Bl/6/129 for the knockout versus C3He/SnJ for *coloboma*) did not result in a hyperactive phenotype, thereby pointing to the contribution of genetic variation at other loci, in addition to *Snap25*, in producing the phenotype (Washbourne et al. 2002).

In our initial study of the *SNAP25* gene, we identified six polymorphisms and used four of these to test for association with ADHD (Barr et al. 2000a). Our first studies showed weak evidence for association with one of the haplotypes of two 3′ polymorphisms. In our latest study of this gene, in an expanded sample of families and with additional markers, we identified significant evidence for association with four markers (Feng et al. 2005). The finding of association between *SNAP25* and ADHD has now been replicated in eight clinical samples (Brophy et al. 2002; Mill et al. 2002b, 2004b; Kustanovich et al. 2003; Hawi et al. 2005; Brookes et al. 2006b; Choi et al. 2007)(MC O'Donovan, personal communication) and one population-based sample (Mill et al. 2005b). However, the markers studied and associated alleles differ between studies.

Based on the findings for *SNAP25*, additional genes involved in the process of neurotransmitter release were investigated, namely, syntaxin 1A (*STX1A*), vesicle associated membrane protein 2 (*VAMP2*), synaptophysin (*SYP*), synaptotagmin (*SYT1*), and vesicular monoamine transporter 2 (*VMAT2*) (Brookes et al. 2005a). Using a DNA pooling approach, association was only identified for one marker in the synaptophysin gene in that study. The reported association was a trend in the case-control sample and significant in the family-based TDT sample; however, the number of informative transmissions was low (20 in total). While interesting, further studies with additional markers are now required.

Based on our positive findings for the *DRD1* and *DRD5* genes and ADHD, we investigated the calcyon gene (*DRD1IP*), which had been reported to encode a brain-specific D1-interacting protein involved in D1/D5 receptor-mediated calcium

Molecular Genetics of ADHD

signaling (Lezcano et al. 2000). However, contrary to original reports, recent evidence indicates that calcyon does not physically interact with the D1 receptor or enhance D1-stimulated intracellular calcium release (Lezcano et al. 2006), but rather is involved in clathrin mediated endocytosis, a process contributing to synaptic plasticity (Xiao et al. 2006; Heijtz et al. 2007). Clathrin mediated endocytosis is involved in the recycling of synaptic vesicles, a process critical in the optimization of levels of releasable pools of neurotransmitters during synaptic activity (Dickman et al. 2005). Interestingly, expression analyses in 3-, 5-, and 10-week-old spontaneously hypertensive rats have found higher expression of *calycon* in the medial, prefrontal and orbital frontal cortices compared to the progenitor strain Wistar-Kyoto (Heijtz et al. 2007). The spontaneously hypertensive rat has been considered a reasonable animal model of ADHD, based on early hyperactivity, motor impulsiveness, behavioral variability and deficits in sustained attention (Sagvolden 2005). Interestingly, both rat strains showed a decrease in *calycon* expression with age.

The possibility of a role of the calcyon gene in the aetiology of ADHD was also supported by the results of a genome-wide scan for ADHD (Fisher et al. 2002b). The location of this gene coincides with one of the highest positive linkage sites identified by that screen, at chromosome 10q26. In a study of multiple polymorphisms of the calcyon gene in our sample of ADHD families, we found that the most common haplotype, designated C1, demonstrated significant evidence for excess transmission (Laurin et al. 2005). In addition, quantitative trait analyses of this haplotype showed significant relationships with both the inattentive (parent's rating, $p = 0.006$; teacher's rating, $p = 0.003$) and hyperactive/impulsive (parent's rating, $p = 0.004$) dimensions of the disorder. At the current time, no other studies have been published on this gene in relation to ADHD. If replicated, this finding would indicate that other genes involved in clathrin mediated endocytosis should also be considered as candidates for involvement in ADHD.

4.1.9 Dopamine Receptor Signaling

Based on the association of *DRD1* and *DRD5*, the gene for $G\alpha_{olf}$ (*GNAL*) was supported as a strong candidate gene for ADHD, owing to the role of this G protein subunit in D1 and D5 signaling, and further, because the G(*olf*) knockout mouse is hyperactive (Belluscio et al. 1998). Among 12 *GNAL* SNPs analysed in our study sample, one in the third intron of the gene, rs2161961, showed evidence for association with ADHD. Further analyses of maternal and paternal transmssions separately, also indicated a parent of origin effect, with biased transmission found for maternal, but not paternal, transmissions (Laurin et al. 2008).

Another study has examined the genes for the G protein α subunits $G\alpha_{T2}$, $G\alpha_{o}$, and $G\alpha_{z}$ (Turic et al. 2004a), on the basis of their involvement in D4 signalling. The exons of these 3 genes were first screened for sequence variation, followed by testing of the DNA variants for association with ADHD. No evidence for association was detected with any of 13 variants tested.

4.1.10 Nicotinic System

Evidence from pharmacological, clinical and animal studies has suggested that the nicotinic system could be involved in ADHD. A particularly strong case for this can be made based on the expression of nicotinic receptors in regions innervated by dopaminergic neurons (Arroyo-Jimenez et al. 1999; Klink et al. 2001; Gotti et al. 2006) and the role of the nicotinic system in modulating dopamine neurotransmission (Picciotto et al. 1998; Grady et al. 2002). Nicotinic receptor signaling also regulates dopamine transporter gene transcription and function (Li et al. 2004; Parish et al. 2005), potentially affecting dopamine uptake.

Further support for involvement of the nicotinic system in ADHD is the documented increased risk of early initiation (Milberger et al. 1997a), higher prevalence (Lambert and Hartsough 1998; Biederman et al. 2006), and lower quit ratios (percentage of ever-smokers who were ex-smokers) of cigarette smoking in ADHD individuals, compared to the general population (Pomerleau et al. 1995). Further, maternal smoking during pregnancy may influence the expression and activation of nicotinic receptors in utero, with subsequent influences on neuronal pathfinding affecting region-specific brain development and subsequent neurotransmission (Ernst et al. 2001). Nicotine and nicotinic agonists have also been shown to improve attention in adult smokers and nonsmokers without attention deficits, as well as in adults with ADHD (Levin et al. 1998; Mancuso et al. 1999; Wilens et al. 1999, 2006).

Studies have implicated the nicotinic acetylcholine receptor alpha 4 subunit gene, *CHRNA4*, in ADHD (Todd et al. 2003). In particular, a polymorphism at the exon 2-intron 2 junction of *CHRNA4* has been associated with severe inattention defined by latent class analysis (Todd et al. 2003). Our study, testing the associated markers in the Todd et al. study, did not confirm the association with the inattentive subtype as defined by the DSM-IV, or with inattention as measured as a quantitative trait. However, we did observe nominally significant evidence of association of individual markers with the combined subtype and with teacher-rated hyperactivity-impulsivity scores, as well as undertransmission of one haplotype. An additional study of the *CHRNA4* gene did not find evidence for association; however, the sample size in that study was relatively small and only a single polymorphism was tested (Kent et al. 2001b) The $\alpha7$ subunit gene, *CHRNA7*, has also been tested in one study, using 3 microsatellite markers, however no association was found (Kent et al. 2001a).

4.2 Molecular Genetic Studies Using Population-Based Samples

As indicated in an earlier section, it is thought that ADHD represents the extreme of the normal distribution of symptoms and that genes contributing to ADHD as a clinical diagnosis will contribute to ADHD symptoms as quantitative trait loci in the population. This is based on the finding of similar heritability for ADHD as a diagnosis, compared to the heritability of the entire distribution of ADHD

Molecular Genetics of ADHD

symptoms. This has led investigators to study the contribution of ADHD-associated genes to ADHD symptoms in population-based samples. Only a few such studies have been published thus far (Payton et al. 2001b; Todd et al. 2001a, b; Mill et al. 2002a, 2005b). The study by Todd et al. (2001b) used a sample of 7- to 19-year-old twins and two polymorphisms of *DRD4* to test for association with DSM-IV subtypes and with ADHD subtypes derived from latent-class analysis. That study found no evidence for association with *DRD4* (Todd et al. 2001b). The study by Mill et al. used an unselected birth cohort to investigate the contribution of *DRD4* to ADHD diagnosis, as well as to hyperactivity/impulsivity as a continuous trait measured at ages 7, 9, 11, and 13 (Mill et al. 2002a). No evidence for association was identified at any age. In their most recent study, 5 genes, *DRD4*, *DAT1*, *DRD5*, *SNAP25* and *HTR1B*, were tested in a twin sample of 329 male dizygotic twins (Mill et al. 2005b). A composite score of ADHD symptoms was used, taking average counts of parent-reported ADHD symptoms at 2, 3, 4, and 7 years of age. They also examined ADHD scores separately at each time point. The authors found no evidence of association for *DRD4*, *HTR1B* or *DRD5*. However, they did find evidence for association of *DAT1* and *SNAP25* with symptom scores.

Similar to studies of these genes in clinical samples, the results from population-based samples are mixed, with some supporting the contribution of these genes to ADHD symptoms, while others not. The conclusions that can be drawn from these studies are presently limited because the method of ascertainment, and the selection of phenotypes for analysis are not standardized across studies. These factors will likely influence the composition of the sample, making comparisons across studies difficult. Further, for the majority of studies, only one or a few polymorphisms per gene were tested, and as discussed above, in the case of negative results, this is unlikely to be sufficient to rule out these genes as contributors to symptoms in the population.

4.3 Genome Scans for ADHD

The first genome scan of 126 affected sibling pairs identified four chromosomal regions (5p12, 10q26, 12q23, and 16p13) with a LOD score greater than 1.5 that, while not significant, suggested the possible locations of susceptibility genes (Fisher et al. 2002b). The 16p13 region was further studied by expanding the sample by 151 sibling pairs and genotyping additional markers across the region. Results from this analysis yielded a multi-point maximum LOD score of 4.2, thus continuing to support this region (Smalley et al. 2002). The power for the genome scan was then increased further, by genotyping an additional 144 affected sibling pairs (Ogdie et al. 2003). The results identified an additional region at 17p11 with suggestive evidence for linkage (LOD 2.98); however, the regions at 10q26 and 12q23 became less significant. The results on 5p continued to show some evidence for linkage, and regions on 6q14, 11q25, and 20q13 showed some hints of linkage, with LOD scores greater than 1.

An independent genome scan in 164 Dutch sibling pairs identified the 15q15 region with significant evidence for linkage (Bakker et al. 2003). Other regions with a suggestion of linkage were 7p13 and 9q33.3. Interestingly, the 15q15 region overlaps a region reported to be linked/associated with reading disabilities (Smith et al. 1983; Grigorenko et al. 1997; Morris et al. 2000; Taipale et al. 2003; Chapman et al. 2004; Wigg et al. 2005; Bates et al. 2007).

A genome scan of Colombian families (Arcos-Burgos et al. 2004) identified novel regions, 4q13.2 (LOD 2.7), 5q33.3 (LOD 1.6) and 8q11.23 (LOD 3.22), as well as the previously identified 17p11 (LOD 3.0) region and 11q23.1–24.3 (LOD 4.0) (reported as11q22), a region partially overlapping the previously identified 11q region (Ogdie et al. 2003). The most recent genome scan found evidence for linkage over a wide region of 5p (LOD 4.75), overlapping the region from previous studies, as well as nominal evidence for linkage to 12q (LOD 2.1) and 17p (LOD 1.39) (Hebebrand et al. 2006).

Thus far, the regions that overlap between the published genome scans are 5p12–13 (max LOD scores 3.24, 1.43, 4.75), 17p11 (max LOD scores 3.63, 3.0, 1.39) and 11q24 (max LOD score 4.0, 1.19). The 6q region (max LOD score 3.3) overlaps with linkage findings in RD (Petryshen et al. 2001; Bates et al. 2007), and the 5p and 11q regions overlap with regions linked to Tourette Syndrome (Simonic et al. 1998; Barr et al. 1999b; Merette et al. 2000; Simonic et al. 2001).

The differences in the regions identified between samples may reflect differences in the clinical characteristics of the samples or may reflect ethnic differences. Some of these weak findings may indicate genes of low relative risk. Alternatively, some of these findings may be false positives that can only be clarified with additional studies. We note that whole genome association studies and additional linkage genome scans are in progress, and the results of these may provide additional support for these regions and/or identify new regions.

The genome scan results have led to the testing of key candidate genes within the linkage peak regions. For example, the glutamate receptor, ionotropic, *N*-methyl D-aspartate 2A gene (*GRIN2A*) that encodes the 2A subunit of the NMDA receptor, maps to the 16p13 region. *GRIN2A* is a plausible candidate for involvement in ADHD, given that the glutamatergic system is involved in both cognitive and motor function, and modulates the dopamine and serotonin systems (Miyamoto et al. 2001). Further, mice lacking the NMDAR 2A subunit gene, GluRε1 (the mouse homolog of the human *GRIN2A* gene), show increased spontaneous loco-motor activity in a novel environment, impaired latent learning associated with selective attention (Miyamoto et al. 2001), and deficits in spatial learning (Sakimura et al. 1995). Support for *GRIN2A* as the susceptibility gene in this region was obtained in a family-based study in which significant association between ADHD and the Grin2a-5 polymorphism was observed (Turic et al. 2004b). However, in our sample of families, we did not find any evidence of *GRIN2A* association with ADHD or with the phenotypes of verbal short term memory or inhibitory control, using the same polymorphism (Adams et al. 2004). We have also tested the gene for the 2B receptor subunit (*GRIN2B*), which is located at 12p13, another region that showed suggestive evidence for linkage in a

Molecular Genetics of ADHD 133

genome scan (Fisher et al. 2002b). Here, we were more successful and found evidence for an association (Dorval et al. 2007).

A number of candidate genes initially chosen for study based solely on biological plausibility, were also supported by their locations identified in subsequent genome scans. These include the calcyon gene on 10q26 (Fisher et al. 2002b), the *DAT1* gene on 5p (Hebebrand et al. 2006) and the *HTR1B* gene on 6q14.1 (Ogdie et al. 2004). Whether these genes are in fact the risk genes identified by linkage, or are fortuitously located within the linkage peak, remains to be determined.

5 Relationship of the Molecular Genetic Findings in ADHD to Reading Disabilities

Linkage/association studies have only recently begun to investigate the overlap in the genetic susceptibility of ADHD and comorbid disorders. Particular focus has been on reading disabilities (RD, also called developmental dyslexia) because twin studies have supported common genetic influences (Willcutt et al. 2000, 2007). Further support for the existence of shared genetic factors between ADHD and RD is found in the overlap in regions identified in linkage/association studies and genome scans, notably, 6p22, 15q, 11p15.5 and 6q12–14 (Fisher et al. 2002a; Willcutt et al. 2002; Bakker et al. 2003; Marlow et al. 2003; Loo et al. 2004; Ogdie et al. 2004; Tzenova et al. 2004).

Gene identification has progressed for RD in the linked regions, and some strong candidates have emerged. These include *DCDC2* and *KIAA0319* on chromosome 6p (Meng et al. 2005; Paracchini et al. 2006), *EKN1* and *Protogenin* on chromosome 15q (Taipale et al. 2003; Wigg et al. 2004; Wigg et al. in press), *ROBO1* on chromosome 3 (Hannula-Jouppi et al. 2005) and *KIAA0319L* on chromosome 1p (Couto et al. in press). At the present time, these genes are associated with RD but the evidence that these are indeed the risk genes is not yet conclusive. In both the 6p and 15q region, more than one gene is supported (Taipale et al. 2003; Meng et al. 2005; Paracchini et al. 2006; Couto et al. submitted; Wigg et al. in press; Wigg et al. submitted).

All of the candidate genes associated with RD, thus far, are known or predicted to be involved in neuronal migration or axonal pathfinding (Fisher and Francks 2006; Galaburda et al. 2006; McGrath et al. 2006). Postmortem neuroanatomical studies have indicated subtle cortical neuronal migration abnormalities (ectopias, dysplasia, vascular micromalformations) in RD (Galaburda et al. 1985; Humphreys et al. 1990). Based on these postmortem findings, Galaburda speculated, over 10 years ago, that genetic risks for dyslexia would contribute to dysregulation of neuronal migration, disrupting neuronal circuits involved in language function (Galaburda et al. 1985). For *DCDC2*, *KIAA0319*, and *DYX1C1* (*EKN1*), *in utero* electroporation of RNA interference (RNAi) into rats disrupts neuronal migration in the developing cortex, resulting in migration abnormalities (Meng et al. 2005; Paracchini et al. 2006; Wang et al. 2006a). Interestingly, the malformations from these RNAi experiments resemble the ectopias identified in the postmortem studies of individuals with dyslexia (Galaburda et al. 2006). Further, recent studies using RNAi to *DYX1C1* indicate that reduced expression of this gene results in impairments in auditory processing and spatial learning, depending on the location of the malformation (cortex or hippocampus, respectively) (Threlkeld et al. 2007). This provides an interesting parallel with certain phenotypes associated with dyslexia, in particular, the difficulties in temporal auditory processing of speech sounds and complex nonspeech sounds that are observed in some individuals with the disorder (Tallal et al. 1996; Schulte-Korne et al. 1999; Kujala et al. 2001; Ramus 2006; Strehlow et al. 2006).

Our study of *EKN1* (*DYX1C1*), a gene located in the 15q region and originally reported to be the susceptibility gene for RD at this locus (Taipale et al. 2003), is also associated with ADHD in our sample (Wigg et al. 2005). However, our most significant results implicate a neighbouring gene, *Protogenin*, as associated with ADHD in our sample (Wigg et al., in press). *Protogenin* is an interesting candidate based on the predicted structure of its protein product. The protogenin protein is structurally similar to the DCC/Neogenin family of proteins that are essential in guidance systems controlling cell migration and axonal pathfinding during development of the vertebrate nervous system (Toyoda et al. 2005).

The genetic overlap between RD and ADHD now leads us to the question of how the genes involved could contribute to these diverse phenotypes. For children with RD, with or without ADHD, treatment with methylphenidate improves reading performance but does not improve the basic phonological deficits (Keulers et al. 2007). Phonological-based training methods are necessary to improve reading ability (Lovett et al. 2000). Thus, to envision a common genetic pathway, one must reconcile these dichotomous phenotypes as well as treatments. Based on the current gene findings, we can speculate that the genetic relationship of ADHD symptoms to RD may be explained by either: (1) changes in neuronal migration secondarily changing innervation of critical brain regions influencing neurotransmission, or (2) a dual role for these genes in neuronal migration in development and neurotransmission in the mature brain. Genes that have a fundamental role in neuronal migration in embryos often have additional functions in the mature brain in relation to dendritic growth, synaptogenesis, synaptic plasticity and neurotransmission (Gao et al. 1998; McAllister 2002). Further, a number of neurotransmitters are known to function in neuronal pathfinding and target selection in the developing brain, with disruption during development altering future neurotransmission (Ivgy-May et al. 1994; Landmesser 1994; Whitaker-Azmitia et al. 1996; Ernst et al. 2001; Jassen et al. 2006; Bonnin et al. 2007). For example, the serotonin receptor 1B gene, which is associated with ADHD (described above) (Quist et al. 1999, 2003; Hawi et al. 2002; Li et al. 2005) and located in a region linked to RD (Petryshen et al. 2001), modulates responsiveness to axon guidance cues (Bonnin et al. 2007). In addition, the genes associated with RD, thus far (*DCDC2, EKN1, KIAA0319, ROBO1*), are also expressed in adult brain tissue (Anthoni et al. 2007), indicating the likelihood of additional functions beyond a role in development.

6 Interpretation of the Molecular Genetic Studies

As we have reviewed, a number of studies have reported linkage or association findings for ADHD, while other studies have not been able to confirm these findings. The conflicting reports at this point should not be discouraging, and are to be expected, given the clinical complexity of the disorder. A recurring issue is that few studies have comprehensively assessed the polymorphic content of the genes of interest, with the majority of studies having investigated only one, or at best a few, polymorphism(s), only some of which are predicted to be functional variants. In addition, many of the studies have used relatively small sample sizes for a complex trait and, therefore, some of the findings may simply be chance false positives. On the other hand, in some cases, false negative results may have arisen, due to insufficient power owing to the small sample size. Discrepancies between studies may also result from a number of other factors, including different inclusion and exclusion criteria (e.g., DSM-IV subtypes, comorbidities, IQ), different ascertainment strategies (e.g., clinical versus epidemiological), and/or different ethnic compositions of the samples that may exhibit different LD patterns between the markers studied and the causal variants. Developmental issues, related to differences in age at ascertainment, may further cloud the interpretation of results. As discussed above, symptoms change over time, and genetic and environmental influences may also change during development. Genetic studies, thus far, have approached these issues differently. For example, regarding the issue of DSM-IV subtypes, studies have used multiple methods, including categorical analysis of DSM-IV subtypes (Lowe et al. 2004a; Feng et al. 2005a; Schmitz et al. 2006; Wang et al. 2006b), analysis of population-based subtypes (Todd et al. 2003, 2005), quantitative analysis of symptom counts (Waldman et al. 1998; Misener et al. 2004; Laurin et al. 2005, 2008), and restricting the collection of subjects to the most common DSM-IV subtype of ADHD, the combined subtype (Swanson et al. 1998; Brookes et al. 2005). Currently, there is no consensus as to how this complex clinical phenotype should be treated for molecular genetic studies, and further attention should be focused on these differences when considering the issue of replication across studies. Cognitive phenotypes associated with ADHD (Arnsten and Li 2005; Lijffijt et al. 2005; Luman et al. 2005; Martinussen et al. 2005; Willcutt et al. 2005) are increasingly the focus of genetic studies, and these may identify genes contributing to specific aspects of the risk (Crosbie et al. 2008). Also, given the possibility that certain genes may be contributing to ADHD in children, but not to persistent ADHD in adults, some groups have begun to investigate candidate genes in adult ADHD.

7 After the Linkage or Association Finding

The finding of linkage or association with a candidate gene is actually just the first step in the understanding of its contribution to susceptibility to the development of the disorder. As noted earlier, in the majority of studies thus far, the polymorphic

DNA changes used to test for linkage or association are not predicted to change the function of the gene. These DNA changes are used simply to investigate the sharing of alleles in ADHD families or sharing among ADHD individuals selected from the population. To further investigate how the gene contributes to the disorder, the gene must be comprehensively screened to identify the DNA change(s) that result in a change or loss of function of the gene, i.e., variants that are likely to play a causal role in the phenotype. Although DNA sequence screening is technically straightforward, this is not generally a trivial task, as many genes may be very large. For example, the coding region of the dopamine transporter gene spans over 55 kb of genomic DNA. Further, the DNA code that regulates appropriate temporal expression and tissue specificity of the gene may extend for thousands or hundreds of thousands of base pairs outside of the coding region (Steidl et al. 2007), and may even be present within neighboring genes (Lettice et al. 2003).

Current theory posits that complex traits are more likely to result from genetic variation that results in changes in gene expression rather than variation in protein coding regions (Rockman and Wray 2002; Buckland et al. 2004, 2005; Jais 2005; Knight 2005). This is particularly appealing for psychiatric disorders, as genetic changes in regulatory regions may result in differential responses to environmental cues. However, one of the limitations of identifying regulatory regions outside the core promoter is that these regions can be a long distance from the coding region and it is labor-intensive to screen these large regions for regulatory elements using functional studies. Thus, when screening of the coding region fails to find potential functional changes, further molecular characterization of the gene is often neglected.

In addition to potential challenges posed by regulatory elements (e.g., distal promoter elements, enhancers, silencers, locus control regions, and insulators) being situated long distances away from the gene, the actual identification of key nucleotides within these and other types of regulatory elements, and determinations as to whether/how genetic variation within them influences gene expression, are not trivial. Transcription factor binding sites, for example, are small (~5–8 bp) and degenerate, making it difficult to exploit bioinformatics alone to predict regulatory elements (Wasserman and Sandelin 2004). Redundancy in the sequence of transcription factor binding sites makes it difficult to predict the impact of a single nucleotide change, and because of the modularity of the binding sites, a combination of changes may be required to result in a change of function (Wasserman and Sandelin 2004). Further, in a recent large-scale study that identified genetic variation in promoters correlated with changes in gene expression, only 33% of the DNA variants were found in known consensus transcription factor binding sites, suggesting the existence of many unknown binding motifs (Buckland et al. 2005).

Conserved regions between species flag some regulatory regions, and multispecies comparisons have been used successfully in identifying several of these, but principally for highly conserved genes that are expressed early in development with likely strict constraints on temporal and spatial gene expression (Boffelli et al. 2004). However, for other genes with only small stretches of conserved sequence, this approach may miss key regions. Further, conservation in function can be maintained even without conservation of sequence, due to compensatory changes at other sites (Ludwig et al. 2000). Even genetic changes in nonconserved regions can result in dysregulation of gene regulation, as was recently shown for a single base pair change that was found to create a new promoter-like element that interferes with expression from the correct downstream promoters of the α-like globin genes, resulting in α thalassemia (De Gobbi et al. 2006). DNA changes between species may also be critical, as such changes may result in new functions with evolutionary advantages (Dermitzakis and Clark 2002; Rockman et al. 2003), including changes that are human specific (King and Wilson 1975; Prabhakar et al. 2006). For example, a comparison of known functional elements in humans indicated that only 60–68% of these are functional in rodents (Dermitzakis and Clark 2002). Specifically relevant to the study of ADHD and RD is that nonconserved regions may be relevant to genes contributing to human higher order complex cognitive traits such as verbal working memory and reading ability.

Recently, methods that exploit the location of modified histones as a marker for regulatory regions have been used successfully to identify remote regulatory regions (i.e., distant from gene promoters) across the genome (Heintzman et al. 2007; Ni et al., 2008). These techniques allow for the efficient identification of regulatory regions that can then be targeted for functional studies and screened for risk alleles. We have recently used these techniques to identify a 2.5 kb region marked by acetylated histones in the 5′ region of the *KIAA0319* gene (Couto et al., submitted). Markers in this 2.5 kb region, which spans the promoter region, first untranslated exon and part of the first intron, have been associated with RD in six independent studies (Kaplan et al. 2002; Deffenbacher et al. 2004; Francks et al. 2004; Cope et al. 2005; Harold et al. 2006; Luciano et al. 2007). Thus, the identification of this putative 5′ regulatory region suggests a possible location of functional risk alleles.

Some of the DNA changes used for genetic studies of ADHD, thus far, are indicated or predicted to change the function of the gene, based on in vivo or *in vitro* studies. However, there may be multiple DNA changes in a gene that are predicted to cause a change in function, and only some of these may cause a change in phenotype in the organism. Further, it may require a combination of alleles to produce the phenotype (Drysdale et al. 2000; Bray et al. 2003; Craddock et al. 2006). For example, as noted in the section on *DRD4*, for the 48 bp repeat in the third exon this gene, the number of repeat units has been reported to change the affinity and intracellular signaling properties of the protein (Asghari et al. 1994, 1995), leading to the hypothesis that the 7-repeat allele is subsensitive to dopamine. However, there are additional polymorphic DNA changes that are predicted to change the function of the *DRD4* gene, and determining which of these variants and/or which combination of variants is related to the phenotype will be challenging. In some cases, animal models may be helpful to examine the relationship between the DNA changes and the phenotype, however, the function of the gene may differ somewhat between species, and not all behavioral and cognitive aspects of ADHD can be modeled in animals. Further, the genetic background of the animal model can also be a factor that influences the phenotype, as in the example of the *SNAP25* gene, described in an earlier section.

8 Implications of the Molecular Genetic Findings for Diagnosis and Treatment

ADHD can be debilitating, resulting in significant academic underachievement and underemployment (Mannuzza et al. 1993). ADHD in childhood is a significant risk factor for later substance abuse, early and continued smoking, as well as automobile accidents in adolescents (Gittelman et al. 1985; Mannuzza et al. 1993; Barkley et al. 2002b; Flory et al. 2003). For example, studies have shown that children with ADHD symptoms are more likely to use illicit drugs earlier and smoke at an early age, with 1/3 of children indicating that they are daily smokers by the age of 16 (Molina and Pelham 2003). ADHD symptoms often continue into adulthood (Barkley et al. 2002a; Kessler et al. 2005). ADHD is associated with comorbid psychiatric disorders both in childhood and later in life (Biederman et al. 1991a, b, 1992), and with increased incidence of psychiatric illness in family members (Schachar and Wachsmuth 1990). Commonly occurring comorbid disorders, including conduct disorder, anxiety disorder, mood disorder, learning disabilities (including RD), and oppositional defiant disorder, often lead to more severe outcomes. Thus, insight into the biological bases of these conditions derived from genetic findings for ADHD and comorbid disorders could make a very significant contribution to the theoretical understanding and practical treatment of a wide spectrum of psychiatric disorders, with the long-term goal of improved pharmacological and behavioral treatments. Considering the prevalence of some of the comorbid disorders, and the effects they have on children and adults, and on their families, this would make a substantial contribution to health care.

While the argument to identify the biological bases for these disorders is rather obvious, a number of questions arise specifically around the issue of genetics. Mainly, what are the ramifications of the genetic findings for diagnosis and treatment of ADHD? Will there be a genetic test that can be used to predict the development of ADHD with a reasonable degree of accuracy, such that it could be clinically useful? Will the findings of genes related to susceptibility relate to predictions of treatment response? These questions cannot be answered at the present time. For the genes that have been studied thus far, the genetic risks conferred by alleles at these genes are estimated to be small (Faraone et al. 2005). Therefore, these genes are clearly not useful for predicting genetic susceptibility. Although the possibility of genes of major effect cannot be completely ruled out at this time, current data do not support this. The first genome scan to be completed for ADHD was able to detect a region with significant evidence for linkage to the 16p13 region using a sample of 270 affected sibling pairs (Smalley et al. 2002), suggesting the existence of genes of major effect. However, further studies did not confirm this particular region and other loci were supported, with little overlap between studies thus far. Thus, current available data does not support the existence of genes of major effect that could be used for predictive tests. Despite the improbability of a predictive genetic test for ADHD and comorbid disorders, concerns have arisen as to the ethics of genetic testing, particularly in regards to aggressive behaviors and substance

Molecular Genetics of ADHD   139

abuse. Thus, treating physicians should be educated such that the implications of new genetic findings can be accurately and effectively explained to families.

The most likely immediate positive outcome of gene-finding efforts in ADHD is that the gene identification studies may increase public awareness of the biological basis of the disorder. Through education of parents, teachers and physicians, early recognition and treatment has the potential to prevent the sequelae associated with poor academic performance and low self-esteem commonly experienced by children with ADHD.

References

Adams J, Crosbie J, Wigg K, Ickowicz A, Pathare T, Roberts W, Malone M, Schachar R, Tannock R, Kennedy JL, Barr CL (2004) Glutamate receptor, ionotropic, N-methyl D-aspartate 2A (GRIN2A) gene as a positional candidate for attention-deficit/hyperactivity disorder in the 16p13 region. Mol Psychiatry 9(5): 494–499

American Psychiatric Association (1994) American psychiatric association: diagnostic and statistical manual of mental disorders, 4th edn. American Psychiatric Association, Washington, DC

Anthoni H, Zucchelli M, Matsson H, Muller-Myhsok B, Fransson I, Schumacher J, Massinen S, Onkamo P, Warnke A, Griesemann H, Hoffmann P, Nopola-Hemmi J, Lyytinen H, Schulte-Korne G, Kere J, Nothen MM, Peyrard-Janvid M (2007) A locus on 2p12 containing the co-regulated MRPL19 and C2ORF3 genes is associated to dyslexia. Hum Mol Genet 16(6): 667–677

Arcos-Burgos M, Castellanos FX, Pineda D, Lopera F, Palacio JD, Palacio LG, Rapoport JL, Berg K, Bailey-Wilson JE, Muenke M (2004) Attention-deficit/hyperactivity disorder in a population isolate: linkage to loci at 4q13.2, 5q33.3, 11q22, and 17p11. Am J Hum Genet 75(6): 998–1014. Epub 20 Oct 2004

Arnsten AF, Li BM (2005) Neurobiology of executive functions: catecholamine influences on prefrontal cortical functions. Biol Psychiatry 57(11): 1377–1384

Arnsten AF, Steere JC, Hunt RD (1996) The contribution of alpha 2-noradrenergic mechanisms of prefrontal cortical cognitive function. Potential significance for attention- deficit hyperactivity disorder. Arch Gen Psychiatry 53(5): 448–455

Arroyo-Jimenez MM, Bourgeois JP, Marubio LM, Le Sourd AM, Ottersen OP, Rinvik E, Fairen A, Changeux JP (1999) Ultrastructural localization of the alpha4-subunit of the neuronal acetylcholine nicotinic receptor in the rat substantia nigra. J Neurosci 19(15): 6475–6487

Asghari V, Schoots O, van Kats S, Ohara K, Jovanovic V, Guan HC, Bunzow JR, Petronis A, Van Tol HH (1994) Dopamine D4 receptor repeat: analysis of different native and mutant forms of the human and rat genes. Mol Pharmacol 46(2): 364–373

Asghari V, Sanyal S, Buchwaldt S, Paterson A, Jovanovic V, Van Tol HH (1995) Modulation of intracellular cyclic AMP levels by different human dopamine D4 receptor variants. J Neurochem 65(3): 1157–1165

Asherson P, Brookes K, Franke B, Chen W, Gill M, Ebstein RP, Buitelaar J, Banaschewski T, Sonuga-Barke E, Eisenberg J, Manor I, Miranda A, Oades RD, Roeyers H, Rothenberger A, Sergeant J, Steinhausen HC, Faraone SV (2007) Confirmation that a specific haplotype of the dopamine transporter gene is associated with combined-type ADHD. Am J Psychiatry 164(4): 674–677

Bakker SC, van der Meulen EM, Buitelaar JK, Sandkuijl LA, Pauls DL, Monsuur AJ, van 't Slot R, Minderaa RB, Gunning WB, Pearson PL, Sinke RJ (2003) A whole-genome scan in 164 Dutch sib pairs with attention-deficit/hyperactivity disorder: suggestive evidence for linkage on chromosomes 7p and 15q. Am J Hum Genet 72(5): 1251–1260

Bakker SC, van der Meulen EM, Oteman N, Schelleman H, Pearson PL, Buitelaar JK, Sinke RJ (2005) DAT1, DRD4, and DRD5 polymorphisms are not associated with ADHD in Dutch families. Am J Med Genet B Neuropsychiatr Genet 132(1): 50–52

Banerjee E, Sinha S, Chatterjee A, Gangopadhyay PK, Singh M, Nandagopal K (2006) A family-based study of Indian subjects from Kolkata reveals allelic association of the serotonin transporter intron-2 (STin2) polymorphism and attention-deficit-hyperactivity disorder (ADHD). Am J Med Genet B Neuropsychiatr Genet 141(4): 361–366

Barkley RA, Fischer M, Smallish L, Fletcher K (2002a) The persistence of attention-deficit/hyperactivity disorder into young adulthood as a function of reporting source and definition of disorder. J Abnorm Psychol 111(2): 279–289

Barkley RA, Murphy KR, Dupaul GI, Bush T (2002b) Driving in young adults with attention deficit hyperactivity disorder: knowledge, performance, adverse outcomes, and the role of executive functioning. J Int Neuropsychol Soc 8(5): 655–672

Barman SK, Pulkkinen L, Kaprio J, Rose RJ (2004) Inattentiveness, parental smoking and adolescent smoking initiation. Addiction 99(8): 1049–1061

Barr CL, Kidd KK (1993) Population frequencies of the A1 allele at the dopamine D2 receptor locus. Biol Psychiatry 34(4): 204–209

Barr CL, Wigg K, Malone M, Schachar R, Tannock R, Roberts W, Kennedy JL (1999a) Linkage study of Catechol-O-Methyltransferase and attention-deficit hyperactivity disorder. Am J Med Genet 88(6): 710–713

Barr CL, Wigg KG, Pakstis AJ, Kurlan R, Pauls D, Kidd KK, Tsui LC, Sandor P (1999b) Genome scan for linkage to Gilles de la Tourette syndrome. Am J Med Genet 88(4): 437–445

Barr CL, Feng Y, Wigg K, Bloom S, Roberts W, Malone M, Schachar R, Tannock R, Kennedy JL (2000a) Identification of DNA variants in the SNAP-25 gene and linkage study of these polymorphisms and attention-deficit hyperactivity disorder. Mol Psychiatry 5(4): 405–409

Barr CL, Wigg KG, Bloom S, Schachar R, Tannock R, Roberts W, Malone M, Kennedy JL (2000b) Further evidence from haplotype analysis for linkage of the dopamine D4 receptor gene and attention-deficit hyperactivity disorder. Am J Med Genet 96(3): 262–267

Barr CL, Wigg KG, Feng Y, Zai G, Malone M, Roberts W, Schachar R, Tannock R, Kennedy JL (2000c) Attention-deficit hyperactivity disorder and the gene for the dopamine D5 receptor. Mol Psychiatry 5(5): 548–551

Barr CL, Wigg KG, Wu J, Zai C, Bloom S, Tannock R, Roberts W, Malone M, Schachar R, Kennedy JL (2000d) Linkage study of two polymorphisms at the dopamine D3 receptor gene and attention-deficit hyperactivity disorder. Am J Med Genet 96(1): 114–117

Barr CL, Feng Y, Wigg KG, Schachar R, Tannock R, Roberts W, Malone M, Kennedy JL (2001a) 5′ untranslated region of the dopamine D4 receptor gene and attention-deficit hyperactivity disorder. Am J Med Genet (Neuropsychiatr Genet) 105: 84–90

Barr CL, Wigg K, Zai G, Roberts W, Malone M, Schachar R, Tannock R, Kennedy JL (2001b) Attention-deficit hyperactivity disorder and the adrenergic receptors alpha1C and alpha2C. Mol Psychiatry 6(3): 334–337

Barr CL, Xu C, Kroft J, Feng Y, Wigg K, Zai G, Tannock R, Schachar R, Malone M, Roberts W, Nothen MM, Grunhage F, Vandenbergh DJ, Uhl G, Sunohara G, King N, Kennedy JL (2001c) Haplotype study of three polymorphisms at the dopamine transporter locus confirm linkage to attention-deficit/hyperactivity disorder. Biol Psychiatry 49(4): 333–339

Barr CL, Kroft J, Feng Y, Wigg K, Roberts W, Malone M, Ickowicz A, Schachar R, Tannock R, Kennedy JL (2002) The norepinephrine transporter gene and attention-deficit hyperactivity disorder. Am J Med Genet 114(3): 255–259

Bates TC, Luciano M, Castles A, Coltheart M, Wright MJ, Martin NG (2007) Replication of reported linkages for dyslexia and spelling and suggestive evidence for novel regions on chromosomes 4 and 17. Eur J Hum Genet 15(2): 194–203

Batstra L, Hadders-Algra M, Neeleman J (2003) Effect of antenatal exposure to maternal smoking on behavioural problems and academic achievement in childhood: prospective evidence from a Dutch birth cohort. Early Hum Dev 75(1–2): 21–33

Beitchman JH, Davidge KM, Kennedy JL, Atkinson L, Lee V, Shapiro S, Douglas L (2003) The serotonin transporter gene in aggressive children with and without ADHD and nonaggressive matched controls. Ann N Y Acad Sci 1008: 248–251

Molecular Genetics of ADHD

Bellgrove MA, Domschke K, Hawi Z, Kirley A, Mullins C, Robertson IH, Gill M (2005) The methionine allele of the COMT polymorphism impairs prefrontal cognition in children and adolescents with ADHD. Exp Brain Res 163(3): 352–360. Epub 15 Jan 2005

Belluscio L, Gold GH, Nemes A, Axel R (1998) Mice deficient in G(olf) are anosmic. Neuron 20(1): 69–81

Bhaduri N, Mukhopadhyay K (2006) Lack of significant association between −1021C–>T polymorphism in the dopamine beta hydroxylase gene and attention deficit hyperactivity disorder. Neurosci Lett 402(1–2): 12–16

Bhutta AT, Cleves MA, Casey PH, Cradock MM, Anand KJ (2002) Cognitive and behavioral outcomes of school-aged children who were born preterm: a meta-analysis. Jama 288(6): 728–737

Biederman J, Spencer T (1999) Attention-deficit/hyperactivity disorder (ADHD) as a noradrenergic disorder. Biol Psychiatry 46(9): 1234–1242

Biederman J, Faraone SV, Keenan K, Steingard R, Tsuang MT (1991a) Familial association between attention deficit disorder and anxiety disorders. Am J Psychiatry 148(2): 251–256

Biederman J, Faraone SV, Keenan K, Tsuang MT (1991b) Evidence of familial association between attention deficit disorder and major affective disorders. Arch Gen Psychiatry 48(7): 633–642

Biederman J, Faraone SV, Keenan K, Benjamin J, Krifcher B, Moore C, Sprich-Buckminster S, Ugaglia K, Jellinek MS, Steingard R, Spencer T, Norman D, Kolodny R, Kraus I, Perrin J, Keller MB, Tsuang MT (1992) Further evidence for family-genetic risk factors in attention deficit hyperactivity disorder. Patterns of comorbidity in probands and relatives psychiatrically and pediatrically referred samples. Arch Gen Psychiatry 49(9): 728–738

Biederman J, Milberger S, Faraone SV, Kiely K, Guite J, Mick E, Ablon S, Warburton R, Reed E (1995) Family-environment risk factors for attention-deficit hyperactivity disorder. A test of Rutter's indicators of adversity. Arch Gen Psychiatry 52(6): 464–470

Biederman J, Mick E, Faraone SV (2000) Age-dependent decline of symptoms of attention deficit hyperactivity disorder: impact of remission definition and symptom type. Am J Psychiatry 157(5): 816–818

Biederman J, Faraone SV, Monuteaux MC (2002) Differential effect of environmental adversity by gender: Rutter's index of adversity in a group of boys and girls with and without ADHD. Am J Psychiatry 159(9): 1556–1562

Biederman J, Monuteaux MC, Mick E, Spencer T, Wilens TE, Silva JM, Snyder LE, Faraone SV (2006) Young adult outcome of attention deficit hyperactivity disorder: a controlled 10-year follow-up study. Psychol Med 36(2): 167–179

Biegon A, Greuner N (1992) Age-related changes in serotonin 5HT2 receptors on human blood platelets. Psychopharmacology (Berl) 108(1–2): 210–212

Blum K, Sheridan PJ, Wood RC, Braverman ER, Chen TJ, Comings DE (1995) Dopamine D2 receptor gene variants: association and linkage studies in impulsive-addictive-compulsive behaviour. Pharmacogenetics 5(3): 121–141

Bobb AJ, Addington AM, Sidransky E, Gornick MC, Lerch JP, Greenstein DK, Clasen LS, Sharp WS, Inoff-Germain G, Wavrant-De Vrieze F, Arcos-Burgos M, Straub RE, Hardy JA, Castellanos FX, Rapoport JL (2005) Support for association between ADHD and two candidate genes: NET1 and DRD1. Am J Med Genet B Neuropsychiatr Genet 134(1): 67–72

Boffelli D, Nobrega MA, Rubin EM (2004) Comparative genomics at the vertebrate extremes. Nat Rev Genet 5(6): 456–465

Bonhomme N, De Deurwaerdere P, Le Moal M, Spampinato U (1995) Evidence for 5-HT4 receptor subtype involvement in the enhancement of striatal dopamine release induced by serotonin: a microdialysis study in the halothane-anesthetized rat. Neuropharmacology 34(3): 269–279

Bonnin A, Torii M, Wang L, Rakic P, Levitt P (2007) Serotonin modulates the response of embryonic thalamocortical axons to netrin-1. Nat Neurosci 10(5): 588–597 Epub 22 Apr 2007

Botting N, Powls A, Cooke RW, Marlow N (1997) Attention deficit hyperactivity disorders and other psychiatric outcomes in very low birthweight children at 12 years. J Child Psychol Psychiatry 38(8): 931–941

Braun JM, Kahn RS, Froehlich T, Auinger P, Lanphear BP (2006) Exposures to environmental toxicants and attention deficit hyperactivity disorder in U.S. children Environ Health Perspect 114(12): 1904–1909

Bray NJ, Buckland PR, Williams NM, Williams HJ, Norton N, Owen MJ, O'Donovan MC (2003) A haplotype implicated in schizophrenia susceptibility is associated with reduced COMT expression in human brain. Am J Hum Genet 73(1): 152–161 Epub 11 June 2003

Bray NJ, Buckland PR, Hall H, Owen MJ, O'Donovan MC (2004) The serotonin-2A receptor gene locus does not contain common polymorphism affecting mRNA levels in adult brain. Mol Psychiatry 9(1): 109–114

Breslau N, Brown GG, DelDotto JE, Kumar S, Ezhuthachan S, Andreski P, Hufnagle KG (1996) Psychiatric sequelae of low birth weight at 6 years of age. J Abnorm Child Psychol 24(3): 385–400

Brookes KJ, Knight J, Xu X, Asherson P (2005a) DNA pooling analysis of ADHD and genes regulating vesicle release of neurotransmitters. Am J Med Genet B Neuropsychiatr Genet 139(1): 33–37

Brookes KJ, Xu X, Chen CK, Huang YS, Wu YY, Asherson P (2005b) No evidence for the association of DRD4 with ADHD in a Taiwanese population within-family study. BMC Med Genet 6: 31

Brookes KJ, Mill J, Guindalini C, Curran S, Xu X, Knight J, Chen CK, Huang YS, Sethna V, Taylor E, Chen W, Breen G, Asherson P (2006a) A common haplotype of the dopamine transporter gene associated with attention-deficit/hyperactivity disorder and interacting with maternal use of alcohol during pregnancy. Arch Gen Psychiatry 63(1): 74–81

Brookes K, Xu X, Chen W, Zhou K, Neale B, Lowe N, Aneey R, Franke B, Gill M, Ebstein R, Buitelaar J, Sham P, Campbell D, Knight J, Andreou P, Altink M, Arnold R, Boer F, Buschgens C, Butler L, Christiansen H, Feldman L, Fleischman K, Fliers E, Howe-Forbes R, Goldfarb A, Heise A, Gabriels I, Korn-Lubetzki I, Marco R, Medad S, Minderaa R, Mulas F, Muller U, Mulligan A, Rabin K, Rommelse N, Sethna V, Sorohan J, Uebel H, Psychogiou L, Weeks A, Barrett R, Craig I, Banaschewski T, Sonuga-Barke E, Eisenberg J, Kuntsi J, Manor I, McGuffin P, Miranda A, Oades RD, Plomin R, Roeyers H, Rothenberger A, Sergeant J, Steinhausen HC, Taylor E, Thompson M, Faraone SV, Asherson P, Johansson L (2006b) The analysis of 51 genes in DSM-IV combined type attention deficit hyperactivity disorder: association signals in DRD4, DAT1 and 16 other genes. Mol Psychiatry 11: 934–953

Brophy K, Hawi Z, Kirley A, Fitzgerald M, Gill M (2002) Synaptosomal-associated protein 25 (SNAP-25) and attention deficit hyperactivity disorder (ADHD): evidence of linkage and association in the Irish population. Mol Psychiatry 7(8): 913–917

Buckland PR, Hoogendoorn B, Guy CA, Coleman SL, Smith SK, Buxbaum JD, Haroutunian V, O'Donovan MC (2004) A high proportion of polymorphisms in the promoters of brain expressed genes influences transcriptional activity. Biochim Biophys Acta 1690(3): 238–249

Buckland PR, Hoogendoorn B, Coleman SL, Guy CA, Smith SK, O'Donovan MC (2005) Strong bias in the location of functional promoter polymorphisms. Hum Mutat 26(3): 214–223

Bunzel R, Blumcke I, Cichon S, Normann S, Schramm J, Propping P, Nothen MM (1998) Polymorphic imprinting of the serotonin-2A (5-HT2A) receptor gene in human adult brain. Brain Res Mol Brain Res 59(1): 90–92

Caspi A, McClay J, Moffitt TE, Mill J, Martin J, Craig IW, Taylor A, Poulton R (2002) Role of genotype in the cycle of violence in maltreated children. Science 297(5582): 851–854

Caspi A, Sugden K, Moffitt TE, Taylor A, Craig IW, Harrington H, McClay J, Mill J, Martin J, Braithwaite A, Poulton R (2003) Influence of life stress on depression: moderation by a polymorphism in the 5-HTT gene. Science 301(5631): 386–389

Castellanos FX, Lau E, Tayebi N, Lee P, Long RE, Giedd JN, Sharp W, Marsh WL, Walter JM, Hamburger SD, Ginns EI, Rapoport JL, Sidransky E (1998) Lack of an association between a dopamine-4 receptor polymorphism and attention-deficit/hyperactivity disorder: genetic and brain morphometric analyses. Mol Psychiatry 3(5): 431–434

Molecular Genetics of ADHD 143

Catalano M, Nobile M, Novelli E, Nothen MM, Smeraldi E (1993) Distribution of a novel mutation in the first exon of the human dopamine D4 receptor gene in psychotic patients. Biol Psychiatry 34(7): 459–464

Chandola CA, Robling MR, Peters TJ, Melville-Thomas G, McGuffin P (1992) Pre- and perinatal factors and the risk of subsequent referral for hyperactivity. J Child Psychol Psychiatry 33(6): 1077–1090

Chang FM, Kidd JR, Livak KJ, Pakstis AJ, Kidd KK (1996) The world-wide distribution of allele frequencies at the human dopamine D4 receptor locus. Hum Genet 98(1): 91–101

Chapman NH, Igo RP, Thomson JB, Matsushita M, Brkanac Z, Holzman T, Berninger VW, Wijsman EM, Raskind WH (2004) Linkage analyses of four regions previously implicated in dyslexia: confirmation of a locus on chromosome 15q. Am J Med Genet B Neuropsychiatr Genet 131(1): 67–75

Chen CK, Chen SL, Mill J, Huang YS, Lin SK, Curran S, Purcell S, Sham P, Asherson P (2003) The dopamine transporter gene is associated with attention deficit hyperactivity disorder in a Taiwanese sample. Mol Psychiatry 8(4): 393–396

Cheuk DK, Li SY, Wong V (2006a) Exon 3 polymorphisms of dopamine D4 receptor (DRD4) gene and attention deficit hyperactivity disorder in Chinese children. Am J Med Genet B Neuropsychiatr Genet 141(8): 907–911

Cheuk DK, Li SY, Wong V (2006b) No association between VNTR polymorphisms of dopamine transporter gene and attention deficit hyperactivity disorder in Chinese children. Am J Med Genet B Neuropsychiatr Genet 141(2): 123–125

Cheuk DK, Wong V (2006c) Meta-analysis of association between a catechol-O-methyltransferase gene polymorphism and attention deficit hyperactivity disorder. Behav Genet 36(5): 651–659

Choi TK, Lee HS, Kim JW, Park TW, Song DH, Yook KW, Lee SH, Kim JI, Suh SY (2007) Support for the MnlI polymorphism of SNAP25; a Korean ADHD case-control study. Mol Psychiatry 12(3): 224–226

Cichon S, Nothen MM, Catalano M, Di Bella D, Maier W, Lichtermann D, Minges J, Albus M, Borrmann M, Franzek E, Stober G, Weigelt B, Korner J, Rietschel M, Propping P (1995) Identification of two novel polymorphisms and a rare deletion variant in the human dopamine D4 receptor gene. Psychiatr Genet 5(3): 97–103

Clifford JJ, Tighe O, Croke DT, Sibley DR, Drago J, Waddington JL (1998) Topographical evaluation of the phenotype of spontaneous behaviour in mice with targeted gene deletion of the D1A dopamine receptor: paradoxical elevation of grooming syntax. Neuropharmacology 37(12): 1595–1602

Comings DE, Comings BG, Muhleman D, Dietz G, Shahbahrami B, Tast D, Knell E, Kocsis P, Baumgarten R, Kovacs BW, Levy DL, Smith M, Borison RL, Evans D, Klein DN, MacMurray J, Tosk JM, Sverd J, Gysin R, Flanagan SD (1991) The dopamine D2 receptor locus as a modifying gene in neuropsychiatric disorders. Jama 266(13): 1793–1800

Comings DE, Gade-Andavolu R, Gonzalez N, Blake H, Wu S, MacMurray JP (1999) Additive effect of three noradrenergic genes (ADRA2a, ADRA2C, DBH) on attention-deficit hyperactivity disorder and learning disabilities in Tourette syndrome subjects. Clin Genet 55(3): 160–172

Cook EH, Jr, Stein MA, Krasowski MD, Cox NJ, Olkon DM, Kieffer JE, Leventhal BL (1995) Association of attention-deficit disorder and the dopamine transporter gene. Am J Hum Genet 56(4): 993–998

Cope N, Harold D, Hill G, Moskvina V, Stevenson J, Holmans P, Owen MJ, O'Donovan MC, Williams J (2005) Strong evidence that KIAA0319 on chromosome 6p is a susceptibility gene for developmental dyslexia. Am J Hum Genet 76(4): 581–591

Couto JM, Gomez L, Wigg K, Cate-Carter T, Archibald J, Anderson B, Tannock R, Kerr EN, Lovett MW, Humphries T, Barr CL. The KIAA0319-Like (*KIAA0319L*) Gene on Chromosome 1p34 as a Candidate for Reading Disabilities. Neurogenetics

Couto JM, Livne-Bar I, Xu Z, Cate-Carter T, Nathaniel A, Anderson B, Tannock R, Kerr EN, Lovett MW, Humphries T, Bremner R, Barr CL (submitted) Association of Reading Disabilites to a Region Marked by Acetylated H3 Histones in *KIAA0319*

Craddock N, Owen MJ, O'Donovan MC (2006) The catechol-*O*-methyl transferase (COMT) gene as a candidate for psychiatric phenotypes: evidence and lessons. Mol Psychiatry 11(5): 446–458

Crosbie J, Perusse D, Barr CL, Schachar RJ (2008) Validating psychiatric endophenotypes: Inhibitory control and attention deficit hyperactivity disorder. Neurosci Biobehav Rev 32: 40–55

Cubells JF, Zabetian CP (2004) Human genetics of plasma dopamine beta-hydroxylase activity: applications to research in psychiatry and neurology. Psychopharmacology (Berl) 174(4): 463–476

Cubells JF, Kranzler HR, McCance-Katz E, Anderson GM, Malison RT, Price LH, Gelernter J (2000) A haplotype at the DBH locus, associated with low plasma dopamine beta-hydroxylase activity, also associates with cocaine-induced paranoia. Mol Psychiatry 5(1): 56–63

Curran S, Mill J, Tahir E, Kent L, Richards S, Gould A, Huckett L, Sharp J, Batten C, Fernando S, Ozbay F, Yazgan Y, Simonoff E, Thompson M, Taylor E, Asherson P (2001) Association study of a dopamine transporter polymorphism and attention deficit hyperactivity disorder in UK and Turkish samples. Mol Psychiatry 6(4): 425–428

Cyr M, Bosse R, Di Paolo T (1998) Gonadal hormones modulate 5-hydroxytryptamine2A receptors: emphasis on the rat frontal cortex. Neuroscience 83(3): 829–836

Daly G, Hawi Z, Fitzgerald M, Gill M (1999) Mapping susceptibility loci in attention deficit hyperactivity disorder: preferential transmission of parental alleles at DAT1, DBH and DRD5 to affected children. Mol Psychiatry 4(2): 192–196

Das M, Mukhopadhyay K (2007) DAT1 3'-UTR 9R allele: preferential transmission in Indian children with attention deficit hyperactivity disorder. Am J Med Genet B Neuropsychiatr Genet 144(6): 826–829

Das M, Bhowmik AD, Sinha S, Chattopadhyay A, Chaudhuri K, Singh M, Mukhopadhyay K (2006) MAOA promoter polymorphism and attention deficit hyperactivity disorder (ADHD) in indian children. Am J Med Genet B Neuropsychiatr Genet 141(6): 637–642

Deckert J, Catalano M, Syagailo YV, Bosi M, Okladnova O, Di Bella D, Nothen MM, Maffei P, Franke P, Fritze J, Maier W, Propping P, Beckmann H, Bellodi L, Lesch KP (1999) Excess of high activity monoamine oxidase A gene promoter alleles in female patients with panic disorder. Hum Mol Genet 8(4): 621–624

Deffenbacher KE, Kenyon JB, Hoover DM, Olson RK, Pennington BF, DeFries JC, Smith SD (2004) Refinement of the 6p21.3 quantitative trait locus influencing dyslexia: linkage and association analyses. Hum Genet 115(2): 128–138

De Gobbi M, Viprakasit V, Hughes JR, Fisher C, Buckle VJ, Ayyub H, Gibbons RJ, Vernimmen D, Yoshinaga Y, de Jong P, Cheng JF, Rubin EM, Wood WG, Bowden D, Higgs DR (2006) A regulatory SNP causes a human genetic disease by creating a new transcriptional promoter. Science 312(5777): 1215–1217

De Luca V, Muglia P, Jain U, Kennedy JL (2004a) No evidence of linkage or association between the norepinephrine transporter (NET) gene MnlI polymorphism and adult ADHD. Am J Med Genet B Neuropsychiatr Genet 124(1): 38–40

De Luca V, Muglia P, Vincent JB, Lanktree M, Jain U, Kennedy JL (2004b) Adrenergic alpha 2C receptor genomic organization: association study in adult ADHD. Am J Med Genet B Neuropsychiatr Genet 127(1): 65–67

Denney RM, Koch H, Craig IW (1999) Association between monoamine oxidase A activity in human male skin fibroblasts and genotype of the MAOA promoter-associated variable number tandem repeat. Hum Genet 105(6): 542–551

Dermitzakis ET, Clark AG (2002) Evolution of transcription factor binding sites in Mammalian gene regulatory regions: conservation and turnover. Mol Biol Evol 19(7): 1114–1121

Deupree JD, Smith SD, Kratochvil CJ, Bohac D, Ellis CR, Polaha J, Bylund DB (2006) Possible involvement of alpha-2A adrenergic receptors in attention deficit hyperactivity disorder: radioligand binding and polymorphism studies. Am J Med Genet B Neuropsychiatr Genet 141(8): 877–884

Molecular Genetics of ADHD

Dickman DK, Horne JA, Meinertzhagen IA, Schwarz TL (2005) A slowed classical pathway rather than kiss-and-run mediates endocytosis at synapses lacking synaptojanin and endophilin. Cell 123(3): 521–533

DiFranza JR, Lew RA (1995) Effect of maternal cigarette smoking on pregnancy complications and sudden infant death syndrome. J Fam Pract 40(4): 385–394

Dluzen DE (2000) Neuroprotective effects of estrogen upon the nigrostriatal dopaminergic system. J Neurocytol 29(5–6): 387–399

Domschke K, Sheehan K, Lowe N, Kirley A, Mullins C, O'Sullivan R, Freitag C, Becker T, Conroy J, Fitzgerald M, Gill M, Hawi Z (2005) Association analysis of the monoamine oxidase A and B genes with attention deficit hyperactivity disorder (ADHD) in an Irish sample: preferential transmission of the MAO-A 941G allele to affected children. Am J Med Genet B Neuropsychiatr Genet 134(1): 110–114

Dorval KM, Wigg KG, Crosbie J, Tannock R, Kennedy JL, Ickowicz A, Pathare T, Malone M, Schachar R, Barr CL (2007) Association of the glutamate receptor subunit gene GRIN2B with attention-deficit/hyperactivity disorder. Genes Brain Behav 6(5): 444–452

Drysdale CM, McGraw DW, Stack CB, Stephens JC, Judson RS, Nandabalan K, Arnold K, Ruano G, Liggett SB (2000) Complex promoter and coding region beta 2-adrenergic receptor haplotypes alter receptor expression and predict in vivo responsiveness. Proc Natl Acad Sci U S A 97(19): 10483–10488

D'Souza UM, Russ C, Tahir E, Mill J, McGuffin P, Asherson PJ, Craig IW (2004) Functional effects of a tandem duplication polymorphism in the 5′ flanking region of the DRD4 gene. Biol Psychiatry 56(9): 691–697

Duan J, Sanders AR, Molen JE, Martinolich L, Mowry BJ, Levinson DF, Crowe RR, Silverman JM, Gejman PV (2003) Polymorphisms in the 5′-untranslated region of the human serotonin receptor 1B (HTR1B) gene affect gene expression. Mol Psychiatry 8(11): 901–910

Eaves LJ, Silberg JL, Meyer JM, Maes HH, Simonoff E, Pickles A, Rutter M, Neale MC, Reynolds CA, Erikson MT, Heath AC, Loeber R, Truett KR, Hewitt JK (1997) Genetics and developmental psychopathology: 2. The main effects of genes and environment on behavioral problems in the Virginia Twin Study of Adolescent Behavioral Development. J Child Psychol Psychiatry 38(8): 965–980

Eisenberg J, Mei-Tal G, Steinberg A, Tartakovsky E, Zohar A, Gritsenko I, Nemanov L, Ebstein RP (1999) Haplotype relative risk study of catechol-O-methyltransferase (COMT) and attention deficit hyperactivity disorder (ADHD): association of the high-enzyme activity Val allele with ADHD impulsive-hyperactive phenotype. Am J Med Genet 88(5): 497–502

Eisenberg J, Zohar A, Mei-Tal G, Steinberg A, Tartakovsky E, Gritsenko I, Nemanov L, Ebstein RP (2000) A haplotype relative risk study of the dopamine D4 receptor (DRD4) exon III repeat polymorphism and attention deficit hyperactivity disorder (ADHD). Am J Med Genet 96(3): 258–261

Ernst M, Moolchan ET, Robinson ML (2001) Behavioral and neural consequences of prenatal exposure to nicotine. J Am Acad Child Adolesc Psychiatry 40(6): 630–641

Faraone SV, Biederman J (1997) Do attention deficit hyperactivity disorder and major depression share familial risk factors? J Nerv Ment Dis 185(9): 533–541

Faraone SV, Biederman J, Keenan K, Tsuang MT (1991) Separation of DSM-III attention deficit disorder and conduct disorder: evidence from a family-genetic study of American child psychiatric patients. Psychol Med 21(1): 109–121

Faraone S, Biederman J, Chen WJ, Kricher B, Keenan K, Moore C, Sprich S, Tsuang MT (1992) Segregation analysis of attention deficit hyperactivity disorder. Psychiatr Genet 2: 257–275

Faraone SV, Biederman J, Lehman BK, Keenan K, Norman D, Seidman LJ, Kolodny R, Kraus I, Perrin J, Chen WJ (1993) Evidence for the independent familial transmission of attention deficit hyperactivity disorder and learning disabilities: results from a family genetic study. Am J Psychiatry 150(6): 891–895

Faraone SV, Biederman J, Mennin D, Wozniak J, Spencer T (1997) Attention-deficit hyperactivity disorder with bipolar disorder: a familial subtype? J Am Acad Child Adolesc Psychiatry 36(10): 1378–1387; discussion 1387–90

Faraone SV, Biederman J, Mennin D, Russell R (1998) Bipolar and antisocial disorders among relatives of ADHD children: parsing familial subtypes of illness. Am J Med Genet 81(1): 108–116

Faraone SV, Biederman J, Weiffenbach B, Keith T, Chu MP, Weaver A, Spencer TJ, Wilens TE, Frazier J, Cleves M, Sakai J (1999) Dopamine D4 gene 7-repeat allele and attention deficit hyperactivity disorder. Am J Psychiatry 156(5): 768–770

Faraone SV, Biederman J, Monuteaux MC (2000) Toward guidelines for pedigree selection in genetic studies of attention deficit hyperactivity disorder. Genet Epidemiol 18(1): 1–16

Faraone SV, Doyle AE, Mick E, Biederman J (2001) Meta-analysis of the association between the 7-repeat allele of the dopamine D(4) receptor gene and attention deficit hyperactivity disorder. Am J Psychiatry 158(7): 1052–1057

Faraone SV, Perlis RH, Doyle AE, Smoller JW, Goralnick JJ, Holmgren MA, Sklar P (2005) Molecular genetics of attention-deficit/hyperactivity disorder. Biol Psychiatry 57(11): 1313–1323

Feng Y, Crosbie J, Wigg K, Pathare T, Ickowicz A, Schachar R, Tannock R, Roberts W, Malone M, Swanson J, Kennedy JL, Barr CL (2005a) The SNAP25 gene as a susceptibility gene contributing to attention-deficit hyperactivity disorder. Mol Psychiatry 10(11): 998–1005, 973

Feng Y, Wigg KG, Makkar R, Ickowicz A, Pathare T, Tannock R, Roberts W, Malone M, Kennedy JL, Schachar R, Barr CL (2005b) Sequence variation in the 3'-untranslated region of the dopamine transporter gene and attention-deficit hyperactivity disorder (ADHD). Am J Med Genet B Neuropsychiatr Genet 139B(1): 1–6

Fink G, Sumner BE (1996) Oestrogen and mental state. Nature 383(6598): 306

Fisher SE, Francks C (2006) Genes, cognition and dyslexia: learning to read the genome. Trends Cogn Sci 10(6): 250–257

Fisher SE, Francks C, Marlow AJ, MacPhie IL, Newbury DF, Cardon LR, Ishikawa-Brush Y, Richardson AJ, Talcott JB, Gayan J, Olson RK, Pennington BF, Smith SD, DeFries JC, Stein JF, Monaco AP (2002a) Independent genome-wide scans identify a chromosome 18 quantitative-trait locus influencing dyslexia. Nat Genet 30(1): 86–91

Fisher SE, Francks C, McCracken JT, McGough JJ, Marlow AJ, MacPhie IL, Newbury DF, Crawford LR, Palmer CG, Woodward JA, Del'Homme M, Cantwell DP, Nelson SF, Monaco AP, Smalley SL (2002b) A genomewide scan for loci involved in attention-deficit/hyperactivity disorder. Am J Hum Genet 70(5): 1183–1196

Fiskerstrand CE, Lovejoy EA, Quinn JP (1999) An intronic polymorphic domain often associated with susceptibility to affective disorders has allele dependent differential enhancer activity in embryonic stem cells. FEBS Lett 458(2): 171–174

Flory K, Milich R, Lynam DR, Leukefeld C, Clayton R (2003) Relation between childhood disruptive behavior disorders and substance use and dependence symptoms in young adulthood: individuals with symptoms of attention-deficit/hyperactivity disorder and conduct disorder are uniquely at risk. Psychol Addict Behav 17(2): 151–158

Foulder-Hughes LA, Cooke RW (2003) Motor, cognitive, and behavioural disorders in children born very preterm. Dev Med Child Neurol 45(2): 97–103

Fowler JS, Alia-Klein N, Kriplani A, Logan J, Williams B, Zhu W, Craig IW, Telang F, Goldstein R, Volkow ND, Vaska P, Wang GJ (2007) Evidence That Brain MAO A Activity Does Not Correspond to MAO A Genotype in Healthy Male Subjects. Biol Psychiatry 62(4): 355–358

Francks C, Paracchini S, Smith SD, Richardson AJ, Scerri TS, Cardon LR, Marlow AJ, MacPhie IL, Walter J, Pennington BF, Fisher SE, Olson RK, DeFries JC, Stein JF, Monaco AP (2004) A 77-kilobase region of chromosome 6p22.2 is associated with dyslexia in families from the United Kingdom and from the United States. Am J Hum Genet 75(6): 1046–1058

Friedel S, Saar K, Sauer S, Dempfle A, Walitza S, Renner T, Romanos M, Freitag C, Seitz C, Palmason H, Scherag A, Windemuth-Kieselbach C, Schimmelmann BG, Wewetzer C, Meyer J, Warnke A, Lesch KP, Reinhardt R, Herpertz-Dahlmann B, Linder M, Hinney A, Remschmidt H, Schafer H, Konrad K, Hubner N, Hebebrand J (2007) Association and linkage of allelic variants of the dopamine transporter gene in ADHD. Mol Psychiatry 12(10): 923–933

Molecular Genetics of ADHD 147

Fuke S, Suo S, Takahashi N, Koike H, Sasagawa N, Ishiura S (2001) The VNTR polymorphism of the human dopamine transporter (DAT1) gene affects gene expression. Pharmacogenomics J 1(2): 152–156

Gainetdinov RR, Wetsel WC, Jones SR, Levin ED, Jaber M, Caron MG (1999) Role of serotonin in the paradoxical calming effect of psychostimulants on hyperactivity. Science 283(5400): 397–401

Galaburda AM, Sherman GF, Rosen GD, Aboitiz F, Geschwind N (1985) Developmental dyslexia: four consecutive patients with cortical anomalies. Ann Neurol 18(2): 222–233

Galaburda AM, LoTurco J, Ramus F, Fitch RH, Rosen GD (2006) From genes to behavior in developmental dyslexia. Nat Neurosci 9(10): 1213–1217

Galili-Weisstub E, Levy S, Frisch A, Gross-Tsur V, Michaelovsky E, Kosov A, Meltzer A, Goltser T, Serretti A, Cusin C, Darvasi A, Inbar E, Weizman A, Segman RH (2005) Dopamine transporter haplotype and attention-deficit hyperactivity disorder. Mol Psychiatry 10(7): 617–618

Gao WQ, Shinsky N, Armanini MP, Moran P, Zheng JL, Mendoza-Ramirez JL, Phillips HS, Winslow JW, Caras IW (1998) Regulation of hippocampal synaptic plasticity by the tyrosine kinase receptor, REK7/EphA5, and its ligand, AL-1/Ephrin-A5. Mol Cell Neurosci 11(5–6): 247–259

Gelernter J, Cubells JF, Kidd JR, Pakstis AJ, Kidd KK (1999) Population studies of polymorphisms of the serotonin transporter protein gene. Am J Med Genet 88(1): 61–66

Genro JP, Zeni C, Polanczyk GV, Roman T, Rohde LA, Hutz MH (2007) A promoter polymorphism (−839 C > T) at the dopamine transporter gene is associated with attention deficit/hyperactivity disorder in Brazilian children. Am J Med Genet B Neuropsychiatr Genet 144(2): 215–219

Gill M, Daly G, Heron S, Hawi Z, Fitzgerald M (1997) Confirmation of association between attention deficit hyperactivity disorder and a dopamine transporter polymorphism. Mol Psychiatry 2(4): 311–313

Gittelman R, Mannuzza S, Shenker R, Bonagura N (1985) Hyperactive boys almost grown up. 1. Psychiatric status. Arch Gen Psychiatry 42(10): 937–947

Gjone H, Stevenson J, Sundet JM (1996) Genetic influence on parent-reported attention-related problems in a Norwegian general population twin sample. J Am Acad Child Adolesc Psychiatry 35(5): 588–596; discussion 596–8

Gornick MC, Addington A, Shaw P, Bobb AJ, Sharp W, Greenstein D, Arepalli S, Castellanos FX, Rapoport JL (2007) Association of the dopamine receptor D4 (DRD4) gene 7-repeat allele with children with attention-deficit/hyperactivity disorder (ADHD): an update. Am J Med Genet B Neuropsychiatr Genet 144(3): 379–382

Gotti C, Zoli M, Clementi F (2006) Brain nicotinic acetylcholine receptors: native subtypes and their relevance. Trends Pharmacol Sci 27(9): 482–491

Grady SR, Murphy KL, Cao J, Marks MJ, McIntosh JM, Collins AC (2002) Characterization of nicotinic agonist-induced [(3)H]dopamine release from synaptosomes prepared from four mouse brain regions. J Pharmacol Exp Ther 301(2): 651–660

Greenwood TA, Kelsoe JR (2003) Promoter and intronic variants affect the transcriptional regulation of the human dopamine transporter gene. Genomics 82(5): 511–520

Grigorenko EL, Wood FB, Meyer MS, Hart LA, Speed WC, Shuster A, Pauls DL (1997) Susceptibility loci for distinct components of developmental dyslexia on chromosomes 6 and 15. Am J Hum Genet 60(1): 27–39

Gross J, Muller I, Chen Y, Elizalde M, Leclere N, Herrera-Marschitz M, Andersson K (2000) Perinatal asphyxia induces region-specific long-term changes in mRNA levels of tyrosine hydroxylase and dopamine D(1) and D(2) receptors in rat brain. Brain Res Mol Brain Res 79(1–2): 110–117

Guimaraes AP, Zeni C, Polanczyk GV, Genro JP, Roman T, Rohde LA, Hutz MH (2007) Serotonin genes and attention deficit/hyperactivity disorder in a Brazilian sample: preferential transmission of the HTR2A 452His allele to affected boys. Am J Med Genet B Neuropsychiatr Genet 144(1): 69–73

Halperin JM, Newcorn JH, Koda VH, Pick L, McKay KE, Knott P (1997) Noradrenergic mechanisms in ADHD children with and without reading disabilities: a replication and extension. J Am Acad Child Adolesc Psychiatry 36(12): 1688–1697

Hannula-Jouppi K, Kaminen-Ahola N, Taipale M, Eklund R, Nopola-Hemmi J, Kaariainen H, Kere J (2005) The Axon Guidance Receptor Gene ROBO1 Is a Candidate Gene for Developmental Dyslexia. PLoS Genet 1(4): e50

Harold D, Paracchini S, Scerri T, Dennis M, Cope N, Hill G, Moskvina V, Walter J, Richardson AJ, Owen MJ, Stein JF, Green ED, O'Donovan MC, Williams J, Monaco AP (2006) Further evidence that the KIAA0319 gene confers susceptibility to developmental dyslexia. Mol Psychiatry 11(12): 1085–1091, 1061

Hart EL, Lahey BB, Loeber R, Applegate B, Frick PJ (1995) Developmental change in attention-deficit hyperactivity disorder in boys: a four-year longitudinal study. J Abnorm Child Psychol 23(6): 729–749

Hawi Z, McCarron M, Kirley A, Daly G, Fitzgerald M, Gill M (2000a) No association of the dopamine DRD4 receptor (DRD4) gene polymorphism with attention deficit hyperactivity disorder (ADHD) in the Irish population. Am J Med Genet 96(3): 268–272

Hawi Z, Millar N, Daly G, Fitzgerald M, Gill M (2000b) No association between catechol-O-methyltransferase (COMT) gene polymorphism and attention deficit hyperactivity disorder (ADHD) in an Irish sample. Am J Med Genet 96(3): 282–284

Hawi Z, Foley D, Kirley A, McCarron M, Fitzgerald M, Gill M (2001) Dopa decarboxylase gene polymorphisms and attention deficit hyperactivity disorder (ADHD): no evidence for association in the Irish population. Mol Psychiatry 6(4): 420–424

Hawi Z, Dring M, Kirley A, Foley D, Kent L, Craddock N, Asherson P, Curran S, Gould A, Richards S, Lawson D, Pay H, Turic D, Langley K, Owen M, O'Donovan M, Thapar A, Fitzgerald M, Gill M (2002) Serotonergic system and attention deficit hyperactivity disorder (ADHD): a potential susceptibility locus at the 5-HT(1B) receptor gene in 273 nuclear families from a multi-centre sample. Mol Psychiatry 7(7): 718–725

Hawi Z, Lowe N, Kirley A, Gruenhage F, Nothen M, Greenwood T, Kelsoe J, Fitzgerald M, Gill M (2003) Linkage disequilibrium mapping at DAT1, DRD5 and DBH narrows the search for ADHD susceptibility alleles at these loci. Mol Psychiatry 8(3): 299–308

Hawi Z, Segurado R, Conroy J, Sheehan K, Lowe N, Kirley A, Shields D, Fitzgerald M, Gallagher L, Gill M (2005) Preferential transmission of paternal alleles at risk genes in attention-deficit/hyperactivity disorder. Am J Hum Genet 77(6): 958–965

Hazelwood LA, Sanders-Bush E (2004) His452Tyr polymorphism in the human 5-HT2A receptor destabilizes the signaling conformation. Mol Pharmacol 66(5): 1293–1300

Hebebrand J, Dempfle A, Saar K, Thiele H, Herpertz-Dahlmann B, Linder M, Kiefl H, Remschmidt H, Hemminger U, Warnke A, Knolker U, Heiser P, Friedel S, Hinney A, Schafer H, Nurnberg P, Konrad K (2006) A genome-wide scan for attention-deficit/hyperactivity disorder in 155 German sib-pairs. Mol Psychiatry 11(2): 196–205

Heijtz RD, Alexeyenko A, Castellanos FX (2007) Calcyon mRNA expression in the frontal-striatal circuitry and its relationship to vesicular processes and ADHD. Behav Brain Funct 3: 33

Heintzman ND, Stuart RK, Hon G, Fu Y, Ching CW, Hawkins RD, Barrera LO, Van Calcar S, Qu C, Ching KA, Wang W, Weng Z, Green RD, Crawford GE, Ren B (2007) Distinct and predictive chromatin signatures of transcriptional promoters and enhancers in the human genome. Nat Genet 39(3): 311–318

Heinz A, Goldman D, Jones DW, Palmour R, Hommer D, Gorey JG, Lee KS, Linnoila M, Weinberger DR (2000) Genotype influences in vivo dopamine transporter availability in human striatum. Neuropsychopharmacology 22(2): 133–139

Heiser P, Dempfle A, Friedel S, Konrad K, Hinney A, Kiefl H, Walitza S, Bettecken T, Saar K, Linder M, Warnke A, Herpertz-Dahlmann B, Schafer H, Remschmidt H, Hebebrand J (2007) Family-based association study of serotonergic candidate genes and attention-deficit/hyperactivity disorder in a German sample. J Neural Transm 114(4): 513–521

Herbison AE, Simonian SX, Thanky NR, Bicknell RJ (2000) Oestrogen modulation of noradrenaline neurotransmission. Novartis Found Symp 230: 74–85; discussion 85–93

Hess EJ, Collins KA, Wilson MC (1996) Mouse model of hyperkinesis implicates SNAP-25 in behavioral regulation. J Neurosci 16(9): 3104–3111

Molecular Genetics of ADHD

Heyser CJ, Wilson MC, Gold LH (1995) Coloboma hyperactive mutant exhibits delayed neurobehavioral developmental milestones. Brain Res Dev Brain Res 89(2): 264–269

Holmes J, Payton A, Barrett JH, Hever T, Fitzpatrick H, Trumper AL, Harrington R, McGuffin P, Owen M, Ollier W, Worthington J, Thapar A (2000) A family-based and case-control association study of the dopamine D4 receptor gene and dopamine transporter gene in attention deficit hyperactivity disorder. Mol Psychiatry 5(5): 523–530

Hotamisligil GS, Breakefield XO (1991) Human monoamine oxidase A gene determines levels of enzyme activity. Am J Hum Genet 49(2): 383–392

Hu XZ, Lipsky RH, Zhu G, Akhtar LA, Taubman J, Greenberg BD, Xu K, Arnold PD, Richter MA, Kennedy JL, Murphy DL, Goldman D (2006) Serotonin transporter promoter gain-of-function genotypes are linked to obessive-compulsive disorder. Am J Hum Genet 78(5): 815–826

Hudziak JJ, Derks EM, Althoff RR, Rettew DC, Boomsma DI (2005) The genetic and environmental contributions to attention deficit hyperactivity disorder as measured by the Conners' Rating Scales – Revised. Am J Psychiatry 162(9): 1614–1620

Hultman CM, Torrang A, Tuvblad C, Cnattingius S, Larsson JO, Lichtenstein P (2007) Birth weight and attention-deficit/hyperactivity symptoms in childhood and early adolescence: a prospective Swedish twin study. J Am Acad Child Adolesc Psychiatry 46(3): 370–377

Humphreys P, Kaufmann WE, Galaburda AM (1990) Developmental dyslexia in women: neuropathological findings in three patients. Ann Neurol 28(6): 727–738

Ickowicz A, Feng Y, Wigg K, Quist J, Pathare T, Roberts W, Malone M, Schachar R, Tannock R, Kennedy JL, Barr CL (2007) The serotonin receptor HTR1B: Gene polymorphisms in attention deficit hyperactivity disorder. Am J Med Genet B Neuropsychiatr Genet 144: 121–125

Inkster B, Muglia P, Jain U, Kennedy JL (2004) Linkage disequilibrium analysis of the dopamine beta-hydroxylase gene in persistent attention deficit hyperactivity disorder. Psychiatr Genet 14(2): 117–120

Ivgy-May N, Tamir H, Gershon MD (1994) Synaptic properties of serotonergic growth cones in developing rat brain. J Neurosci 14(3 Pt 1): 1011–1029

Jackson DM, Westlind-Danielsson A (1994) Dopamine receptors: molecular biology, biochemistry and behavioural aspects. Pharmacol Ther 64(2): 291–370

Jacobsen LK, Staley JK, Zoghbi SS, Seibyl JP, Kosten TR, Innis RB, Gelernter J (2000) Prediction of dopamine transporter binding availability by genotype: a preliminary report. Am J Psychiatry 157(10): 1700–1703

Jais PH (2005) How frequent is altered gene expression among susceptibility genes to human complex disorders? Genet Med 7(2): 83–96

Jassen AK, Yang H, Miller GM, Calder E, Madras BK (2006) Receptor regulation of gene expression of axon guidance molecules: implications for adaptation. Mol Pharmacol 70(1): 71–77

Jiang S, Xin R, Wu X, Lin S, Qian Y, Ren D, Tang G, Wang D (2000) Association between attention deficit hyperactivity disorder and the DXS7 locus. Am J Med Genet 96(3): 289–292

Jiang S, Xin R, Lin S, Qian Y, Tang G, Wang D, Wu X (2001) Linkage studies between attention-deficit hyperactivity disorder and the monoamine oxidase genes. Am J Med Genet 105(8): 783–788

Jin LQ, Wang HY, Friedman E (2001) Stimulated D(1) dopamine receptors couple to multiple Galpha proteins in different brain regions. J Neurochem 78(5): 981–990

Jovanovic V, Guan HC, Van Tol HH (1999) Comparative pharmacological and functional analysis of the human dopamine D4.2 and D4.10 receptor variants. Pharmacogenetics 9(5): 561–568

Kahn RS, Khoury J, Nichols WC, Lanphear BP (2003) Role of dopamine transporter genotype and maternal prenatal smoking in childhood hyperactive-impulsive, inattentive, and oppositional behaviors. J Pediatr 143(1): 104–110

Kaplan DE, Gayan J, Ahn J, Won TW, Pauls D, Olson RK, DeFries JC, Wood F, Pennington BF, Page GP, Smith SD, Gruen JR (2002) Evidence for linkage and association with reading disability on 6p21.3–22. Am J Hum Genet 70(5): 1287–1298

Kelada SN, Costa-Mallen P, Checkoway H, Carlson CS, Weller TS, Swanson PD, Franklin GM, Longstreth WT, Jr, Afsharinejad Z, Costa LG (2005) Dopamine transporter (SLC6A3) 5 region haplotypes significantly affect transcriptional activity in vitro but are not associated with Parkinson's disease. Pharmacogenet Genomics 15(9): 659–668

Kent L, Green E, Holmes J, Thapar A, Gill M, Hawi Z, Fitzgerald M, Asherson P, Curran S, Mills J, Payton A, Craddock N (2001a) No association between CHRNA7 microsatellite markers and attention-deficit hyperactivity disorder. Am J Med Genet 105(8): 686–689

Kent L, Middle F, Hawi Z, Fitzgerald M, Gill M, Feehan C, Craddock N (2001b) Nicotinic acetylcholine receptor alpha4 subunit gene polymorphism and attention deficit hyperactivity disorder. Psychiatr Genet 11(1): 37–40

Kent L, Doerry U, Hardy E, Parmar R, Gingell K, Hawi Z, Kirley A, Lowe N, Fitzgerald M, Gill M, Craddock N (2002) Evidence that variation at the serotonin transporter gene influences susceptibility to attention deficit hyperactivity disorder (ADHD): analysis and pooled analysis. Mol Psychiatry 7(8): 908–912

Kereszturi E, Kiraly O, Csapo Z, Tarnok Z, Gadoros J, Sasvari-Szekely M, Nemoda Z (2007) Association between the 120-bp duplication of the dopamine D4 receptor gene and attention deficit hyperactivity disorder: genetic and molecular analyses. Am J Med Genet B Neuropsychiatr Genet 144(2): 231–236

Kessler RC, Adler LA, Barkley R, Biederman J, Conners CK, Faraone SV, Greenhill LL, Jaeger S, Secnik K, Spencer T, Ustun TB, Zaslavsky AM (2005) Patterns and predictors of attention-deficit/hyperactivity disorder persistence into adulthood: results from the national comorbidity survey replication. Biol Psychiatry 57(11): 1442–1451

Keulers EH, Hendriksen JG, Feron FJ, Wassenberg R, Wuisman-Frerker MG, Jolles J, Vles JS (2007) Methylphenidate improves reading performance in children with attention deficit hyperactivity disorder and comorbid dyslexia: an unblinded clinical trial. Eur J Paediatr Neurol 11(1): 21–28

Kim SJ, Badner J, Cheon KA, Kim BN, Yoo HJ, Cook E, Jr, Leventhal BL, Kim YS (2005a) Family-based association study of the serotonin transporter gene polymorphisms in Korean ADHD trios. Am J Med Genet B Neuropsychiatr Genet 139(1): 14–18

Kim YS, Leventhal BL, Kim SJ, Kim BN, Cheon KA, Yoo HJ, Badner J, Cook EH (2005b) Family-based association study of DAT1 and DRD4 polymorphism in Korean children with ADHD. Neurosci Lett 390(3): 176–181

Kim CH, Hahn MK, Joung Y, Anderson SL, Steele AH, Mazei-Robinson MS, Gizer I, Teicher MH, Cohen BM, Robertson D, Waldman ID, Blakely RD, Kim KS (2006) A polymorphism in the norepinephrine transporter gene alters promoter activity and is associated with attention-deficit hyperactivity disorder. Proc Natl Acad Sci U S A 103(50): 19164–19169

King MC, Wilson AC (1975) Evolution at two levels in humans and chimpanzees. Science 188(4184): 107–116

Kirley A, Hawi Z, Daly G, McCarron M, Mullins C, Millar N, Waldman I, Fitzgerald M, Gill M (2002) Dopaminergic system genes in ADHD: toward a biological hypothesis. Neuropsychopharmacology 27(4): 607–619

Kirley A, Lowe N, Hawi Z, Mullins C, Daly G, Waldman I, McCarron M, O'Donnell D, Fitzgerald M, Gill M (2003) Association of the 480bp DAT1 allele with methylphenidate response in a sample of Irish children with ADHD. Am J Med Genet 121B(1): 50–54

Klink R, de Kerchove d'Exaerde A, Zoli M, Changeux JP (2001) Molecular and physiological diversity of nicotinic acetylcholine receptors in the midbrain dopaminergic nuclei. J Neurosci 21(5): 1452–1463

Knight JC (2005) Regulatory polymorphisms underlying complex disease traits. J Mol Med 83(2): 97–109

Knopik VS, Sparrow EP, Madden PA, Bucholz KK, Hudziak JJ, Reich W, Slutske WS, Grant JD, McLaughlin TL, Todorov A, Todd RD, Heath AC (2005) Contributions of parental alcoholism, prenatal substance exposure, and genetic transmission to child ADHD risk: a female twin study. Psychol Med 35(5): 625–635

Kotimaa AJ, Moilanen I, Taanila A, Ebeling H, Smalley SL, McGough JJ, Hartikainen AL, Jarvelin MR (2003) Maternal smoking and hyperactivity in 8-year-old children. J Am Acad Child Adolesc Psychiatry 42(7): 826–833

Kotler M, Manor I, Sever Y, Eisenberg J, Cohen H, Ebstein RP, Tyano S (2000) Failure to replicate an excess of the long dopamine D4 exon III repeat polymorphism in ADHD in a family-based study. Am J Med Genet 96(3): 278–281

Kujala T, Karma K, Ceponiene R, Belitz S, Turkkila P, Tervaniemi M, Naatanen R (2001) Plastic neural changes and reading improvement caused by audiovisual training in reading-impaired children. Proc Natl Acad Sci U S A 98(18): 10509–10514

Kuntsi J, Rijsdijk F, Ronald A, Asherson P, Plomin R (2005) Genetic influences on the stability of attention-deficit/hyperactivity disorder symptoms from early to middle childhood. Biol Psychiatry 57(6): 647–654

Kustanovich V, Merriman B, McGough J, McCracken JT, Smalley SL, Nelson SF (2003) Biased paternal transmission of SNAP-25 risk alleles in attention-deficit hyperactivity disorder. Mol Psychiatry 8(3): 309–315

Kustanovich V, Ishii J, Crawford L, Yang M, McGough JJ, McCracken JT, Smalley SL, Nelson SF (2004) Transmission disequilibrium testing of dopamine-related candidate gene polymorphisms in ADHD: confirmation of association of ADHD with DRD4 and DRD5. Mol Psychiatry 9: 711–717

LaHoste GJ, Swanson JM, Wigal SB, Glabe C, Wigal T, King N, Kennedy JL (1996) Dopamine D4 receptor gene polymorphism is associated with attention deficit hyperactivity disorder. Mol Psychiatry 1(2): 121–124

Lambert NM, Hartsough CS (1998) Prospective study of tobacco smoking and substance dependencies among samples of ADHD and non-ADHD participants. J Learn Disabil 31(6): 533–544

Landmesser L (1994) Axonal outgrowth and pathfinding. Prog Brain Res 103: 67–73

Langley K, Payton A, Hamshere ML, Pay HM, Lawson DC, Turic D, Ollier W, Worthington J, Owen MJ, O'Donovan MC, Thapar A (2003) No evidence of association of two 5HT transporter gene polymorphisms and attention deficit hyperactivity disorder. Psychiatr Genet 13(2): 107–110

Langley K, Turic D, Peirce TR, Mills S, Van Den Bree MB, Owen MJ, O'Donovan MC, Thapar A (2005) No support for association between the dopamine transporter (DAT1) gene and ADHD. Am J Med Genet B Neuropsychiatr Genet 139(1): 7–10

Larsson JO, Larsson H, Lichtenstein P (2004) Genetic and environmental contributions to stability and change of ADHD symptoms between 8 and 13 years of age: a longitudinal twin study. J Am Acad Child Adolesc Psychiatry 43(10): 1267–1275

Laucht M, Skowronek MH, Becker K, Schmidt MH, Esser G, Schulze TG, Rietschel M (2007) Interacting effects of the dopamine transporter gene and psychosocial adversity on attention-deficit/hyperactivity disorder symptoms among 15-year-olds from a high-risk community sample. Arch Gen Psychiatry 64(5): 585–590

Laurin N, Misener VL, Crosbie J, Ickowicz A, Pathare T, Roberts W, Malone M, Tannock R, Schachar R, Kennedy JL, Barr CL (2005) Association of the calcyon gene (DRD1IP) with attention deficit/hyperactivity disorder. Mol Psychiatry 10(12): 1117–1125

Laurin N, Ickowicz A, Pathare T, Malone M, Tannock R, Schachar R, Kennedy JL, Barr CL (2008) Investigation of the G protein subunit Galpha(olf) gene (GNAL) in attention deficit/ hyperactivity disorder. J Psychiatr Res 42: 117–124

Lawson DC, Turic D, Langley K, Pay HM, Govan CF, Norton N, Hamshere ML, Owen MJ, O'Donovan MC, Thapar A (2003) Association analysis of monoamine oxidase A and attention deficit hyperactivity disorder. Am J Med Genet B Neuropsychiatr Genet 116(1): 84–89

Lee SH, Mouradian MM (1999) Up-regulation of D1A dopamine receptor gene transcription by estrogen. Mol Cell Endocrinol 156(1–2): 151–157

Lesch KP, Bengel D, Heils A, Sabol SZ, Greenberg BD, Petri S, Benjamin J, Muller CR, Hamer DH, Murphy DL (1996) Association of anxiety-related traits with a polymorphism in the serotonin transporter gene regulatory region [see comments]. Science 274(5292): 1527–1531

Lettice LA, Heaney SJ, Purdie LA, Li L, de Beer P, Oostra BA, Goode D, Elgar G, Hill RE, de Graaff E (2003) A long-range Shh enhancer regulates expression in the developing limb and fin and is associated with preaxial polydactyly. Hum Mol Genet 12(14): 1725–1735

Leung PW, Lee CC, Hung SF, Ho TP, Tang CP, Kwong SL, Leung SY, Yuen ST, Lieh-Mak F, Oosterlaan J, Grady D, Harxhi A, Ding YC, Chi HC, Flodman P, Schuck S, Spence MA, Moyzis R, Swanson J (2005) Dopamine receptor D4 (DRD4) gene in Han Chinese children with attention-deficit/hyperactivity disorder (ADHD): increased prevalence of the 2-repeat allele. Am J Med Genet B Neuropsychiatr Genet 133(1): 54–56

Levin ED, Conners CK, Silva D, Hinton SC, Meck WH, March J, Rose JE (1998) Transdermal nicotine effects on attention. Psychopharmacology (Berl) 140(2): 135–141

Levy F, Swanson JM (2001) Timing, space and ADHD: the dopamine theory revisited. Aust N Z J Psychiatry 35(4): 504–511

Levy F, Hay DA, McStephen M, Wood C, Waldman I (1997) Attention-deficit hyperactivity disorder: a category or a continuum? Genetic analysis of a large-scale twin study. J Am Acad Child Adolesc Psychiatry 36(6): 737–744

Levy F, McStephen M, Hay DA (2001) The diagnostic genetics of ADHD symptoms and subtypes. Attention-Genes and ADHD. Levy F, Hay D (eds). Brunner-Routledge, Hove, UK

Levy F, Hay DA, Bennett KS, McStephen M (2005) Gender differences in ADHD subtype comorbidity. J Am Acad Child Adolesc Psychiatry 44(4): 368–376

Lezcano N, Mrzljak L, Eubanks S, Levenson R, Goldman-Rakic P, Bergson C (2000) Dual signaling regulated by calcyon, a D1 dopamine receptor interacting protein. Science 287(5458): 1660–1664

Lezcano N, Mrzljak L, Levenson R, Bergson C (2006) Retraction. Science 314(5806): 1681

Li S, Kim KY, Kim JH, Park MS, Bahk JY, Kim MO (2004) Chronic nicotine and smoking treatment increases dopamine transporter mRNA expression in the rat midbrain. Neurosci Lett 363(1): 29–32

Li J, Wang Y, Zhou R, Zhang H, Yang L, Wang B, Khan S, Faraone SV (2005) Serotonin 5-HT1B receptor gene and attention deficit hyperactivity disorder in Chinese Han subjects. Am J Med Genet B Neuropsychiatr Genet 132(1): 59–63

Li D, Sham PC, Owen MJ, He L (2006a) Meta-analysis shows significant association between dopamine system genes and attention deficit hyperactivity disorder (ADHD). Hum Mol Genet 15(14): 2276–2284

Li J, Kang C, Wang Y, Zhou R, Wang B, Guan L, Yang L, Faraone SV (2006b) Contribution of 5-HT2A receptor gene -1438A>G polymorphism to outcome of attention-deficit/hyperactivity disorder in adolescents. Am J Med Genet B Neuropsychiatr Genet 141(5): 473–476

Li J, Wang Y, Zhou R, Wang B, Zhang H, Yang L, Faraone SV (2006c) Association of attention-deficit/hyperactivity disorder with serotonin 4 receptor gene polymorphisms in Han Chinese subjects. Neurosci Lett 401(1–2): 6–9

Li J, Wang Y, Zhou R, Wang B, Zhang H, Yang L, Faraone SV (2006d) No association of attention-deficit/hyperactivity disorder with genes of the serotonergic pathway in Han Chinese subjects. Neurosci Lett 403(1–2): 172–175

Li J, Wang Y, Zhou R, Zhang H, Yang L, Wang B, Faraone SV (2006e) Association between polymorphisms in serotonin 2C receptor gene and attention-deficit/hyperactivity disorder in Han Chinese subjects. Neurosci Lett 407(2): 107–111

Li J, Wang Y, Zhou R, Zhang H, Yang L, Wang B, Faraone SV (2006f) Association between tryptophan hydroxylase gene polymorphisms and attention deficit hyperactivity disorder in Chinese Han population. Am J Med Genet B Neuropsychiatr Genet 141(2): 126–129

Li J, Zhang X, Wang Y, Zhou R, Zhang H, Yang L, Wang B, Faraone SV (2006g) The serotonin 5-HT1D receptor gene and attention-deficit/hyperactivity disorder in Chinese Han subjects. Am J Med Genet B Neuropsychiatr Genet 141(8): 874–876

Li J, Wang Y, Zhou R, Zhang H, Yang L, Wang B, Faraone SV (2007) Association between polymorphisms in serotonin transporter gene and attention deficit hyperactivity disorder in Chinese Han subjects. Am J Med Genet B Neuropsychiatr Genet 144(1): 14–19

Lichter JB, Barr CL, Kennedy JL, Van Tol HH, Kidd KK, Livak KJ (1993) A hypervariable segment in the human dopamine receptor D4 (DRD4) gene. Hum Mol Genet 2(6): 767–773

Lijffijt M, Kenemans JL, Verbaten MN, van Engeland H (2005) A meta-analytic review of stopping performance in attention-deficit/hyperactivity disorder: deficient inhibitory motor control? J Abnorm Psychol 114(2): 216–222

Molecular Genetics of ADHD

Lim MH, Kim HW, Paik KC, Cho SC, Yoon do Y, Lee HJ (2006) Association of the DAT1 polymorphism with attention deficit hyperactivity disorder (ADHD): a family-based approach. Am J Med Genet B Neuropsychiatr Genet 141(3): 309–311

Linnet KM, Dalsgaard S, Obel C, Wisborg K, Henriksen TB, Rodriguez A, Kotimaa A, Moilanen I, Thomsen PH, Olsen J, Jarvelin MR (2003) Maternal lifestyle factors in pregnancy risk of attention deficit hyperactivity disorder and associated behaviors: review of the current evidence. Am J Psychiatry 160(6): 1028–1040

Loo SK, Fisher SE, Francks C, Ogdie MN, MacPhie IL, Yang M, McCracken JT, McGough JJ, Nelson SF, Monaco AP, Smalley SL (2004) Genome-wide scan of reading ability in affected sibling pairs with attention-deficit/hyperactivity disorder: unique and shared genetic effects. Mol Psychiatry 9(5): 485–493

Lovett MW, Steinbach KA, Frijters JC (2000) Remediating the core deficits of developmental reading disability: A double deficit perspective. J Learn Disabil 33(4): 334–358

Lowe N, Hawi Z, Fitzgerald M, Gill M (2001) No evidence of linkage or association between ADHD and DXS7 locus in Irish population. Am J Med Genet 105(4): 394–395

Lowe N, Kirley A, Hawi Z, Sham P, Wickham H, Kratochvil CJ, Smith SD, Lee SY, Levy F, Kent L, Middle F, Rohde LA, Roman T, Tahir E, Yazgan Y, Asherson P, Mill J, Thapar A, Payton A, Todd RD, Stephens T, Ebstein RP, Manor I, Barr CL, Wigg KG, Sinke RJ, Buitelaar JK, Smalley SL, Nelson SF, Biederman J, Faraone SV, Gill M (2004a) Joint analysis of the DRD5 marker concludes association with attention-deficit/hyperactivity disorder confined to the predominantly inattentive and combined subtypes. Am J Hum Genet 74(2): 348–356

Lowe N, Kirley A, Mullins C, Fitzgerald M, Gill M, Hawi Z (2004b) Multiple marker analysis at the promoter region of the DRD4 gene and ADHD: evidence of linkage and association with the SNP -616. Am J Med Genet B Neuropsychiatr Genet 131(1): 33–37

Luca P, Laurin N, Misener VL, Wigg KG, Anderson B, Cate-Carter T, Tannock R, Humphries T, Lovett MW, Barr CL (2007) Association of the dopamine receptor D1 gene, DRD1, with inattention symptoms in families selected for reading problems. Mol Psychiatry 12(8): 776–785

Luciano M, Lind PA, Duffy DL, Castles A, Wright MJ, Montgomery GW, Martin NG, Bates TC (2007) A Haplotype Spanning KIAA0319 and TTRAP Is Associated with Normal Variation in Reading and Spelling Ability. Biol Psychiatry 62(7): 811–817

Ludwig MZ, Bergman C, Patel NH, Kreitman M (2000) Evidence for stabilizing selection in a eukaryotic enhancer element. Nature 403(6769): 564–567

Luman M, Oosterlaan J, Sergeant JA (2005) The impact of reinforcement contingencies on AD/HD: a review and theoretical appraisal. Clin Psychol Rev 25(2): 183–213

MacKenzie A, Quinn J (1999) A serotonin transporter gene intron 2 polymorphic region, correlated with affective disorders, has allele-dependent differential enhancer-like properties in the mouse embryo. Proc Natl Acad Sci U S A 96(26): 15251–15255

Maher BS, Marazita ML, Ferrell RE, Vanyukov MM (2002) Dopamine system genes and attention deficit hyperactivity disorder: a meta-analysis. Psychiatr Genet 12(4): 207–215

Mancuso G, Warburton DM, Melen M, Sherwood N, Tirelli E (1999) Selective effects of nicotine on attentional processes. Psychopharmacology (Berl) 146(2): 199–204

Mannuzza S, Klein RG, Bessler A, Malloy P, LaPadula M (1993) Adult outcome of hyperactive boys. Educational achievement, occupational rank, and psychiatric status. Arch Gen Psychiatry 50(7): 565–576

Manor I, Kotler M, Sever Y, Eisenberg J, Cohen H, Ebstein RP, Tyano S (2000) Failure to replicate an association between the catechol-O-methyltransferase polymorphism and attention deficit hyperactivity disorder in a second, independently recruited Israeli cohort. Am J Med Genet 96(6): 858–860

Manor I, Eisenberg J, Tyano S, Sever Y, Cohen H, Ebstein RP, Kotler M (2001) Family-based association study of the serotonin transporter promoter region polymorphism (5-HTTLPR) in attention deficit hyperactivity disorder. Am J Med Genet 105(1): 91–95

Manor I, Tyano S, Eisenberg J, Bachner-Melman R, Kotler M, Ebstein RP (2002a) The short DRD4 repeats confer risk to attention deficit hyperactivity disorder in a family-based design and impair performance on a continuous performance test (TOVA). Mol Psychiatry 7(7): 790–794

Manor I, Tyano S, Mel E, Eisenberg J, Bachner-Melman R, Kotler M, Ebstein RP (2002b) Family-based and association studies of monoamine oxidase A and attention deficit hyperactivity disorder (ADHD): preferential transmission of the long promoter-region repeat and its association with impaired performance on a continuous performance test (TOVA). Mol Psychiatry 7(6): 626–632

Manor I, Corbex M, Eisenberg J, Gritsenkso I, Bachner-Melman R, Tyano S, Ebstein RP (2004) Association of the dopamine D5 receptor with attention deficit hyperactivity disorder (ADHD) and scores on a continuous performance test (TOVA). Am J Med Genet B Neuropsychiatr Genet 127(1): 73–77

Manuck SB, Flory JD, Ferrell RE, Mann JJ, Muldoon MF (2000) A regulatory polymorphism of the monoamine oxidase-A gene may be associated with variability in aggression, impulsivity, and central nervous system serotonergic responsivity. Psychiatry Res 95(1): 9–23

Marlow AJ, Fisher SE, Francks C, MacPhie IL, Cherny SS, Richardson AJ, Talcott JB, Stein JF, Monaco AP, Cardon LR (2003) Use of multivariate linkage analysis for dissection of a complex cognitive trait. Am J Hum Genet 72(3): 561–570

Martin N, Scourfield J, McGuffin P (2002) Observer effects and heritability of childhood attention-deficit hyperactivity disorder symptoms. Br J Psychiatry 180: 260–265

Martin NC, Piek JP, Hay D (2006) DCD and ADHD: a genetic study of their shared aetiology. Hum Mov Sci 25(1): 110–124

Martinez D, Gelernter J, Abi-Dargham A, van Dyck CH, Kegeles L, Innis RB, Laruelle M (2001) The variable number of tandem repeats polymorphism of the dopamine transporter gene is not associated with significant change in dopamine transporter phenotype in humans. Neuropsychopharmacology 24(5): 553–560

Martinussen R, Hayden J, Hogg-Johnson S, Tannock R (2005) A meta-analysis of working memory impairments in children with attention-deficit/hyperactivity disorder. J Am Acad Child Adolesc Psychiatry 44(4): 377–384

McAllister AK (2002) Conserved cues for axon and dendrite growth in the developing cortex. Neuron 33(1): 2–4

McCracken JT, Smalley SL, McGough JJ, Crawford L, Del'Homme M, Cantor RM, Liu A, Nelson SF (2000) Evidence for linkage of a tandem duplication polymorphism upstream of the dopamine D4 receptor gene (DRD4) with attention deficit hyperactivity disorder (ADHD). Mol Psychiatry 5(5): 531–536

McEvoy B, Hawi Z, Fitzgerald M, Gill M (2002) No evidence of linkage or association between the norepinephrine transporter (NET) gene polymorphisms and ADHD in the Irish population. Am J Med Genet 114(6): 665–666

McEwen BS, Alves SE (1999) Estrogen actions in the central nervous system. Endocr Rev 20(3): 279–307

McGrath MM, Sullivan MC, Lester BM, Oh W (2000) Longitudinal neurologic follow-up in neonatal intensive care unit survivors with various neonatal morbidities. Pediatrics 106(6): 1397–1405

McGrath LM, Smith SD, Pennington BF (2006) Breakthroughs in the search for dyslexia candidate genes. Trends Mol Med 12(7): 333–341

McLoughlin G, Ronald A, Kuntsi J, Asherson P, Plomin R (2007) Genetic support for the dual nature of attention deficit hyperactivity disorder: substantial genetic overlap between the inattentive and hyperactive-impulsive components. J Abnorm Child Psychol 35(6): 999–1008

Meng SZ, Ozawa Y, Itoh M, Takashima S (1999) Developmental and age-related changes of dopamine transporter, and dopamine D1 and D2 receptors in human basal ganglia. Brain Res 843(1–2): 136–144

Meng H, Smith SD, Hager K, Held M, Liu J, Olson RK, Pennington BF, DeFries JC, Gelernter J, O'Reilly-Pol T, Somlo S, Skudlarski P, Shaywitz SE, Shaywitz BA, Marchione K, Wang Y, Paramasivam M, LoTurco JJ, Page GP, Gruen JR (2005) DCDC2 is associated with reading disability and modulates neuronal development in the brain. Proc Natl Acad Sci U S A 102(47): 17053–17058

Merette C, Brassard A, Potvin A, Bouvier H, Rousseau F, Emond C, Bissonnette L, Roy MA, Maziade M, Ott J, Caron C (2000) Significant linkage for Tourette syndrome in a large French Canadian family Am J Hum Genet 67(4): 1008–1013

Molecular Genetics of ADHD

Mick E, Biederman J, Faraone SV, Sayer J, Kleinman S (2002) Case-control study of attention-deficit hyperactivity disorder and maternal smoking, alcohol use, and drug use during pregnancy. J Am Acad Child Adolesc Psychiatry 41(4): 378–385

Milberger S, Biederman J, Faraone SV, Chen L, Jones J (1997a) ADHD is associated with early initiation of cigarette smoking in children and adolescents. J Am Acad Child Adolesc Psychiatry 36(1): 37–44

Milberger S, Biederman J, Faraone SV, Guite J, Tsuang MT (1997b) Pregnancy, delivery and infancy complications and attention deficit hyperactivity disorder: issues of gene-environment interaction. Biol Psychiatry 41(1): 65–75

Mill J, Curran S, Kent L, Richards S, Gould A, Virdee V, Huckett L, Sharp J, Batten C, Fernando S, Simanoff E, Thompson M, Zhao J, Sham P, Taylor E, Asherson P (2001) Attention deficit hyperactivity disorder (ADHD) and the dopamine D4 receptor gene: evidence of association but no linkage in a UK sample. Mol Psychiatry 6(4): 440–444

Mill JS, Caspi A, McClay J, Sugden K, Purcell S, Asherson P, Craig I, McGuffin P, Braithwaite A, Poulton R, Moffitt TE (2002a) The dopamine D4 receptor and the hyperactivity phenotype: a developmental-epidemiological study. Mol Psychiatry 7(4): 383–391

Mill J, Curran S, Kent L, Gould A, Huckett L, Richards S, Taylor E, Asherson P (2002b) Association study of a SNAP-25 microsatellite and attention deficit hyperactivity disorder. Am J Med Genet 114(3): 269–271

Mill J, Fisher N, Curran S, Richards S, Taylor E, Asherson P (2003) Polymorphisms in the dopamine D4 receptor gene and attention-deficit hyperactivity disorder. Neuroreport 14(11): 1463–1466

Mill J, Curran S, Richards S, Taylor E, Asherson P (2004a) Polymorphisms in the dopamine D5 receptor (DRD5) gene and ADHD. Am J Med Genet B Neuropsychiatr Genet 125B(1): 38–42

Mill J, Richards S, Knight J, Curran S, Taylor E, Asherson P (2004b) Haplotype analysis of SNAP-25 suggests a role in the aetiology of ADHD. Mol Psychiatry 9(8): 801–810

Mill J, Asherson P, Craig I, D'Souza UM (2005a) Transient expression analysis of allelic variants of a VNTR in the dopamine transporter gene (DAT1). BMC Genet 6(1): 3

Mill J, Xu X, Ronald A, Curran S, Price T, Knight J, Craig I, Sham P, Plomin R, Asherson P (2005b) Quantitative trait locus analysis of candidate gene alleles associated with attention deficit hyperactivity disorder (ADHD) in five genes: DRD4, DAT1, DRD5, SNAP-25, and 5HT1B. Am J Med Genet B Neuropsychiatr Genet 133(1): 68–73

Miller GM, Madras BK (2002) Polymorphisms in the 3'-untranslated region of human and monkey dopamine transporter genes affect reporter gene expression. Mol Psychiatry 7(1): 44–55

Misener V, Luca P, Azeke O, Crosbie J, Waldman I, Tannock R, Roberts W, Malone M, Schachar R, Ickowicz A, Kennedy J, Barr C (2004) Linkage of the dopamine receptor D1 gene to attention-deficit/hyperactivity disorder. Mol Psychiatry 9(5): 500–509

Misener VL, Wigg K, Couto JM, So C, Zai C, Shulman R, Crosbie J, Tannock R, Malone M, Schachar R, Ickowicz A, Kennedy JL, Cl B (submitted) Family-based association study of attention-deficit/hyperactivity disorder and the gene for dopamine receptor D2

Missale C, Nash SR, Robinson SW, Jaber M, Caron MG (1998) Dopamine receptors: from structure to function. Physiol Rev 78(1): 189–225

Miyamoto Y, Yamada K, Noda Y, Mori H, Mishina M, Nabeshima T (2001) Hyperfunction of dopaminergic and serotonergic neuronal systems in mice lacking the NMDA receptor epsilon1 subunit. J Neurosci 21(2): 750–7

Moffitt TE, Caspi A, Rutter M (2005) Strategy for investigating interactions between measured genes and measured environments. Arch Gen Psychiatry 62(5): 473–481

Molina BS, Pelham WE, Jr (2003) Childhood predictors of adolescent substance use in a longitudinal study of children with ADHD. J Abnorm Psychol 112(3): 497–507

Morris DW, Robinson L, Turic D, Duke M, Webb V, Milham C, Hopkin E, Pound K, Fernando S, Easton M, Hamshere M, Williams N, McGuffin P, Stevenson J, Krawczak M, Owen MJ, O'Donovan MC, Williams J (2000) Family-based association mapping provides evidence for a gene for reading disability on chromosome 15q. Hum Mol Genet 9(5): 843–848

Muglia P, Jain U, Inkster B, Kennedy JL (2002a) A quantitative trait locus analysis of the dopamine transporter gene in adults with ADHD. Neuropsychopharmacology 27(4): 655–662

Muglia P, Jain U, Kennedy JL (2002b) A transmission disequilibrium test of the Ser9/Gly dopamine D3 receptor gene polymorphism in adult attention-deficit hyperactivity disorder. Behav Brain Res 130(1–2): 91–95

Nadder TS, Silberg JL, Eaves LJ, Maes HH, Meyer JM (1998) Genetic effects on ADHD symptomatology in 7- to 13-year-old twins: results from a telephone survey. Behav Genet 28(2): 83–99

Nadder TS, Rutter M, Silberg JL, Maes HH, Eaves LJ (2002) Genetic effects on the variation and covariation of attention deficit-hyperactivity disorder (ADHD) and oppositional-defiant disorder/conduct disorder (Odd/CD) symptomatologies across informant and occasion of measurement. Psychol Med 32(1): 39–53

Nadeau L, Boivin M, Tessier R, Lefebvre F, Robaey P (2001) Mediators of behavioral problems in 7-year-old children born after 24 to 28 weeks of gestation. J Dev Behav Pediatr 22(1): 1–10

Neuman RJ, Lobos E, Reich W, Henderson CA, Sun LW, Todd RD (2007) Prenatal smoking exposure and dopaminergic genotypes interact to cause a severe ADHD subtype. Biol Psychiatry 61(12): 1320–1328

Neville MJ, Johnstone EC, Walton RT (2004) Identification and characterization of ANKK1: a novel kinase gene closely linked to DRD2 on chromosome band 11q23.1. Hum Mutat 23(6): 540–545

Nguyen T, Bard J, Jin H, Taruscio D, Ward DC, Kennedy JL, Weinshank R, Seeman P, O'Dowd BF (1991) Human dopamine D5 receptor pseudogenes. Gene 109(2): 211–218

Ni Z, Abou El Hassan M, Xu Z, Yu T, Bremner R (2008) The chromatin-remodeling anzyme BRG1 coordinates CIITA induction through many interdependent distal enhancers. Nat Immunol 9(7): 785–793

Nothen MM, Cichon S, Hemmer S, Hebebrand J, Remschmidt H, Lehmkuhl G, Poustka F, Schmidt M, Catalano M, Fimmers R, Korner J, Rietschel M, Propping P (1994) Human dopamine D4 receptor gene: frequent occurrence of a null allele and observation of homozygosity. Hum Mol Genet 3(12): 2207–2212

Ogdie MN, Macphie IL, Minassian SL, Yang M, Fisher SE, Francks C, Cantor RM, McCracken JT, McGough JJ, Nelson SF, Monaco AP, Smalley SL (2003) A genomewide scan for attention-deficit/hyperactivity disorder in an extended sample: suggestive linkage on 17p11. Am J Hum Genet 72(5): 1268–1279

Ogdie MN, Fisher SE, Yang M, Ishii J, Francks C, Loo SK, Cantor RM, McCracken JT, McGough JJ, Smalley SL, Nelson SF (2004) Attention deficit hyperactivity disorder: fine mapping supports linkage to 5p13, 6q12, 16p13, and 17p11. Am J Hum Genet 75(4): 661–668

Ohadi M, Shirazi E, Tehranidoosti M, Moghimi N, Keikhaee MR, Ehssani S, Aghajani A, Najmabadi H (2006) Attention-deficit/hyperactivity disorder (ADHD) association with the DAT1 core promoter -67 T allele. Brain Res 1101(1): 1–4

Okuyama Y, Ishiguro H, Toru M, Arinami T (1999) A genetic polymorphism in the promoter region of DRD4 associated with expression and schizophrenia. Biochem Biophys Res Commun 258(2): 292–295

O'Neill MF, Heron-Maxwell CL, Shaw G (1999) 5-HT2 receptor antagonism reduces hyperactivity induced by amphetamine, cocaine, and MK-801 but not D1 agonist C-APB. Pharmacol Biochem Behav 63(2): 237–243

Osen-Sand A, Catsicas M, Staple JK, Jones KA, Ayala G, Knowles J, Grenningloh G, Catsicas S (1993) Inhibition of axonal growth by SNAP-25 antisense oligonucleotides in vitro and in vivo. Nature 364(6436): 445–448

Ouellet-Morrin I, Wigg KG, Feng Y, Dionne G, Robaey P, Brendgen M, Vitaro F, Simard L, Schachar R, Tremblay RE, Pacrusse D, Boivin M, Barr CL (2007) Association of the dopamine transporter gene and ADHD symptoms in a Canadian population-based sample of same-age twins. Am J Med Genet B Neuropsychiatr Genet (Epub ahead of print)

Ozaki N, Manji H, Lubierman V, Lu SJ, Lappalainen J, Rosenthal NE, Goldman D (1997) A naturally occurring amino acid substitution of the human serotonin 5-HT2A receptor influences amplitude and timing of intracellular calcium mobilization. J Neurochem 68(5): 2186–2193

Molecular Genetics of ADHD

Ozaki N, Goldman D, Kaye WH, Plotnicov K, Greenberg BD, Lappalainen J, Rudnick G, Murphy DL (2003) Serotonin transporter missense mutation associated with a complex neuropsychiatric phenotype. Mol Psychiatry 8(11): 895, 933–936

Palmer CG, Bailey JN, Ramsey C, Cantwell D, Sinsheimer JS, Del'Homme M, McGough J, Woodward JA, Asarnow R, Asarnow J, Nelson S, Smalley SL (1999) No evidence of linkage or linkage disequilibrium between DAT1 and attention deficit hyperactivity disorder in a large sample. Psychiatr Genet 9(3): 157–160

Paracchini S, Thomas A, Castro S, Lai C, Paramasivam M, Wang Y, Keating BJ, Taylor JM, Hacking DF, Scerri T, Francks C, Richardson AJ, Wade-Martins R, Stein JF, Knight JC, Copp AJ, Loturco J, Monaco AP (2006) The chromosome 6p22 haplotype associated with dyslexia reduces the expression of KIAA0319, a novel gene involved in neuronal migration. Hum Mol Genet 15(10): 1659–1666

Parish CL, Nunan J, Finkelstein DI, McNamara FN, Wong JY, Waddington JL, Brown RM, Lawrence AJ, Horne MK, Drago J (2005) Mice lacking the alpha4 nicotinic receptor subunit fail to modulate dopaminergic neuronal arbors and possess impaired dopamine transporter function. Mol Pharmacol 68(5): 1376–1386

Park L, Nigg JT, Waldman ID, Nummy KA, Huang-Pollock C, Rappley M, Friderici KH (2005) Association and linkage of alpha-2A adrenergic receptor gene polymorphisms with childhood ADHD. Mol Psychiatry 10(6): 572–580

Pastinen T, Sladek R, Gurd S, Sammak A, Ge B, Lepage P, Lavergne K, Villeneuve A, Gaudin T, Brandstrom H, Beck A, Verner A, Kingsley J, Harmsen E, Labuda D, Morgan K, Vohl MC, Naumova AK, Sinnett D, Hudson TJ (2004) A survey of genetic and epigenetic variation affecting human gene expression. Physiol Genomics 16(2): 184–193

Payton A, Holmes J, Barrett JH, Hever T, Fitzpatrick H, Trumper AL, Harrington R, McGuffin P, O'Donovan M, Owen M, Ollier W, Worthington J, Thapar A (2001a) Examining for association between candidate gene polymorphisms in the dopamine pathway and attention-deficit hyperactivity disorder: a family-based study. Am J Med Genet 105(5): 464–470

Payton A, Holmes J, Barrett JH, Sham P, Harrington R, McGuffin P, Owen M, Ollier W, Worthington J, Thapar A (2001b) Susceptibility genes for a trait measure of attention deficit hyperactivity disorder: a pilot study in a non-clinical sample of twins. Psychiatry Res 105(3): 273–278

Petronis A, O'Hara K, Barr CL, Kennedy JL, Van Tol HH (1994) (G)n-mononucleotide polymorphism in the human D4 dopamine receptor (DRD4) gene. Hum Genet 93(6): 719

Petryshen TL, Kaplan BJ, Liu MF, Schmill de French N, Tobias R, Hughes ML, Field LL (2001) Evidence for a susceptibility locus on chromosome 6q influencing phonological coding dyslexia. Am J Med Genet (Neuropsychiatr Genet) 105: 507–517

Picciotto MR, Zoli M, Rimondini R, Lena C, Marubio LM, Pich EM, Fuxe K, Changeux JP (1998) Acetylcholine receptors containing the beta2 subunit are involved in the reinforcing properties of nicotine. Nature 391(6663): 173–177

Pick LH, Halperin JM, Schwartz ST, Newcorn JH (1999) A longitudinal study of neurobiological mechanisms in boys with attention-deficit hyperactivity disorder: preliminary findings. Biol Psychiatry 45(3): 371–373

Pliszka SR, McCracken JT, Maas JW (1996) Catecholamines in attention-deficit hyperactivity disorder: current perspectives. J Am Acad Child Adolesc Psychiatry 35(3): 264–272

Pomerleau OF, Downey KK, Stelson FW, Pomerleau CS (1995) Cigarette smoking in adult patients diagnosed with attention deficit hyperactivity disorder. J Subst Abuse 7(3): 373–378

Porras G, Di Matteo V, De Deurwaerdere P, Esposito E, Spampinato U (2002a) Central serotonin4 receptors selectively regulate the impulse-dependent exocytosis of dopamine in the rat striatum: in vivo studies with morphine, amphetamine and cocaine. Neuropharmacology 43(7): 1099–1109

Porras G, Di Matteo V, Fracasso C, Lucas G, De Deurwaerdere P, Caccia S, Esposito E, Spampinato U (2002b) 5-HT2A and 5-HT2C/2B receptor subtypes modulate dopamine release induced in vivo by amphetamine and morphine in both the rat nucleus accumbens and striatum. Neuropsychopharmacology 26(3): 311–324

Prabhakar S, Noonan JP, Paabo S, Rubin EM (2006) Accelerated evolution of conserved noncoding sequences in humans. Science 314(5800): 786

Price TS, Simonoff E, Waldman I, Asherson P, Plomin R (2001) Hyperactivity in preschool children is highly heritable. J Am Acad Child Adolesc Psychiatry 40(12): 1362–1364

Purper-Ouakil D, Wohl M, Mouren MC, Verpillat P, Ades J, Gorwood P (2005) Meta-analysis of family-based association studies between the dopamine transporter gene and attention deficit hyperactivity disorder. Psychiatr Genet 15(1): 53–59

Qian Q, Wang Y, Zhou R, Li J, Wang B, Glatt S, Faraone SV (2003) Family-based and case-control association studies of catechol-O-methyltransferase in attention deficit hyperactivity disorder suggest genetic sexual dimorphism. Am J Med Genet B Neuropsychiatr Genet 118(1): 103–109

Qian Q, Wang Y, Zhou R, Yang L, Faraone SV (2004) Family-based and case-control association studies of DRD4 and DAT1 polymorphisms in Chinese attention deficit hyperactivity disorder patients suggest long repeats contribute to genetic risk for the disorder. Am J Med Genet B Neuropsychiatr Genet 128(1): 84–89

Qian Q, Wang Y, Li J, Yang L, Wang B, Zhou R, Glatt SJ, Faraone SV (2007) Evaluation of potential gene-gene interactions for attention deficit hyperactivity disorder in the Han Chinese population. Am J Med Genet B Neuropsychiatr Genet 144(2): 200–206

Quist JF, Kennedy JL (2001) Genetics of childhood disorders: XXIII. ADHD, Part 7: The serotonin system. J Am Acad Child Adolesc Psychiatry 40(2): 253–256

Quist JF, Barr CL, Shachar R, Roberts W, Malone M, Tannock R, Bloom S, Basile VS, Kennedy JL (1999) Evidence for an association of the serotonin 5-HT1B G861C polymorphism with ADHD. Mol Psychiatry 4(S1): S93

Quist JF, Barr CL, Schachar R, Roberts W, Malone M, Tannock R, Basile VS, Beitchman J, Kennedy JL (2000) Evidence for the serotonin HTR2A receptor gene as a susceptibility factor in attention deficit hyperactivity disorder (ADHD). Mol Psychiatry 5(5): 537–541

Quist JF, Barr CL, Schachar R, Roberts W, Malone M, Tannock R, Basile VS, Beitchman J, Kennedy JL (2003) The serotonin 5-HT1B receptor gene and attention deficit hyperactivity disorder. Mol Psychiatry 8(1): 98–102

Ramus F (2006) Genes, brain, and cognition: a roadmap for the cognitive scientist. Cognition 101(2): 247–269

Rasmussen ER, Neuman RJ, Heath AC, Levy F, Hay DA, Todd RD (2002) Replication of the latent class structure of Attention-Deficit/Hyperactivity Disorder (ADHD) subtypes in a sample of Australian twins. J Child Psychol Psychiatry 43(8): 1018–1028

Rhee SH, Waldman ID, Hay DA, Levy F (2001) Aetiology of the sex difference in the prevalence of DSM-III-R ADHD: A comparison of two models. Attention, Genes and ADHD. Hay DA, Levy F (eds) Brunner-Routledge, East Sussex, pp 139–156

Rietveld MJ, Hudziak JJ, Bartels M, van Beijsterveldt CE, Boomsma DI (2004) Heritability of attention problems in children: longitudinal results from a study of twins, age 3 to 12. J Child Psychol Psychiatry 45(3): 577–588

Rockman MV, Wray GA (2002) Abundant raw material for cis-regulatory evolution in humans. Mol Biol Evol 19(11): 1991–2004

Rockman MV, Hahn MW, Soranzo N, Goldstein DB, Wray GA (2003) Positive selection on a human-specific transcription factor binding site regulating IL4 expression. Curr Biol 13(23): 2118–2123

Roman T, Schmitz M, Polanczyk G, Eizirik M, Rohde LA, Hutz MH (2001) Attention-deficit hyperactivity disorder: a study of association with both the dopamine transporter gene and the dopamine D4 receptor gene. Am J Med Genet 105(5): 471–478

Roman T, Schmitz M, Polanczyk GV, Eizirik M, Rohde LA, Hutz MH (2002a) Further evidence for the association between attention-deficit/hyperactivity disorder and the dopamine-beta-hydroxylase gene. Am J Med Genet 114(2): 154–158

Roman T, Szobot C, Martins S, Biederman J, Rohde LA, Hutz MH (2002b) Dopamine transporter gene and response to methylphenidate in attention-deficit/hyperactivity disorder. Pharmacogenetics 12(6): 497–499

Roman T, Schmitz M, Polanczyk GV, Eizirik M, Rohde LA, Hutz MH (2003) Is the alpha-2A adrenergic receptor gene (ADRA2A) associated with attention-deficit/hyperactivity disorder? Am J Med Genet B Neuropsychiatr Genet 120(1): 116–120

Roman T, Polanczyk GV, Zeni C, Genro JP, Rohde LA, Hutz MH (2006) Further evidence of the involvement of alpha-2A-adrenergic receptor gene (ADRA2A) in inattentive dimensional scores of attention-deficit/hyperactivity disorder. Mol Psychiatry 11(1): 8–10

Rowe DC, Stever C, Giedinghagen LN, Gard JM, Cleveland HH, Terris ST, Mohr JH, Sherman S, Abramowitz A, Waldman ID (1998) Dopamine DRD4 receptor polymorphism and attention deficit hyperactivity disorder. Mol Psychiatry 3(5): 419–426

Rowe DC, Van den Oord EJ, Stever C, Giedinghagen LN, Gard JM, Cleveland HH, Gilson M, Terris ST, Mohr JH, Sherman S, Abramowitz A, Waldman ID (1999) The DRD2 TaqI polymorphism and symptoms of attention deficit hyperactivity disorder. Mol Psychiatry 4(6): 580–586

Rutter M, Silberg J (2002) Gene-environment interplay in relation to emotional and behavioral disturbance. Annu Rev Psychol 53: 463–490

Sabol SZ, Hu S, Hamer D (1998) A functional polymorphism in the monoamine oxidase A gene promoter. Hum Genet 103(3): 273–279

Sagvolden T, Russell VA, Aase H, Johansen EB, Farshbaf M (2005) Rodent models of attention-deficit/hyperactivity disorder. Biol Psychiatry 57(11): 1239–1247

Sakimura K, Kutsuwada T, Ito I, Manabe T, Takayama C, Kushiya E, Yagi T, Aizawa S, Inoue Y, Sugiyama H, et al (1995) Reduced hippocampal LTP and spatial learning in mice lacking NMDA receptor epsilon 1 subunit. Nature 373(6510): 151–155

Schachar R, Wachsmuth R (1990) Hyperactivity and parental psychopathology. J Child Psychol Psychiatry 31(3): 381–392

Schmitz M, Denardin D, Silva TL, Pianca T, Roman T, Hutz MH, Faraone SV, Rohde LA (2006) Association between alpha-2a-adrenergic receptor gene and ADHD inattentive type. Biol Psychiatry 60(10): 1028–1033

Schulte-Korne G, Deimel W, Bartling J, Remschmidt H (1999) Pre-attentive processing of auditory patterns in dyslexic human subjects. Neurosci Lett 276(1): 41–44

Seaman MI, Fisher JB, Chang F, Kidd KK (1999) Tandem duplication polymorphism upstream of the dopamine D4 receptor gene (DRD4). Am J Med Genet 88(6): 705–709

Seeger G, Schloss P, Schmidt MH (2001) Functional polymorphism within the promotor of the serotonin transporter gene is associated with severe hyperkinetic disorders. Mol Psychiatry 6(2): 235–238

Shaltiel G, Shamir A, Agam G, Belmaker RH (2005) Only tryptophan hydroxylase (TPH)-2 is relevant to the CNS. Am J Med Genet B Neuropsychiatr Genet 136(1): 106

Sheehan K, Lowe N, Kirley A, Mullins C, Fitzgerald M, Gill M, Hawi Z (2005) Tryptophan hydroxylase 2 (TPH2) gene variants associated with ADHD. Mol Psychiatry 10(10): 944–949

Sheehan K, Hawi Z, Gill M, Kent L (2007) No association between TPH2 gene polymorphisms and ADHD in a UK sample. Neurosci Lett 412(2): 105–107

Sherman DK, Iacono WG, McGue MK (1997) Attention-deficit hyperactivity disorder dimensions: a twin study of inattention and impulsivity-hyperactivity. J Am Acad Child Adolesc Psychiatry 36(6): 745–753

Sherrington R, Mankoo B, Attwood J, Kalsi G, Curtis D, Buetow K, Povey S, Gurling H (1993) Cloning of the human dopamine D5 receptor gene and identification of a highly polymorphic microsatellite for the DRD5 locus that shows tight linkage to the chromosome 4p reference marker RAF1P1. Genomics 18(2): 423–425

Simonic I, Gericke GS, Ott J, Weber JL (1998) Identification of genetic markers associated with Gilles de la Tourette syndrome in an Afrikaner population. Am J Hum Genet 63(3): 839–846

Simonic I, Nyholt DR, Gericke GS, Gordon D, Matsumoto N, Ledbetter DH, Ott J, Weber JL (2001) Further evidence for linkage of Gilles de la Tourette syndrome (GTS) susceptibility loci on chromosomes 2p11, 8q22 and 11q23–24 in South African Afrikaners. Am J Med Genet 105(2): 163–167

Simsek M, Al-Sharbati M, Al-Adawi S, Ganguly SS, Lawatia K (2005) Association of the risk allele of dopamine transporter gene (DAT1*10) in Omani male children with attention-deficit hyperactivity disorder. Clin Biochem 38(8): 739–742

Simsek M, Al-Sharbati M, Al-Adawi S, Lawatia K (2006) The VNTR polymorphism in the human dopamine transporter gene: improved detection and absence of association of VNTR alleles with attention-deficit hyperactivity disorder. Genet Test 10(1): 31–34

Smalley SL, Bailey JN, Palmer CG, Cantwell DP, McGough JJ, Del'Homme MA, Asarnow JR, Woodward JA, Ramsey C, Nelson SF (1998) Evidence that the dopamine D4 receptor is a susceptibility gene in attention deficit hyperactivity disorder. Mol Psychiatry 3(5): 427–430

Smalley SL, Kustanovich V, Minassian SL, Stone JL, Ogdie MN, McGough JJ, McCracken JT, MacPhie IL, Francks C, Fisher SE, Cantor RM, Monaco AP, Nelson SF (2002) Genetic linkage of attention-deficit/hyperactivity disorder on chromosome 16p13, in a region implicated in autism. Am J Hum Genet 71(4): 959–963

Smith SD, Kimberling WJ, Pennington BF, Lubs HA (1983) Specific reading disability: identification of an inherited form through linkage analysis. Science 219(4590): 1345–1347

Smith KM, Daly M, Fischer M, Yiannoutsos CT, Bauer L, Barkley R, Navia BA (2003) Association of the dopamine beta hydroxylase gene with attention deficit hyperactivity disorder: genetic analysis of the Milwaukee longitudinal study. Am J Med Genet B Neuropsychiatr Genet 119(1): 77–85

Smoller JW, Biederman J, Arbeitman L, Doyle AE, Fagerness J, Perlis RH, Sklar P, Faraone SV (2006) Association Between the 5HT1B Receptor Gene (HTR1B) and the Inattentive Subtype of ADHD. Biol Psychiatry 59(5): 460–467

Solanto MV (1998) Neuropsychopharmacological mechanisms of stimulant drug action in attention-deficit hyperactivity disorder: a review and integration. Behav Brain Res 94(1): 127–152

Sommerfelt K, Troland K, Ellertsen B, Markestad T (1996) Behavioral problems in low-birthweight preschoolers. Dev Med Child Neurol 38(10): 927–940

Spencer T, Biederman J, Wilens T, Prince J, Hatch M, Jones J, Harding M, Faraone SV, Seidman L (1998) Effectiveness and tolerability of tomoxetine in adults with attention deficit hyperactivity disorder. Am J Psychiatry 155(5): 693–695

Spielman RS, Ewens WJ (1996) The TDT and other family-based tests for linkage disequilibrium and association. Am J Hum Genet 59(5): 983–989

Sprich-Buckminster S, Biederman J, Milberger S, Faraone SV, Lehman BK (1993) Are perinatal complications relevant to the manifestation of ADD? Issues of comorbidity and familiality. J Am Acad Child Adolesc Psychiatry 32(5): 1032–1037

Steidl U, Steidl C, Ebralidze A, Chapuy B, Han HJ, Will B, Rosenbauer F, Becker A, Wagner K, Koschmieder S, Kobayashi S, Costa DB, Schulz T, O'Brien KB, Verhaak RG, Delwel R, Haase D, Trumper L, Krauter J, Kohwi-Shigematsu T, Griesinger F, Tenen DG (2007) A distal single nucleotide polymorphism alters long-range regulation of the PU.1 gene in acute myeloid leukemia. J Clin Invest 117(9): 2611–2620

Stevenson J, Langley K, Pay H, Payton A, Worthington J, Ollier W, Thapar A (2005) Attention deficit hyperactivity disorder with reading disabilities: preliminary genetic findings on the involvement of the ADRA2A gene. J Child Psychol Psychiatry 46(10): 1081–1088

Strehlow U, Haffner J, Bischof J, Gratzka V, Parzer P, Resch F (2006) Does successful training of temporal processing of sound and phoneme stimuli improve reading and spelling? Eur Child Adolesc Psychiatry 15(1): 19–29

Sunohara GA, Roberts W, Malone M, Schachar RJ, Tannock R, Basile VS, Wigal T, Wigal SB, Schuck S, Moriarty J, Swanson JM, Kennedy JL, Barr CL (2000) Linkage of the dopamine D4 receptor gene and attention-deficit/hyperactivity disorder. J Am Acad Child Adolesc Psychiatry 39(12): 1537–1542

Swanson JM, Sunohara GA, Kennedy JL, Regino R, Fineberg E, Wigal T, Lerner M, Williams L, LaHoste GJ, Wigal S (1998) Association of the dopamine receptor D4 (DRD4) gene with a refined phenotype of attention deficit hyperactivity disorder (ADHD): a family- based approach. Mol Psychiatry 3(1): 38–41

Szantai E, Szmola R, Sasvari-Szekely M, Guttman A, Ronai Z (2005) The polymorphic nature of the human dopamine D4 receptor gene: a comparative analysis of known variants and a novel 27 bp deletion in the promoter region. BMC Genet 6(1): 39

Szatmari P, Offord DR, Boyle MH (1989) Ontario Child Health Study: prevalence of attention deficit disorder with hyperactivity. J Child Psychol Psychiatry 30(2): 219–230

Szatmari P, Saigal S, Rosenbaum P, Campbell D, King S (1990) Psychiatric disorders at five years among children with birthweights less than 1000g: a regional perspective. Dev Med Child Neurol 32(11): 954–962

Taerk E, Grizenko N, Ben Amor L, Lageix P, Mbekou V, Deguzman R, Torkaman-Zehi A, Ter Stepanian M, Baron C, Joober R (2004) Catechol-O-methyltransferase (COMT) Val108/158 Met polymorphism does not modulate executive function in children with ADHD. BMC Med Genet 5: 30

Tahir E, Yazgan Y, Cirakoglu B, Ozbay F, Waldman I, Asherson PJ (2000) Association and linkage of DRD4 and DRD5 with attention deficit hyperactivity disorder (ADHD) in a sample of Turkish children. Mol Psychiatry 5(4): 396–404

Taipale M, Kaminen N, Nopola-Hemmi J, Haltia T, Myllyluoma B, Lyytinen H, Muller K, Kaaranen M, Lindsberg PJ, Hannula-Jouppi K, Kere J (2003) A candidate gene for developmental dyslexia encodes a nuclear tetratricopeptide repeat domain protein dynamically regulated in brain. Proc Natl Acad Sci U S A 100(20): 11553–11558

Tallal P, Miller SL, Bedi G, Byma G, Wang X, Nagarajan SS, Schreiner C, Jenkins WM, Merzenich MM (1996) Language comprehension in language-learning impaired children improved with acoustically modified speech. Science 271(5245): 81–84

Tang G, Ren D, Xin R, Qian Y, Wang D, Jiang S (2001) Lack of association between the tryptophan hydroxylase gene A218C polymorphism and attention-deficit hyperactivity disorder in Chinese Han population. Am J Med Genet 105(6): 485–488

Tang Y, Buxbaum SG, Waldman I, Anderson GM, Zabetian CP, Kohnke MD, Cubells JF (2006) A single nucleotide polymorphism at DBH, possibly associated with attention-deficit/hyperactivity disorder, associates with lower plasma dopamine beta-hydroxylase activity and is in linkage disequilibrium with two putative functional single nucleotide polymorphisms. Biol Psychiatry 60(10): 1034–1038

Thapar A, Hervas A, McGuffin P (1995) Childhood hyperactivity scores are highly heritable and show sibling competition effects: twin study evidence. Behav Genet 25(6): 537–544

Thapar A, Harrington R, Ross K, McGuffin P (2000) Does the definition of ADHD affect heritability? J Am Acad Child Adolesc Psychiatry 39(12): 1528–1536

Thapar A, Harrington R, McGuffin P (2001) Examining the comorbidity of ADHD-related behaviours and conduct problems using a twin study design. Br J Psychiatry 179: 224–229

Thapar A, Fowler T, Rice F, Scourfield J, van den Bree M, Thomas H, Harold G, Hay D (2003) Maternal smoking during pregnancy and attention deficit hyperactivity disorder symptoms in offspring. Am J Psychiatry 160(11): 1985–1989

Thapar A, Langley K, Fowler T, Rice F, Turic D, Whittinger N, Aggleton J, Van den Bree M, Owen M, O'Donovan M (2005) Catechol O-methyltransferase gene variant and birth weight predict early-onset antisocial behavior in children with attention-deficit/hyperactivity disorder. Arch Gen Psychiatry 62(11): 1275–1278

Threlkeld SW, McClure MM, Bai J, Wang Y, Loturco JJ, Rosen GD, Fitch RH (2007) Developmental disruptions and behavioral impairments in rats following in utero RNAi of Dyx1c1. Brain Res Bull 71(5): 508–514

Todd RD, Jong YJ, Lobos EA, Reich W, Heath AC, Neuman RJ (2001a) No association of the dopamine transporter gene 3' VNTR polymorphism with ADHD subtypes in a population sample of twins. Am J Med Genet 105(8): 745–748

Todd RD, Neuman RJ, Lobos EA, Jong YJ, Reich W, Heath AC (2001b) Lack of association of dopamine D4 receptor gene polymorphisms with ADHD subtypes in a population sample of twins. Am J Med Genet 105(5): 432–438

Todd RD, Rasmussen ER, Neuman RJ, Reich W, Hudziak JJ, Bucholz KK, Madden PA, Heath A (2001c) Familiality and heritability of subtypes of attention deficit hyperactivity disorder in a population sample of adolescent female twins. Am J Psychiatry 158(11): 1891–1898

Todd RD, Lobos EA, Sun LW, Neuman RJ (2003) Mutational analysis of the nicotinic acetylcholine receptor alpha 4 subunit gene in attention deficit/hyperactivity disorder: evidence for association of an intronic polymorphism with attention problems. Mol Psychiatry 8(1): 103–108

Todd RD, Huang H, Smalley SL, Nelson SF, Willcutt EG, Pennington BF, Smith SD, Faraone SV, Neuman RJ (2005) Collaborative analysis of DRD4 and DAT genotypes in population-defined ADHD subtypes. J Child Psychol Psychiatry 46(10): 1067–1073

Toyoda R, Nakamura H, Watanabe Y (2005) Identification of protogenin, a novel immunoglobulin superfamily gene expressed during early chick embryogenesis. Gene Expr Patterns 5(6): 778–785

Turic D, Langley K, Kirov G, Owen MJ, Thapar A, O'Donovan MC (2004a) Direct analysis of the genes encoding G proteins G alpha T2, G alpha o, G alpha Z in ADHD. Am J Med Genet B Neuropsychiatr Genet 127(1): 68–72

Turic D, Langley K, Mills S, Stephens M, Lawson D, Govan C, Williams N, Van Den Bree M, Craddock N, Kent L, Owen M, O'Donovan M, Thapar A (2004b) Follow-up of genetic linkage findings on chromosome 16p13: evidence of association of N-methyl-D aspartate glutamate receptor 2A gene polymorphism with ADHD. Mol Psychiatry 9(2): 169–173

Turic D, Williams H, Langley K, Owen M, Thapar A, O'Donovan MC (2005) A family based study of catechol-O-methyltransferase (COMT) and attention deficit hyperactivity disorder (ADHD). Am J Med Genet B Neuropsychiatr Genet 133(1): 64–67

Tzenova J, Kaplan BJ, Petryshen TL, Field LL (2004) Confirmation of a dyslexia susceptibility locus on chromosome 1p34-p36 in a set of 100 Canadian families. Am J Med Genet B Neuropsychiatr Genet 127(1): 117–124

VanNess SH, Owens MJ, Kilts CD (2005) The variable number of tandem repeats element in DAT1 regulates in vitro dopamine transporter density. BMC Genet 6: 55

Waldman ID, Rowe DC, Abramowitz A, Kozel ST, Mohr JH, Sherman SL, Cleveland HH, Sanders ML, Gard JM, Stever C (1998) Association and Linkage of the Dopamine Transporter Gene and Attention- Deficit Hyperactivity Disorder in Children: Heterogeneity owing to Diagnostic Subtype and Severity. Am J Hum Genet 63(6): 1767–1776

Waldman ID, Robinson BF, Rhee SH (1999) A logistic regression extension of the transmission disequilibrium test for continuous traits: application to linkage disequilibrium between alcoholism and the candidate genes DRD2 and ADH3. Genet Epidemiol 17 Suppl 1: S379–S384

Walitza S, Renner TJ, Dempfle A, Konrad K, Wewetzer C, Halbach A, Herpertz-Dahlmann B, Remschmidt H, Smidt J, Linder M, Flierl L, Knolker U, Friedel S, Schafer H, Gross C, Hebebrand J, Warnke A, Lesch KP (2005) Transmission disequilibrium of polymorphic variants in the tryptophan hydroxylase-2 gene in attention-deficit/hyperactivity disorder. Mol Psychiatry 10(12): 1126–1132

Walther DJ, Peter JU, Bashammakh S, Hortnagl H, Voits M, Fink H, Bader M (2003) Synthesis of serotonin by a second tryptophan hydroxylase isoform. Science 299(5603): 76

Wang HY, Undie AS, Friedman E (1995) Evidence for the coupling of Gq protein to D1-like dopamine sites in rat striatum: possible role in dopamine-mediated inositol phosphate formation. Mol Pharmacol 48(6): 988–994

Wang Y, Paramasivam M, Thomas A, Bai J, Kaminen-Ahola N, Kere J, Voskuil J, Rosen GD, Galaburda AM, Loturco JJ (2006a) DYX1C1 functions in neuronal migration in developing neocortex. Neuroscience 143(2): 515–522

Wang B, Wang Y, Zhou R, Li J, Qian Q, Yang L, Guan L, Faraone SV (2006b) Possible association of the alpha-2A adrenergic receptor gene (ADRA2A) with symptoms of attention-deficit/hyperactivity disorder. Am J Med Genet B Neuropsychiatr Genet 141(2): 130–134

Washbourne P, Thompson PM, Carta M, Costa ET, Mathews JR, Lopez-Bendito G, Molnar Z, Becher MW, Valenzuela CF, Partridge LD, Wilson MC (2002) Genetic ablation of the t-SNARE SNAP-25 distinguishes mechanisms of neuroexocytosis. Nat Neurosci 5(1): 19–26

Wasserman WW, Sandelin A (2004) Applied bioinformatics for the identification of regulatory elements. Nat Rev Genet 5(4): 276–287

Wei J, Ramchand CN, Hemmings GP (1997) Possible control of dopamine beta-hydroxylase via a codominant mechanism associated with the polymorphic (GT)n repeat at its gene locus in healthy individuals. Hum Genet 99(1): 52–55

Molecular Genetics of ADHD

Weissman MM, Warner V, Wickramaratne PJ, Kandel DB (1999) Maternal smoking during pregnancy and psychopathology in offspring followed to adulthood. J Am Acad Child Adolesc Psychiatry 38(7): 892–899

Whitaker-Azmitia PM, Druse M, Walker P, Lauder JM (1996) Serotonin as a developmental signal. Behav Brain Res 73(1–2): 19–29

Wigg K, Feng J, Crosbie J, Tannock R, Kennedy JL, Ickowicz A, Malone M, Schachar R, Barr CL. (in press) Association of ADHD and the Protogenin Gene in the Chrromosome 15q21.3 region. Genes, Brain, and Behavior

Wigg K, Feng Y, Anderson B, Cate-Carter T, Archibald J, Kerr E, Tannock R, Lovett M, Humphries T, Barr C (submitted) Association of Reading Disabilities and the Protogenin Gene in the Chromosome 15q21.3 Region.

Wigg K, Zai G, Schachar R, Tannock R, Roberts W, Malone M, Kennedy JL, Barr CL (2002) Attention deficit hyperactivity disorder and the gene for dopamine Beta-hydroxylase. Am J Psychiatry 159(6): 1046–1048

Wigg KG, Couto JM, Feng Y, Anderson B, Cate-Carter TD, Macciardi F, Tannock R, Lovett MW, Humphries TW, Barr CL (2004) Support for EKN1 as the susceptibility locus for dyslexia on 15q21. Mol Psychiatry 9(12): 1111–1121

Wigg K, Couto J, Feng Y, Crosbie J, Anderson B, Cate-Carter TD, Tannock R, Lovett MW, Humphries T, Kennedy JL, Ickowicz A, Pathare T, Roberts W, Malone M, Schachar R, Barr CL (2005) Investigation of the relationship of attention deficit hyperactivity disorder to the EKN1 gene on chromosome 15q21. Sci Stud Read 9(3): 261–283

Wigg KG, Takhar A, Ickowicz A, Tannock R, Kennedy JL, Pathare T, Malone M, Schachar R, Barr CL (2006) Gene for the serotonin transporter and ADHD: No association with two functional polymorphisms. Am J Med Genet B Neuropsychiatr Genet 141(6): 566–570

Wilens TE, Biederman J, Spencer TJ, Bostic J, Prince J, Monuteaux MC, Soriano J, Fine C, Abrams A, Rater M, Polisner D (1999) A pilot controlled clinical trial of ABT-418, a cholinergic agonist, in the treatment of adults with attention deficit hyperactivity disorder. Am J Psychiatry 156(12): 1931–1937

Wilens TE, Verlinden MH, Adler LA, Wozniak PJ, West SA (2006) ABT-089, a neuronal nicotinic receptor partial agonist, for the treatment of attention-deficit/hyperactivity disorder in adults: results of a pilot study. Biol Psychiatry 59(11): 1065–1070

Willcutt EG, Pennington BF, DeFries JC (2000) Twin study of the etiology of comorbidity between reading disability and attention-deficit/hyperactivity disorder. Am J Med Genet 96(3): 293–301

Willcutt EG, Pennington BF, Smith SD, Cardon LR, Gayan J, Knopik VS, Olson RK, DeFries JC (2002) Quantitative trait locus for reading disability on chromosome 6p is pleiotropic for attention-deficit/hyperactivity disorder. Am J Med Genet 114(3): 260–268

Willcutt EG, Doyle AE, Nigg JT, Faraone SV, Pennington BF (2005) Validity of the executive function theory of attention-deficit/hyperactivity disorder: a meta-analytic review. Biol Psychiatry 57(11): 1336–1346

Willcutt EG, Pennington BF, Olson RK, Defries JC (2007) Understanding comorbidity: A twin study of reading disability and attention-deficit/hyperactivity disorder. Am J Med Genet B Neuropsychiatr Genet 144(6): 709–714

Winsberg BG, Comings DE (1999) Association of the dopamine transporter gene (DAT1) with poor methylphenidate response. J Am Acad Child Adolesc Psychiatry 38(12): 1474–1477

Wozniak J, Biederman J, Mundy E, Mennin D, Faraone SV (1995) A pilot family study of childhood-onset mania. J Am Acad Child Adolesc Psychiatry 34(12): 1577–1583

Xiao J, Dai R, Negyessy L, Bergson C (2006) Calcyon, a novel partner of clathrin light chain, stimulates clathrin-mediated endocytosis. J Biol Chem 281(22): 15182–15193

Xu M, Hu XT, Cooper DC, Moratalla R, Graybiel AM, White FJ, Tonegawa S (1994a) Elimination of cocaine-induced hyperactivity and dopamine-mediated neurophysiological effects in dopamine D1 receptor mutant mice. Cell 79(6): 945–955

Xu M, Moratalla R, Gold LH, Hiroi N, Koob GF, Graybiel AM, Tonegawa S (1994b) Dopamine D1 receptor mutant mice are deficient in striatal expression of dynorphin and in dopamine-mediated behavioral responses. Cell 79(4): 729–742

Xu C, Schachar R, Tannock R, Roberts W, Malone M, Kennedy JL, Barr CL (2001) Linkage study of the alpha2A adrenergic receptor in attention-deficit hyperactivity disorder families. Am J Med Genet 105(2): 159–162

Xu X, Knight J, Brookes K, Mill J, Sham P, Craig I, Taylor E, Asherson P (2005a) DNA pooling analysis of 21 norepinephrine transporter gene SNPs with attention deficit hyperactivity disorder: no evidence for association. Am J Med Genet B Neuropsychiatr Genet 134(1): 115–118

Xu X, Mill J, Chen CK, Brookes K, Taylor E, Asherson P (2005b) Family-based association study of serotonin transporter gene polymorphisms in attention deficit hyperactivity disorder: no evidence for association in UK and Taiwanese samples. Am J Med Genet B Neuropsychiatr Genet 139(1): 11–13

Xu X, Brookes K, Chen CK, Huang YS, Wu YY, Asherson P (2007) Association study between the monoamine oxidase A gene and attention deficit hyperactivity disorder in Taiwanese samples. BMC Psychiatry 7: 10

Yang B, Chan RC, Jing J, Li T, Sham P, Chen RY (2007) A meta-analysis of association studies between the 10-repeat allele of a VNTR polymorphism in the 3′-UTR of dopamine transporter gene and attention deficit hyperactivity disorder. Am J Med Genet B Neuropsychiatr Genet 144(4): 541–550

Yang JW, Jang WS, Hong SD, Ji YI, Kim DH, Park J, Kim SW, Joung YS (2008) A case-control association study of the polymorphism at the promoter region of the DRD4 gene in Korean boys with attention deficit-hyperactivity disorder: Evidence of association with the -521 C/T SNP. Prog Neuropsychopharmacol Biol Psychiatry 32(1): 243–248

Zhao AL, Su LY, Zhang YH, Tang BS, Luo XR, Huang CX, Su QR (2005) Association analysis of serotonin transporter promoter gene polymorphism with ADHD and related symptomatology. Int J Neurosci 115(8): 1183–1191

Zhou W, Cunningham KA, Thomas ML (2002) Estrogen regulation of gene expression in the brain: a possible mechanism altering the response to psychostimulants in female rats. Brain Res Mol Brain Res 100(1–2): 75–83

Zhuang X, Belluscio L, Hen R (2000) G(olf)alpha mediates dopamine D1 receptor signaling. J Neurosci 20(16): RC91

Zill P, Buttner A, Eisenmenger W, Moller HJ, Ackenheil M, Bondy B (2007) Analysis of tryptophan hydroxylase I and II mRNA expression in the human brain: a post-mortem study. J Psychiatr Res 41(1–2): 168–173

Zoroglu SS, Erdal ME, Alasehirli B, Erdal N, Sivasli E, Tutkun H, Savas HA, Herken H (2002) Significance of serotonin transporter gene 5-HTTLPR and variable number of tandem repeat polymorphism in attention deficit hyperactivity disorder. Neuropsychobiology 45(4): 176–181

Zoroglu SS, Erdal ME, Erdal N, Ozen S, Alasehirli B, Sivasli E (2003) No evidence for an association between the T102C and 1438 G/A polymorphisms of the serotonin 2A receptor gene in attention deficit/hyperactivity disorder in a Turkish population. Neuropsychobiology 47(1): 17–20

The Genetics of Anxiety Disorders

Steven P. Hamilton

Contents

1 Anxiety Disorders 165
2 Panic Disorder 166
3 Phobias 169
4 Generalized Anxiety Disorder 170
5 Obsessive Compulsive Disorder 171
6 Genetic Analysis of Anxious Personality Traits 174
7 Conclusion 175
References 176

Abstract Anxiety disorders are common and chronic disorder characterized by inappropriate fear responses to known or unknown stimuli. These disorders have prominent familial and genetic components, based on family and twin data. Linkage studies have provided provisional findings supporting chromosomal intervals containing anxiety disorder loci, although no convincing and replicated findings have been reported. Candidate gene studies also provide conflicting findings. Advances in genome screening technology will allow genetic researchers to test the hypothesis that common DNA variation will be associated with one or more anxiety disorders.

1 Anxiety Disorders

During the majority of the previous century, the biological roots of neuroses (later reconceptualized as anxiety disorders) were minimized, with etiologic explanations relying upon processes relating to intra-psychic conflict. More recent observations derived from clinical psychopharmacology (Klein 1964) and animal models (Shekhar et al. 2001) have led to the development of a perspective incorporating the neural systems underlying fear processing and emotional regulation. An elaboration of this view would lead to the hypothesis that a heritable biological susceptibility

---

S.P. Hamilton
University of California, San Francisco, Department of Psychiatry, 401 Parnassus Avenue, Box NGL, San Francisco, CA 94143-0984, USA
steveh@lppi.ucsf.edu

D.B. Wildenauer (ed.), *Molecular Biology of Neuropsychiatric Disorders*,
Nucleic Acids and Molecular Biology, © Springer-Verlag Berlin Heidelberg 2009

to anxiety disorders may account for a substantial proportion of the likelihood for their development.

The lifetime incidence of DSM-IV anxiety disorders was estimated to be 28.8% among 9,282 participants in the National Comorbidity Survey Replication (Kessler et al. 2005a), making them the most common type of mental disorder. The 12-month prevalence of DSM-IV anxiety disorders is estimated to be 18.1%, nearly twice that of mood disorders (Kessler et al. 2005b). The individual lifetime prevalence rates were 4.7% for panic disorder, 12.5% for specific phobia, 12.1% for social anxiety disorder, 5.7% for generalized anxiety disorder, and 1.6% for obsessive-compulsive disorder. Twelve-month prevalence rates for the specific disorders were 2.7% for panic disorder, 8.7% for specific phobia, 6.8% for social anxiety disorder, 3.1% for generalized anxiety disorder, and 1.0% for obsessive-compulsive disorder, suggesting that anxiety disorders are typically chronic diseases. The average age of onset of anxiety disorders was at age 11 years, and a striking imbalance in the ratio of female to male cases of these disorders was observed. Females show a 1.6-fold higher risk for an anxiety disorder than do males (Kessler et al. 2005a).

Genetic epidemiological research has consistently documented familial aggregation of anxiety disorders, and that genetic factors are probably behind these observations. A meta-analytic survey of family studies focusing on anxiety disorders showed that panic disorder, specific phobia, social anxiety disorder, generalized anxiety disorder, and obsessive-compulsive disorder all showed significant levels of familial aggregation (Hettema et al. 2001a). By incorporating twin study data, a prominent contribution from genes for panic disorder, specific phobia, social anxiety disorder, and generalized anxiety disorder was observed. This meta-analysis was unable to provide insight into obsessive-compulsive disorder, given the paucity of twin studies for this disorder (Hettema et al. 2001a). Multivariate structural equation modeling of twin data was subsequently used to show that, despite prevalence differences between males and females, it appears that genetic and environmental risk factors underlying anxiety disorders do not differ by gender (Hettema et al. 2005).

2 Panic Disorder

The diagnosis of panic disorder requires recurrent panic attacks, accompanied by anticipatory anxiety, worry about the implications of the attacks, or significant attack-related behavior changes for at least 1 month (American Psychiatric Association 1994). Panic attacks are characterized as the experience of spontaneous intense anxiety accompanied by a number of somatic and psychological symptoms. Panic disorder as a clinical syndrome was conceptualized relatively recently, based on clinical observations regarding the pharmacologic response of anxiety syndromes (Klein 1964) followed by the development of operational criteria for the syndrome in DSM-III (American Psychiatric Association 1980).

Females are affected twice as often as males (Eaton et al. 1994), and the mean age of onset is 24, with the period between 25 and 44 years being the period of highest risk for the disorder (Robins et al. 1984). The 12-month prevalence of panic disorder is 2.7% (Kessler et al. 2005b), an estimate largely concordant with previous epidemiologic data derived from 40,000 subjects, where lifetime prevalence was observed to be between 1.4 and 2.9% (Weissman et al. 1997).

Family and twin studies support a modest genetic contribution to panic disorder. Rigorously designed family studies, involving direct interviews of probands and their relatives using operationalized criteria, have been used to assess familial aggregation of panic disorder. An early review of adequately designed studies reports that familial aggregation studies demonstrate a relative risk of 7.8 (range 2.6–20) for first-degree relatives of persons with panic disorder (Knowles and Weissman 1995). Analyses of second-degree relatives of panic disorder probands using the family history approach showed a relative risk of 6.8, with a rate of panic disorder of 9.5% in second-degree relatives of panic probands compared to 1.4% among relatives of control probands (Pauls et al. 1979b). Family studies have also been able to provide further insight, for example showing that first-degree relatives of panic disorder probands with early (<20 years) versus later onset panic have an increased relative risk of panic disorder (Goldstein et al. 1997). Similarly, specific symptom constellations have been found to increase the risk of having panic disorder for first-degree relatives of probands who had panic disorder accompanied by smothering symptoms by 2.7-fold (Horwath et al. 1997).

Early twin studies, which were methodologically challenged by vague diagnostic criteria, showed higher concordance rates in monozygotic twins for "neurosis" when compared to dizygotic twins, suggesting a genetic etiology (Slater and Shields 1969). A series of small studies utilizing DSM diagnoses suggested a genetic influence on panic disorder (Torgersen 1983; Skre et al. 1993; Perna et al. 1997), which was confirmed in much larger twin registry samples (Kendler et al. 1993b; Scherrer et al. 2000). Subsequent meta-analysis of available family and twin data argued that the data best fit to a model consisting of additive genetic factors and individual environmental factors, with a heritability of 0.48 (Hettema et al. 2001a). Early studies attempting to determine the mode of transmission in panic disorder argued variably for a single-gene dominant model or a model in which both dominant or polygenic modes of inheritance could be ruled out (Pauls et al. 1979a, 1980; Crowe et al. 1983). Subsequent segregation analyses contended that dominant and recessive models were equally likely, and predicted known twin concordance rates (Vieland et al. 1993, 1996).

The presence of panic disorder with other psychiatric conditions in individuals or among members of the same family has generated hypotheses about genetic determinants common to more than one disorder. One analysis of pedigrees segregating bipolar disorder resulted in the observation of aggregation of panic disorder in particular families. These investigators observed that 18% of persons with bipolar disorder in their family collection were also diagnosed with comorbid panic disorder, and that just 88% of subjects with panic disorder also had bipolar disorder (MacKinnon et al. 1997). These data suggest that a subset of families share a potentially heritable susceptibility to bipolar disorder and/or panic disorder (MacKinnon et al. 2002), which linkage analysis suggested occurred on chromosome 18 (MacKinnon et al. 1998). Experiments taking the converse approach, carrying out identification of bipolar disorder in pedigrees selected for the segregation of panic disorder, will be necessary to further determine if both disorders share a common genetic mechanism. The presence of nonpsychiatric medical disorders has raised the possibility of pleiotropic genetic effects. One group has described a "panic disorder syndrome" comprised of panic disorder, severe headaches, mitral valve prolapse, thyroid abnormalities, and bladder/renal problems. Observation of cosegregation of this constellation of seemingly unrelated conditions in numerous pedigrees led to linkage analysis supporting linkage to chromosome 13 in a collection of 19 pedigrees (Weissman et al. 2000). When the number of similar families was tripled, the linkage evidence still pointed to the same region (Hamilton et al. 2003). Independent studies of case-control samples, either starting with subjects with the urologic disorder interstitial cystitis (Weissman et al. 2004) or with subjects with panic disorder (Talati et al. 2008), found high rates of the "syndrome" conditions in first-degree relatives when compared to control subjects. Finally, there has long been interest in an "endophenotype" for panic disorder, namely the response to "panicogens"-like inhaled carbon dioxide (CO_2) or infused lactate, which can precipitate panic attacks in subjects with panic disorder, but not in healthy controls (Gorman et al. 1990). Studies of the rates of anxiety disorder in first-degree relatives of subjects with lactate sensitivity are conflicting (Reschke et al. 1995; Balon et al. 1989), leaving this issue unresolved. There is greater concordance of findings in studies of heritability in CO_2 challenges, where panic attacks in response to 35% CO_2 are reported to occur at higher rates in first-degree relatives of panic disorder subjects (Perna et al. 1995; Coryell 1997; van and Griez 2000). The panic response to CO_2 is associated with a higher familial risk for panic disorder (Perna et al. 1996), although a lack of the CO_2 response in children with a parent with panic disorder suggests a more complex picture (Pine et al. 2005). Concordance of panic response after CO_2 inhalation is four times higher in monozygotic twins when compared to dizygotic twins, suggesting a genetic component (Bellodi et al. 1998) with moderate heritability (Battaglia et al. 2007). Further study will be required to determine if the genetic factors underlying lactate and CO_2 sensitivity contribute to the genetic risk of panic disorder.

An early genetic linkage analysis in 26 North American pedigrees segregating panic disorder with polymorphic blood cell markers and DNA markers near those loci were negative (Crowe et al. 1987, 1990). With the advent of genome-wide microsatellite scans, a subset of 23 of these same families was analyzed, with modest evidence for linkage (lod score = 2.23) on chromosome 7p12 (Crowe et al. 2001). Other U.S. researchers ascertained families with multiple affected individuals (Fyer and Weissman 1999), followed by performance of a genome screen in 23 families that also resulted in modest linkage in the chromosome 7p15 region (Knowles et al. 1998). Expansion of this same pedigree set to 120 families showed diminished support for this region, but identified two novel loci, on chromosomes 15q and 2q, exhibiting genome-wide statistical significance (Fyer et al. 2006). An additional North American group performed genetic linkage analysis of a genome scan in 20 pedigrees, observing evidence for linkage on chromosome 1q (Gelernter et al. 2001). A scan of 25 Icelandic families using a complex phenotype including panic disorder, generalized anxiety disorder, phobias, and somatoform pain, led to a lod score of 4.18 on chromosome 9q31 (Thorgeirsson et al. 2003). The discrepancy between these findings may be explained by genetic heterogeneity due to geographical and population history, as well as by differing ascertainment schemes.

The Genetics of Anxiety Disorders

A large number of candidate genes that may be mechanistically involved in panic disorder susceptibility have been studied in association studies. Most of these findings have remain unreplicated (Gratacos et al. 2007). Genes representing many aspects of fear circuitry biology, including receptors for GABA, monoamines, and neuropeptides, have been examined. The two most thoroughly examined genes are the adenosine 2A receptor (ADORA2A) and catechol-O-methyltransferase (COMT). Two studies in Caucasian populations investigating ADORA2A reported association or linkage (Deckert et al. 1998; Hamilton et al. 2004); however, two Asian studies did not report a similar association (Yamada et al. 2001; Lam et al. 2005). An ADORA2A variant has also been associated with caffeine-induced anxiety (Alsene et al. 2003). The COMT studies are notable in that five studies support the role of this gene in panic disorder (Hamilton et al. 2002; Woo et al. 2002, 2004; Domschke et al. 2004; Rothe et al. 2006) in Caucasian and Asian populations. For much of the panic disorder association literature, negative or conflicting findings are likely derived from low statistical power due to limited sample sizes, heterogeneity in phenotype, low prior probability compounded by multiple comparisons issues, and population heterogeneity at the genetic level.

In aggregate, linkage studies have not provided consistent support for chromosomal regions related to panic disorder. This may be due to the decreased performance of parametric linkage studies in diseases of uncertain mode of inheritance in the context of polygenic inheritance. Candidate gene association studies have tried to remedy this situation, although the low likelihood of these being truly related to panic disorder biology largely makes selection of such genes guesswork. Genomewide association methods in large clinical samples may provide novel directions for panic disorder genetics.

3 Phobias

The most commonly encountered anxiety disorders are specific phobia and social anxiety disorder. Specific phobia is notable for an persistent and unreasonable fear of situations or objects. The presence of a phobic stimulus immediately elicits intense anxiety associated with physiological arousal symptoms, resulting in avoidance of the stimulus. Social anxiety disorder (formerly social phobia) is characterized by marked fear of performance or social situations, which can provoke intense anxiety. Patients often experience panic attacks, and they avoid the feared situations. A small number of phobia family studies support familial aggregation of these disorders. For specific phobia, there is a four-fold increase in risk for the disorder among first-degree relatives of probands compared to relatives of nonphobic controls (Fyer et al. 1995). Similarly, elevated risk is seen with generalized social phobia (Fyer et al. 1993; Mannuzza et al. 1995; Stein et al. 1998b). Twin studies of phobias suggest moderate genetic heritability, with residual risk derived from individual-specific environmental factors (Hettema et al. 2001a). Large twin studies focusing on social anxiety disorder and the situational and animal subtypes of specific phobia support a shared genetic liability between phobias, amounting to

30–40% of the risk (Kendler et al. 1992b, 1993a; Hettema et al. 2005). Twin studies of pre-adolescent children using parental reporting on fears and phobias in their children showed high heritability, but with a larger contribution from shared (i.e., family) environment (Bolton et al. 2006; Lichtenstein and Annas, 2000). Orthogonal studies of irrational fears in twins, regardless of clinical disorder status, also support the role of genetic determinants for fears often found in specific phobia such as situations (enclosed spaces, heights, water) or small animals, with modest estimates of heritability (Sundet et al. 2003; Phillips et al. 1987).

Despite the high prevalence, no pedigrees identified through specific phobia probands have been subjected to linkage analysis. One group focused on specific phobia as a phenotype by identifying specific phobia cases in pedigrees ascertained through panic disorder probands (Gelernter et al. 2003). Fourteen families contained more than one subject with DSM-III-R specific phobia. Parametric analysis of a microsatellite genomic screen showed a peak on chromosome 14 at 36.7 cM, with a lod score = 3.17 under a dominant genetic model, with an enrichment in linked families of non-situational phobia subtypes.

Using the same panic disorder pedigrees, Gelernter and colleagues carried out analyses focusing on social anxiety disorder. A parametric lod score of 2.22 was observed at 71 cM on chromosome 16, while a non-parametric linkage score of 3.4 was observed at 62 cM on chromosome 16 (Gelernter et al. 2004). Subsequent data-mining in these families showed that a broader phenotype of specific phobia, social anxiety disorder, panic disorder, and agoraphobia could be clustered in families, with linkage analysis showing genome-wide significant linkage to chromosome 4q31-q34, near the neuropeptide Y1 receptor gene (NPY1R), a plausible functional candidate gene. A parallel approach was performed in Icelandic pedigrees by combining anxiety disorder phenotypes in 62 pedigrees with high rates of specific phobia and social anxiety disorder. Chromosome 9q revealed a maximum allele-sharing lod score of 2.0 at 104 cM for the broad anxiety phenotype. When a subset of 25 families containing individuals with panic disorder were examined, the lod score increased to 4.18 (Thorgeirsson et al. 2003). Two negative candidate gene studies of social anxiety disorder, looking at dopamine and serotonin genes (Stein et al. 1998a; Kennedy et al. 2001), have been reported, while there are no reported candidate gene studies of specific phobia.

4 Generalized Anxiety Disorder

Generalized anxiety disorder (GAD) is characterized by excessive an persistent worry about several aspects of everyday life (e.g., work, relationships) present for "more days than not" over at least 6 months (DSM-IV). GAD is accompanied by two or more symptoms of arousal (e.g., muscle tension, irritability, sleep disturbance). The lifetime prevalence of GAD is estimated at 5–6% (Kessler et al. 2005a) when using the 6-month DSM-IV duration criterion (Ruscio et al. 2007). GAD occurs more commonly in women, and as many as 80% of individuals have comorbid anxiety disorders or depression (Kessler et al. 2005b). Much of the genetic epidemiology of GAD has focused on untangling its relation to depression,

The Genetics of Anxiety Disorders   171

with the goal of also clarifying the diagnosis. It is not surprising then that diagnostic criteria have continually evolved over the past three decades, so that the inferences about any previous definition of the disorder require validation with currently accepted constructs.

The results of existing direct interview family studies of GAD are generally consistent with familial aggregation, although the magnitude varies greatly between studies (Noyes, Jr. et al. 1987; Mendlewicz et al. 1993; Newman and Bland, 2006). For example, the population-based study of Newman and Bland (2006) reported that relatives of GAD probands were 1.5–2 times more likely to have GAD as relatives of control probands. Large registry-based twin studies suggest modest to moderate heritabilities, ranging from 0.15 to 0.40 (Kendler 1996; Roy et al. 1995; Hettema et al. 2001b; Mackintosh et al. 2006). As with phobias, residual GAD heritability is derived from environmental events specific to the individual. These twin analyses are complicated by the observation that there is complete overlap in the genetic factors that contribute to risk for both GAD and major depression. Within this context, it is argued that individual-specific environmental factors establish whether the clinical syndrome is expressed as MDD or GAD (Kendler et al. 1992a, 2007; Kendler 1996; Roy et al. 1995). There are no published linkage studies or replicated candidate gene findings for GAD. Two studies with small samples comprised of a mixture of anxiety disorder subjects reported association between GAD and the monoamine oxidase A gene (MAOA), although using different genetic markers (Tadic et al. 2003; Samochowiec et al. 2004).

5 Obsessive Compulsive Disorder

Obsessive compulsive disorder (OCD) involves the experience of persistent intrusive thoughts, or obsessions, typically centered on symmetry, checking, or contamination. Ritualistic tasks interfering with normal functioning, or compulsions, are also characteristic of OCD. Early estimates of OCD prevalence characterized the disorder as rare (0.05%). Rigorous epidemiologic studies of the 1980s estimated lifetime prevalence of DSM-III OCD ranging from 1.3 to 3.0%, (Robins et al. 1984; Karno et al. 1988), which is similar to later estimates of 1.9–2.5% seen in the Cross National Collaborative Group (Weissman et al. 1994) and the report of lifetime prevalence of 1.6% in the National Comorbidity Survey replication (Kessler et al. 2005a).

OCD is more common in females, with recent studies suggesting a female to male ratio of about 1.2–1.6 to 1 (Weissman et al. 1994). The typical age at onset of OCD is 22.0 years for females and 19.5 years for males, which represents a significant difference (Rasmussen and Eisen 1992). This difference in age of onset contributes to the prominent differences in OCD gender ratios for prepubertal childhood-onset (3:1, male:female) and postpubertal adolescent-onset (~1:1, male:female). The sex ratio observed in adolescent OCD is more characteristic of that in adults (Flament et al. 1988). OCD has been closely studied in children and adolescents, where the disorder can be particularly impairing. Birth complications have been raised as potential risk factor for early-onset OCD in males, as well as for more severe and chronic symptomatology (Flament et al. 1990; Lensi et al. 1996).

Early family studies, while methodologically flawed, suggested familial aggregation of obsessional syndromes (Lewis 1936; Brown 1942). Subsequent larger studies also report higher risk of OCD in first-degree relatives of OCD probands. A review of 11 OCD family studies found rates of OCD in first-degree relatives to be 9.5–25% for child probands and 0–20% in adult probands (Sobin and Karayiorgou 2000), raising the possibility that early onset OCD may represent a more heritable form of the disorder (Pauls et al. 1995; Hanna et al. 2005b; Nestadt et al. 2000b). This observation is not universal (Black et al. 1992; Fyer et al. 2005), suggesting substantial heterogeneity regarding familial aggregation.

There are fewer twin studies of OCD compared with other anxiety disorders. One example involves assessment at age 6 for DSM-IV OCD in a large collection of twins and supports a more substantial genetic influence (Bolton et al. 2007). Usage of twin samples to estimate the heritability of obsessive-compulsive symptoms have been productive. One study of 10,110 Dutch and U.S. twins, composed only of children, reported a robust additive genetic influence (~55%) for obsessive compulsive symptoms, with residual influence from unique environmental factors (Hudziak et al. 2004). Smaller studies in adults typically support this estimate (Clifford et al. 1984; Jonnal et al. 2000; van Grootheest et al. 2007). Without a collection of twin studies, meta-analysis has not been possible (Hettema et al. 2001a), but the theme of the extant work argues that OCD has a genetic component, particularly when assessed in childhood (van Grootheest et al. 2005). While some studies of segregation patterns suggest a simple genetic model for OCD (Cavallini et al. 1999; Nestadt et al. 2000a; Hanna et al. 2005a), other investigators have not made this observation (Alsobrook II et al. 1999). With the known phenotypic heterogeneity of OCD, it is likely that multiple genes may be involved.

Clinical observation of early-onset OCD has led to interesting hypotheses about co-occurring conditions (Leonard et al. 1999). Obsessive-compulsive symptoms in disorders like autism have been used to stratify families for genetic analyses (McCauley et al. 2004; Buxbaum et al. 2004). Early-onset OCD is frequently associated with tics and Tourette's disorder (Pauls et al. 1986, 1995; Leonard et al. 1992). Tics are more seen commonly in families segregating OCD (Grados et al. 2001). Another example of this approach involves the frequent occurrence of obsessive-compulsive symptoms in children with the poststreptococcal autoimmune syndrome Sydenham's chorea (Swedo et al. 1989). This led to characterization of the syndrome called pediatric autoimmune neuropsychiatric disorder associated with streptococcal infections (PANDAS) (Swedo et al. 1998). Further investigation indicated that children with PANDAS, Sydenham's chorea, and Tourette's disorder were positive for D8/17, a monoclonal antibody, significantly more often than seen in control subjects (Swedo et al. 1997; Murphy et al. 1997). First-degree relatives of PANDAS probands have higher rates of OCD when compared to the population, which suggests a potential genetic component to this interesting phenotype (Lougee et al. 2000).

An early genome scan of OCD included seven pedigrees identified through pediatric-onset probands (Hanna et al. 2002). The authors observed a maximum

The Genetics of Anxiety Disorders

multipoint linkage score of 2.25 on chromosome 9p, with somewhat decreased support when more samples and markers were added. Subsequently, 26 pedigrees were added, but support for linkage in this region dissipated, with the rise of a separate peak on chromosome 10p (Hanna et al. 2007). Using 41 affected sibling pairs, another group observed linkage in the 9p region described above, with a parametric heterogeneity lod score of 2.26 (Willour et al. 2004). Unfortunately, the addition of 178 families led to diminished evidence in this region. A novel locus was observed with a multipoint nonparametric analysis ($p = 0.0003$) on chromosome 3q, with less support for regions on chromosomes 1, 7, and 15 (Shugart et al. 2006). The investigators have explored endophenotypes involving obsessive-compulsive symptoms with the same pedigrees, reporting modest evidence for linkage on chromosome 14 for hoarding symptoms in families with two or more relatives with hoarding obsessions or compulsions (Samuels et al. 2007). Interestingly, this trait clusters in families (Hasler et al. 2007) and shows moderate heritability (Mathews et al. 2007).

Candidate gene association studies in OCD are motivated by pharmacologic data (Goodman et al. 1990), with serotonin pathway genes driving the initial focus. For example, serotonin transporter function in blood and brain in OCD subjects is altered (Delorme et al. 2005; Hesse et al. 2005; Hasselbalch et al. 2007). Numerous family-based and case-control association studies have been reported, with conflicting results. Some groups report associations with particular variants within the serotonin transporter gene (Ozaki et al. 2003; Hu et al. 2006), and meta-analysis suggests very minor association (Dickel et al. 2007; Lin 2007). Additionally, larger studies do provide association to the repeat polymorphism (5-HTTLPR) in the promoter region of the gene (Wendland et al. 2007). Candidate gene studies involving serotonin receptors and dopamine system proteins have likewise proven inconclusive (Hemmings and Stein 2006). Much effort has focused on COMT, which carries out the enzymatic metabolism of monoamine neurotransmitters. The functional valine to methionine variant initially showed an association between the lower activity allele in male subjects with OCD (Karayiorgou et al. 1997), a finding replicated using family samples (Karayiorgou et al. 1999). However, later studies and a meta-analysis (Azzam and Mathews 2003) argue for a more limited role in OCD, possibly limited to males (Denys et al. 2006; Pooley et al. 2007). Positional cloning efforts in OCD described in the previous section have informed candidate gene studies. One reasonable candidate gene in the chromosome 9p region, SLC1A1, encodes the neuronal and epithelial high-affinity glutamate transporter EAAC1. While early mutation detection approaches found no functional variants (Veenstra-VanderWeele et al. 2001), two later analyses with a joint total of 228 families used family-based analyses of nine markers in each study, with five overlapping between studies. Three SNPs were positive in one study, with the signal derived from males (Arnold et al. 2006). The second study showed OCD being associated with two SNPs, one of which was in male samples (Dickel et al. 2006). Of note, these two SNPs were also genotyped in the first study, but were not associated with OCD.

6 Genetic Analysis of Anxious Personality Traits

Neuroticism is a personality trait included in most models of personality (Cloninger 1987), and may be a construct that underlies traits of anxiety, depression, impulsiveness, and self-consciousness (Ebstein 2006). Twin data argues for a simple genetic model that is sufficient to explain familial correlation of neuroticism, and suggests moderate heritability on the order of 0.5–0.6 (Jang et al. 1996; Lake et al. 2000; Rettew et al. 2006).

There is a strong association between neuroticism and anxiety and depression. High neuroticism scores are correlated with levels of depression and anxiety symptoms as measured by the Eysenck Personality Inventory and the Beck Depression Scale (Jylha and Isometsa 2006). Elevated neuroticism was associated with the categorical diagnosis of major depression, as well as all anxiety disorders that were examined, including specific phobia, social anxiety disorder, agoraphobia, panic disorder, GAD, and OCD (Bienvenu et al. 2004), and is associated with comorbidity between many of these same disorders (Cuijpers et al. 2005; Bienvenu et al. 2001). Structural equation modeling was used in a sample of 9,270 twins to investigate the correlation between neuroticism and a group of "internalizing" disorders, including depression, panic disorder, agoraphobia, social anxiety disorder, GAD, and two subtypes of specific phobias (situational, animal). The investigators determined that neuroticism shared significant genetic risk with these disorders (Hettema et al. 2006). The genetic overlap between GAD alone and neuroticism was nearly complete, with a correlation of 0.8 (Hettema et al. 2004). Twin studies also suggest a similar substantial genetic correlation between major depression and neuroticism (Kendler et al. 2006).

The ability to collect personality trait information from large nonclinical samples greatly facilitates genetic studies. One research group utilized data from 35,000 sibling pairs, providing the opportunity to stratify by extreme discordance of neuroticism scores within pairs, as well as identify sibling pairs that are highly concordant. A genome screen of 629 pairs reported significant linkage to neuroticism, or quantitative trait loci (QTL), on chromosomes 1q, 4q, 7p, 12q, and 13q, with female-specific findings on chromosomes 1, 12, and 13 (Fullerton et al. 2003). A similarly designed study utilizing 757 individuals in 297 families culled from over 34,000 individuals led to the development of a phenotype based on factor analysis involving neuroticism scores combined with subscales of anxiety and mood questionnaires. A genome scan revealed some evidence for linkage on chromosomes 1p and 6p, also with a gender-specific effect (Nash et al. 2004). A final study utilizes a genome-wide association approach, in which hundreds of thousands of SNPs are genotyped simultaneously on a microarray. This approach facilitates a test of association between the marker and a phenotype, typically by comparing the magnitude of allele frequency differences for each marker between cases and matched controls (Christensen and Murray 2007). In the first such study of this kind with neuroticism, 2,054 individuals were selected from over 88,000 subjects for their extreme high and low neuroticism scores (Shifman et al. 2008). The investigators pooled DNA samples for genotyping, and obtained data on about 450,000 SNPs. They found that none of the markers met statistical significance using a threshold taking into account the large number of tests carried out. The 19 highest scoring SNPs were genotyped in a replication sample of 1,534 individuals at the extremes of the neuroticism distribution taken from their original 88,000 subjects. A single SNP, in the cAMP-specific phosphodiesterase 4D gene (PDE4D), met their criteria for replication, but this finding was not seen in three additional independent replication groups the investigators genotyped, totaling some 2,199 individuals. Previously reported linkage regions did not show up as associated regions. These results raise the possibility that much larger samples may be required to detect small genetic influences on neuroticism. More targeted genetic association studies have also been used to look at the possible role of specific genes in neuroticism. The serotonin transporter has been studied, with an association to neuroticism reported in a sample of 505 subjects (Lesch et al. 1996), an association supported in a meta-analysis of a number of small studies (Sen et al. 2004). This finding did not stand up in a larger single study of 4,800 subjects (Willis-Owen et al. 2005), nor in other recent large samples (Middeldorp et al. 2007).

7 Conclusion

The anxiety disorders discussed above are all moderately heritable, with 20–40% of the phenotypic variance explained by additive genetic effects. Linkage analyses, while providing important leads, have not produced replicable findings for anxiety disorders. Several developments in the field have paved the way for a different approach to mapping complex disease genes. The Human Genome Project and International HapMap Project have led to the recognition and quantitation of human genetic diversity, with the subsequent development of collections of SNPs usable for genetic analyses. Technological advances in array-based genotyping have facilitated inexpensive and rapid genotyping of hundreds of thousands of SNPs. Additionally, the consensus that large samples will be needed to recognize subtle genetic effects has led to increased efforts to ascertain sizable (i.e., 1,000–2,000 cases) population-based samples, and has fostered collaboration between diverse research groups. Numerous successful applications of this genome-wide association approach have been published (Klein et al. 2005; Wellcome Trust Case Control Consortium 2007; Scott et al. 2007), with the promise of such studies for psychiatric disorders already a reality (Baum et al. 2008; Lencz et al. 2007). Challenges for genome-wide association methods include the reliance on assumptions regarding the genetic model for anxiety disorders, as well as how to account for large-scale hypothesis testing and its attendant risk of false positive findings. Many of these debates can only be resolved when reported associations are replicated in independent samples or in meta-analyses, or with convincing functional correlates. A promise of this unbiased genome-wide approach is that it will promote discovery of novel genes or known genes not considered to be related to anxiety disorders. Such discoveries may foster targeted genetic analyses in human anxiety disorder populations as well as catalyze hypotheses for in vitro and animal model studies.

References

Alsene K., Deckert J., Sand P., De Wit H. (2003) Association between A(2a) receptor gene polymorphisms and caffeine-induced anxiety. Neuropsychopharmacology 28:1694–1702

Alsobrook J.P., Leckman J.F., Goodman W.K., Rasmussen S.A., Pauls D.L. (1999) Segregation analysis of obsessive-compulsive disorder using symptom- based factor scores. Am J Med Genet 88:669–675

American Psychiatric Association (1980). Diagnostic and statistical manual of mental disorders, 3rd edn. American Psychiatric Association, Washington, D.C

American Psychiatric Association (1994). Diagnostic and Statistical Manual of Mental Disorders, 4th edn. American Psychiatric Association, Washington, D.C

Arnold P.D., Sicard T., Burroughs E., Richter M.A., Kennedy J.L. (2006). Glutamate transporter gene SLC1A1 associated with obsessive-compulsive disorder. Arch Gen Psychiatry 63:769–776

Azzam A. Mathews C.A. (2003). Meta-analysis of the association between the catecholamine-O-methyl-transferase gene and obsessive-compulsive disorder. Am J Med Genet B Neuropsychiatr Genet 123:64–69

Balon R., Jordan M., Pohl R., Yeragani V.K. (1989). Family history of anxiety disorders in control subjects with lactate- induced panic attacks. Am J Psychiatry 146:1304–1306

Battaglia M., Ogliari A., Harris J., Spatola C.A.M., Pesenti-Gritti P., Reichborn-Kjennerud T., Torgersen S., Kringlen E., Tambs K. (2007). A genetic study of the acute anxious response to carbon dioxide stimulation in man. J Psychiatric Res 41:906–917

Baum A.E., Akula N., Cabanero M., Cardona I., Corona W., Klemens B., Schulze T.G., Cichon S., Rietschel M., Nothen M.M., Georgi A., Schumacher J., Schwarz M., Abou J.R., Hofels S., Propping P., Satagopan J., tera-Wadleigh S.D., Hardy J., McMahon F.J. (2008) A genome-wide association study implicates diacylglycerol kinase eta (DGKH) and several other genes in the etiology of bipolar disorder. Mol Psychiatry 13:197–207

Bellodi L., Perna G., Caldirola D., Arancio C., Bertani A., Di Bella D. (1998). CO2-induced panic attacks: a twin study. Am J Psychiatry 155:1184–1188

Bienvenu O.J., Brown C., Samuels J.F., Liang K.Y., Costa P.T., Eaton W.W., Nestadt G. (2001) Normal personality traits and comorbidity among phobic, panic and major depressive disorders. Psychiatry Res 102:73–85

Bienvenu O.J., Samuels J.F., Costa P.T., Reti I.M., Eaton W.W., Nestadt G. (2004) Anxiety and depressive disorders and the five-factor model of personality: a higher- and lower-order personality trait investigation in a community sample. Depress Anxiety 20:92–97

Black D.W., Noyes R., Jr., Goldstein R.B., Blum N. (1992). A family study of obsessive-compulsive disorder. Arch Gen Psychiatry 49:362–368

Bolton D., Eley T.C., O'Connor T.G., Perrin S., Rabe-Hesketh S., Rijsdijk F., Smith P. (2006). Prevalence and genetic and environmental influences on anxiety disorders in 6-year-old twins. Psychol Med 36:335–344

Bolton D., Rijsdijk F., O'Connor T.G., Perrin S., Eley T.C. (2007) Obsessive-compulsive disorder, tics and anxiety in 6-year-old twins. Psychol Med 37:39–48

Brown F.W. (1942) Heredity in the psychoneuroses. Proc R Soc Med 35:785–790

Buxbaum J.D., Silverman J., Keddache M., Smith C.J., Hollander E., Ramoz N., Reichert J.G. (2004) Linkage analysis for autism in a subset families with obsessive-compulsive behaviors: evidence for an autism susceptibility gene on chromosome 1 and further support for susceptibility genes on chromosome 6 and 19. Mol Psychiatry 9:144–150

The Genetics of Anxiety Disorders

Cavallini M.C., Pasquale L., Bellodi L., Smeraldi E. (1999) Complex segregation analysis for obsessive compulsive disorder and related disorders. Am J Med Genet 88:38–43

Christensen K. Murray J.C. (2007) What Genome-wide Association Studies Can Do for Medicine. N Engl J Med 356:1094–1097

Clifford C.A., Murray R.M., Fulker D.W. (1984) Genetic and environmental influences on obsessional traits and symptoms. Psychol Med 14:791–800

Cloninger C.R. (1987) A systematic method for clinical description and classification of personality variants. A proposal. Arch Gen Psychiatry 44:573–588

Coryell W. (1997) Hypersensitivity to carbon dioxide as a disease-specific trait marker. Biol Psychiatry 41:259–263

Crowe R.R., Noyes R., Pauls D.L., Slymen D. (1983) A family study of panic disorder. Arch Gen Psychiatry 40:1065–1069

Crowe R.R., Noyes R., Jr., Wilson A.F., Elston R.C., Ward L.J. (1987) A linkage study of panic disorder. Arch Gen Psychiatry 44:933–937

Crowe R.R., Noyes R., Jr., Samuelson S., Wesner R., Wilson R. (1990) Close linkage between panic disorder and alpha-haptoglobin excluded in 10 families. Arch Gen Psychiatry 47:377–380

Crowe R.R., Goedken R., Wilson R., Samuelson S., Nelson J., Noyes R. (2001) A genome-wide survey of panic disorder. Am J Med Genet (Neuropsychiatr Genet) 105:105–109

Cuijpers P., van Straten A., Donker M. (2005) Personality traits of patients with mood and anxiety disorders. Psychiatry Res 133:229–237

Deckert J., Nothen M.M., Franke P., Delmo C., Fritze J., Knapp M., Maier W., Beckmann H., Propping P. (1998) Systematic mutation screening and association study of the A1 and A2a adenosine receptor genes in panic disorder suggest a contribution of the A2a gene to the development of disease. Mol Psychiatry 3:81–85

Delorme R., Betancur C., Callebert J., Chabane N., Laplanche J.L., Mouren-Simeoni M.C., Launay J.M., Leboyer M. (2005) Platelet serotonergic markers as endophenotypes for obsessive-compulsive disorder. Neuropsychopharmacology 30:1539–1547

Denys D., Van Nieuwerburgh F., Deforce D., Westenberg H. (2006) Association between the dopamine D2 receptor TaqI A2 allele and low activity COMT allele with obsessive-compulsive disorder in males. Eur Neuropsychopharmacol 16:446–450

Dickel D.E., Veenstra-VanderWeele J., Cox N.J., Wu X., Fischer D.J., Van Etten-Lee M., Himle J.A., Leventhal B.L., Cook E.H., Jr., Hanna G.L. (2006) Association testing of the positional and functional candidate gene SLC1A1/EAAC1 in early-onset obsessive-compulsive disorder. Arch Gen Psychiatry 63:778–785

Dickel D.E., Veenstra-VanderWeele J., Bivens N.C., Wu X., Fischer D.J., Van Etten-Lee M., Himle J.A., Leventhal B.L., Cook J., Hanna G.L. (2007) Association studies of serotonin system candidate genes in early-onset obsessive-compulsive disorder. Biol Psychiatry 61:322–329

Domschke K., Freitag C.M., Kuhlenbaumer G., Schirmacher A., Sand P., Nyhuis P., Jacob C., Fritze J., Franke P., Rietschel M., Garritsen H.S., Fimmers R., Nothen M.M., Lesch K.P., Stogbauer F., Deckert J. (2004) Association of the functional V158M catechol-O-methyltransferase polymorphism with panic disorder in women. Int J Neuropsychopharmacol 7:183–188

Eaton W.W., Kessler R.C., Wittchen H.U., Magee W.J. (1994) Panic and panic disorder in the United States. Am J Psychiatry 151:413–420

Ebstein R.P. (2006) The molecular genetic architecture of human personality: beyond self-report questionnaires. Mol Psychiatry 11:427–445

Flament M.F., Whitaker A., Rapoport J.L., Davies M., Berg C.Z., Kalikow K., Sceery W., Shaffer D. (1988) Obsessive compulsive disorder in adolescence: an epidemiological study. J Am Acad Child Adolesc Psychiatry 27:764–771

Flament M.F., Koby E., Rapoport J.L., Berg C.J., Zahn T., Cox C., Denckla M., Lenane M. (1990) Childhood obsessive-compulsive disorder: a prospective follow-up study. J Child Psychol Psychiatry 31:363–380

Fullerton J., Cubin M., Tiwari H., Wang C., Bomhra A., Davidson S., Miller S., Fairburn C., Goodwin G., Neale M.C., Fiddy S., Mott R., Allison D.B., Flint J. (2003) Linkage analysis of extremely discordant and concordant sibling pairs identifies quantitative-trait loci that influence variation in the human personality trait neuroticism. Am J Hum Genet 72:879–890

Fyer A.J., Mannuzza S., Chapman T.F., Liebowitz M.R., Klein D.F. (1993) A direct interview family study of social phobia. Arch Gen Psychiatry 50:286–293

Fyer A.J., Mannuzza S., Chapman T.F., Martin L.Y., Klein D.F. (1995) Specificity in familial aggregation of phobic disorders. Arch Gen Psychiatry 52:564–573

Fyer A.J. Weissman M.M. (1999) Genetic linkage study of panic: clinical methodology and description of pedigrees. Am J Med Genet (Neuropsychiatr Genet) 88:173–181

Fyer A.J., Lipsitz J.D., Mannuzza S., Aronowitz B., Chapman T.F. (2005) A direct interview family study of obsessive-compulsive disorder. I Psychol Med 35:1611–1621

Fyer A.J., Hamilton S.P., Durner M., Haghighi F., Heiman G.A., Costa R., Evgrafov O., Adams P., de Leon A.B., Taveras N., Klein D.F., Hodge S.E., Weissman M.M., Knowles J.A. (2006) A Third-Pass Genome Scan in Panic Disorder: Evidence for Multiple Susceptibility Loci. Biol Psychiatry 60:388–401

Gelernter J., Bonvicini K., Page G., Woods S.W., Goddard A.W., Kruger S., Pauls D.L., Goodson S. (2001) Linkage genome scan for loci predisposing to panic disorder or agoraphobia. Am J Med Genet 105:548–557

Gelernter J., Page G.P., Bonvicini K., Woods S.W., Pauls D.L., Kruger S. (2003) A chromosome 14 risk locus for simple phobia: results from a genomewide linkage scan. Mol Psychiatry 8:71–82

Gelernter J., Page G.P., Stein M.B., Woods S.W. (2004) Genome-wide linkage scan for loci predisposing to social phobia: evidence for a chromosome 16 risk locus. Am J Psychiatry 161:59–66

Goldstein R.B., Wickramaratne P.J., Horwath E., Weissman M.M. (1997) Familial aggregation phenomenology of 'early-onset' (at or before age 20 years) panic disorder. Arch Gen Psychiatry 54:271–278

Goodman W.K., McDougle C.J., Price L.H., Riddle M.A., Pauls D.L., Leckman J.F. (1990) Beyond the serotonin hypothesis: a role for dopamine in some forms of obsessive compulsive disorder? J Clin Psychiatry 51 (Suppl 8):36–43

Gorman J.M., Papp L.A., Martinez J., Goetz R.R., Hollander E., Liebowitz M.R., Jordan F. (1990) High-dose carbon dioxide challenge test in anxiety disorder patients. Biol Psychiatry 28:743–757

Grados M.A., Riddle M.A., Samuels J.F., Liang K.Y., Hoehn-Saric R., Bienvenu O.J., Walkup J.T., Song D., Nestadt G. (2001) The familial phenotype of obsessive-compulsive disorder in relation to tic disorders: the Hopkins OCD family study. Biol Psychiatry 50:559–565

Gratacos M., Sahun I., Gallego X., mador-Arjona A., Estivill X., Dierssen M. (2007) Candidate genes for panic disorder: insight from human and mouse genetic studies. Genes, Brain Behav 6:2–23

Hamilton S.P., Slager S.L., Heiman G.A., Deng Z., Haghighi F., Klein D.F., Hodge S.E., Weissman M.M., Fyer A.J., Knowles J.A. (2002) Evidence for a susceptibility locus for panic disorder near the catechol-O-methyltransferase gene on chromosome 22. Biol Psychiatry 51:591–601

Hamilton S.P., Fyer A.J., Durner M., Heiman G.A., Baisre D.L., Hodge S.E., Knowles J.A., Weissman M.M. (2003) Further genetic evidence for a panic disorder syndrome mapping to chromosome 13q. Proc Natl Acad Sci USA 100:2550–2555

Hamilton S.P., Slager S.L., De Leon A.B., Heiman G.A., Klein D.F., Hodge S.E., Weissman M.M., Fyer A.J., Knowles J.A. (2004) Evidence for genetic linkage between a polymorphism in the adenosine 2A receptor and panic disorder. Neuropsychopharmacology 29:558–565

Hanna G.L., Veenstra-VanderWeele J., Cox N.J., Boehnke M., Himle J.A., Curtis G.C., Leventhal B.L., Cook E.H., Jr. (2002) Genome-wide linkage analysis of families with obsessive-compulsive disorder ascertained through pediatric probands. Am J Med Genet 114:541–552

Hanna G.L., Fingerlin T.E., Himle J.A., Boehnke M. (2005a) Complex segregation analysis of obsessive-compulsive disorder in families with pediatric probands. Hum Hered 60:1–9

The Genetics of Anxiety Disorders

Hanna G.L., Himle J.A., Curtis G.C., Gillespie B.W. (2005b) A family study of obsessive-compulsive disorder with pediatric probands. Am J Med Genet B Neuropsychiatr Genet 134:13–19

Hanna G.L., Veenstra-VanderWeele J., Cox N.J., Van E.M., Fischer D.J., Himle J.A., Bivens N.C., Wu X., Roe C.A., Hennessy K.A., Dickel D.E., Leventhal B.L., Cook E.H., Jr. (2007) Evidence for a susceptibility locus on chromosome 10p15 in early-onset obsessive-compulsive disorder. Biol Psychiatry 62:856–862

Hasler G., Pinto A., Greenberg B.D., Samuels J., Fyer A.J., Pauls D., Knowles J.A., McCracken J.T., Piacentini J., Riddle M.A., Rauch S.L., Rasmussen S.A., Willour V.L., Grados M.A., Cullen B., Bienvenu O.J., Shugart Y.Y., Liang K.Y., Hoehn-Saric R., Wang Y., Ronquillo J., Nestadt G., Murphy D.L. (2007) Familiality of factor analysis-derived YBOCS dimensions in OCD-affected sibling pairs from the OCD collaborative genetics study. Biol Psychiatry 61:617–625

Hasselbalch S.G., Hansen E.S., Jakobsen T.B., Pinborg L.H., Lonborg J.H., Bolwig T.G. (2007) Reduced midbrain-pons serotonin transporter binding in patients with obsessive-compulsive disorder. Acta Psychiatrica Scand 115:388–394

Hemmings S.M. Stein D.J. (2006) The current status of association studies in obsessive-compulsive disorder. Psychiatr Clin North Am 29:411–444

Hesse S., Muller U., Lincke T., Barthel H., Villmann T., Angermeyer M.C., Sabri O., Stengler-Wenzke K. (2005) Serotonin and dopamine transporter imaging in patients with obsessive-compulsive disorder. Psychiatry Res: Neuroimaging 140:63–72

Hettema J.M., Neale M.C., Kendler K.S. (2001a) A review and meta-analysis of the genetic epidemiology of anxiety disorders. Am J Psychiatry 158:1568–1578

Hettema J.M., Prescott C.A., Kendler K.S. (2001b) A population-based twin study of generalized anxiety disorder in men and women. J Nerv Ment Dis 189:413–420

Hettema J.M., Prescott C.A., Kendler K.S. (2004) Genetic and environmental sources of covariation between generalized anxiety disorder and neuroticism. Am J Psychiatry 161:1581–1587

Hettema J.M., Prescott C.A., Myers J.M., Neale M.C., Kendler K.S. (2005) The structure of genetic and environmental risk factors for anxiety disorders in men and women. Arch Gen Psychiatry 62:182–189

Hettema J.M., Neale M.C., Myers J.M., Prescott C.A., Kendler K.S. (2006) A population-based twin study of the relationship between neuroticism and internalizing disorders. Am J Psychiatry 163:857–864

Horwath E., Adams P., Wickramaratne P., Pine D., Weissman M.M. (1997) Panic disorder with smothering symptoms: evidence for increased risk in first-degree relatives. Depress Anxiety 6:147–153

Hu X.Z., Lipsky R.H., Zhu G., Akhtar L.A., Taubman J., Greenberg B.D., Xu K., Arnold P.D., Richter M.A., Kennedy J.L., Murphy D.L., Goldman D. (2006) Serotonin transporter promoter gain-of-function genotypes are linked to obsessive-compulsive disorder. Am J Hum Genet 78:815–826

Hudziak J.J., van Beijsterveldt C.E.M., Althoff R.R., Stanger C., Rettew D.C., Nelson E.C., Todd R.D., Bartels M., Boomsma D.I. (2004) Genetic and environmental contributions to the child behavior checklist obsessive-compulsive scale: a cross-cultural twin study. Arch Gen Psychiatry 61:608–616

Jang K.L., Livesley W.J., Vernon P.A. (1996) Heritability of the big five personality dimensions and their facets: a twin study. J Pers 64:577–591

Jonnal A.H., Gardner C.O., Prescott C.A., Kendler K.S. (2000) Obsessive and compulsive symptoms in a general population sample of female twins. Am J Med Genet 96:791–796

Jylha P. Isometsa E. (2006) The relationship of neuroticism and extraversion to symptoms of anxiety and depression in the general population. Depress Anxiety 23:281–289

Karayiorgou M., Altemus M., Galke B.L., Goldman D., Murphy D.L., Ott J., Gogos J.A. (1997) Genotype determining low catechol-O-methyltransferase activity as a risk factor for obsessive-compulsive disorder. Proc Natl Acad Sci USA 94:4572–4575

Karayiorgou M., Sobin C., Blundell M.L., Galke B.L., Malinova L., Goldberg P., Ott J., Gogos J.A. (1999) Family-based association studies support a sexually dimorphic effect of COMT and MAOA on genetic susceptibility to obsessive-compulsive disorder. Biol Psychiatry 45:1178–1189

Karno M., Golding J.M., Sorenson S.B., Burnam M.A. (1988) The epidemiology of obsessive-compulsive disorder in five US communities. Arch Gen Psychiatry 45:1094–1099

Kendler K.S. (1996) Major depression and generalised anxiety disorder. Same genes, (partly)different environments—revisited. Br J Psychiatry 168 (Suppl 30):68–75

Kendler K.S., Gardner C.O., Gatz M., Pedersen N.L. (2007) The sources of co-morbidity between major depression and generalized anxiety disorder in a Swedish national twin sample. Psychol Med 37:453–462

Kendler K.S., Neale M.C., Kessler R.C., Heath A.C., Eaves L.J. (1992a) Major depression and generalized anxiety disorder. Same genes, (partly) different environments? Arch Gen Psychiatry 49:716–722

Kendler K.S., Neale M.C., Kessler R.C., Heath A.C., Eaves L.J. (1992b) The genetic epidemiology of phobias in women. The interrelationship of agoraphobia, social phobia, situational phobia, and simple phobia. Arch Gen Psychiatry 49:273–281

Kendler K.S., Neale M.C., Kessler R.C., Heath A.C., Eaves L.J. (1993a) Major depression and phobias: the genetic and environmental sources of comorbidity. Psychol Med 23:361–371

Kendler K.S., Neale M.C., Kessler R.C., Heath A.C., Eaves L.J. (1993b) Panic disorder in women: a population-based twin study. Psychol Med 23:397–406

Kendler K.S., Gatz M., Gardner C.O., Pedersen N.L. (2006) Personality and major depression: a swedish longitudinal, population-based twin study. Arch Gen Psychiatry 63:1113–1120

Kennedy J.L., Neves-Pereira M., King N., Lizak M.V., Basile V.S., Chartier M.J., Stein M.B. (2001) Dopamine system genes not linked to social phobia. Psychiatr Genet 11:213–217

Kessler R.C., Berglund P., Demler O., Jin R., Merikangas K.R., Walters E.E. (2005a) Lifetime prevalence and age-of-onset distributions of DSM-IV disorders in the national comorbidity survey replication. Arch Gen Psychiatry 62:593–602

Kessler R.C., Chiu W.T., Demler O., Walters E.E. (2005b) Prevalence, severity, and comorbidity of 12-month DSM-IV disorders in the national comorbidity survey replication. Arch Gen Psychiatry 62:617–627

Klein D.F. (1964) Delineation of two drug-responsive anxiety syndromes. Psychopharmacology 5:397–408

Klein R.J., Zeiss C., Chew E.Y., Tsai J.Y., Sackler R.S., Haynes C., Henning A.K., SanGiovanni J.P., Mane S.M., Mayne S.T., Bracken M.B., Ferris F.L., Ott J., Barnstable C., Hoh J. (2005) Complement factor H polymorphism in age-related macular degeneration. Science 308:385–389

Knowles J.A. Weissman M.M. (1995) Panic disorder and agoraphobia. In Oldham J.M., Riba M.B. (eds) Review of psychiatry, vol 14. American Psychiatric Press, Washington, DC, pp. 383–404

Knowles J.A., Fyer A.J., Vieland V.J., Weissman M.M., Hodge S.E., Heiman G.A., Haghighi F., de Jesus G.M., Rassnick H., Preud'homme-Rivelli X., Austin T., Cunjak J., Mick S., Fine L. D., Woodley K.A., Das K., Maier W., Adams P.B., Freimer N.B., Klein D.F., Gilliam T.C. (1998) Results of a genome-wide genetic screen for panic disorder. Am J Med Genet (Neuropsychiatry Genet) 81:139–147

Lake R.I., Eaves L.J., Maes H.H., Heath A.C., Martin N.G. (2000) Further evidence against the environmental transmission of individual differences in neuroticism from a collaborative study of 45, 850 twins and relatives on two continents. Behav Genet 30:223–233

Lam P., Hong C.J., Tsai S.J. (2005) Association study of A2a adenosine receptor genetic polymorphism in panic disorder. Neurosci Lett 378:98–101

Lencz T., Morgan T.V., Athanasiou M., Dain B., Reed C.R., Kane J.M., Kucherlapati R., Malhotra A.K. (2007) Converging evidence for a pseudoautosomal cytokine receptor gene locus in schizophrenia. Mol Psychiatry 12:572–580

Lensi P., Cassano G.B., Correddu G., Ravagli S., Kunovac J.L., Akiskal H.S. (1996) Obsessive-compulsive disorder. Familial-developmental history, symptomatology, comorbidity and course with special reference to gender- related differences. Br. J Psychiatry 169:101–107

Leonard H.L., Lenane M.C., Swedo S.E., Rettew D.C., Gershon E.S., Rapoport J.L. (1992) Tics and Tourette's disorder: a 2- to 7-year follow-up of 54 obsessive- compulsive children. Am J Psychiatry 149:1244–1251

Leonard H.L., Swedo S.E., Garvey M., Beer D., Perlmutter S., Lougee L., Karitani M., Dubbert B. (1999) Postinfectious and other forms of obsessive-compulsive disorder. Child Adolesc Psychiatr Clin N Am 8:497–511

Lesch K.P., Bengel D., Heils A., Sabol S.Z., Greenberg B.D., Petri S., Benjamin J., Müller C.R., Hamer D.H., Murphy D.L. (1996) Association of anxiety-related traits with a polymorphism in the serotonin transporter gene regulatory region. Science 274:1527–1531

Lewis A. (1936) Problems of obsessional illness. Proc R Soc Med 29:325–336

Lichtenstein P. Annas P. (2000) Heritability and prevalence of specific fears and phobias in childhood. J Child Psychol Psychiatry 41:927–937

Lin P.Y. (2007) Meta-analysis of the association of serotonin transporter gene polymorphism with obsessive-compulsive disorder. Prog Neuro-Psychopharmacol Biol Psychiatry 31:683–689

Lougee L., Perlmutter S.J., Nicolson R., Garvey M.A., Swedo S.E. (2000) Psychiatric disorders in first-degree relatives of children with pediatric autoimmune neuropsychiatric disorders associated with streptococcal infections (PANDAS) J Am Acad Child Adolesc Psychiatry 39:1120–1126

MacKinnon D.F., McMahon F.J., Simpson S.G., McInnis M.G., DePaulo J.R. (1997) Panic disorder with familial bipolar disorder. Biol Psychiatry 42:90–95

MacKinnon D.F., Xu J., McMahon F.J., Simpson S.G., Stine O.C., McInnis M.G., DePaulo J.R. (1998) Bipolar disorder and panic disorder in families: an analysis of chromosome 18 data. Am J Psychiatry 155:829–831

MacKinnon D.F., Zandi P.P., Cooper J., Potash J.B., Simpson S.G., Gershon E., Nurnberger J., Reich T., DePaulo J.R. (2002) Comorbid bipolar disorder and panic disorder in families with a high prevalence of bipolar disorder. Am J Psychiatry 159:30–35

Mackintosh M.A., Gatz M., Wetherell J.L., Pedersen N.L. (2006) A twin study of lifetime Generalized Anxiety Disorder (GAD) in older adults: genetic and environmental influences shared by neuroticism and GAD. Twin Res Hum Genet 9:30–37

Mannuzza S., Schneier F.R., Chapman T.F., Liebowitz M.R., Klein D.F., Fyer A.J. (1995) Generalized social phobia. Reliability and validity. Arch Gen Psychiatry 52:230–237

Mathews C.A., Nievergelt C.M., Azzam A., Garrido H., Chavira D.A., Wessel J., Bagnarello M., Reus V.I., Schork N.J. (2007) Heritability and clinical features of multigenerational families with obsessive-compulsive disorder and hoarding. Am J Med Genet B Neuropsychiatry Genet 144:174–182

McCauley J.L., Olson L.M., Dowd M., Amin T., Steele A., Blakely R.D., Folstein S.E., Haines J.L., Sutcliffe J.S. (2004) Linkage and association analysis at the serotonin transporter (SLC6A4) locus in a rigid-compulsive subset of autism. Am J Med Genet B Neuropsychiatr Genet 127:104–112

Mendlewicz J., Papadimitriou G., Wilmotte J. (1993) Family study of panic disorder: comparison with generalized anxiety disorder, major depression and normal subjects. Psychiatry Genet 3:73–78

Middeldorp C.M., de Geus E.J., Beem A.L., Lakenberg N., Hottenga J.J., Slagboom P.E., Boomsma D.I. (2007) Family based association analyses between the serotonin transporter gene polymorphism (5-HTTLPR) and neuroticism, anxiety and depression. Behav Genet 37:294–301

Murphy T.K., Goodman W.K., Fudge M.W., Williams R.C., Jr., Ayoub E.M., Dalal M., Lewis M.H., Zabriskie J.B. (1997) B lymphocyte antigen D8/17: a peripheral marker for childhood-onset obsessive-compulsive disorder and Tourette's syndrome? Am J Psychiatry 154:402–407

Nash M.W., Huezo-Diaz P., Williamson R.J., Sterne A., Purcell S., Hoda F., Cherny S.S., Abecasis G.R., Prince M., Gray J.A., Ball D., Asherson P., Mann A., Goldberg D., McGuffin P., Farmer A., Plomin R., Craig I.W., Sham P.C. (2004) Genome-wide linkage analysis of a composite index of neuroticism and mood-related scales in extreme selected sibships. Hum Mol Genet 13:2173–2182

Nestadt G., Lan T., Samuels J., Riddle M., Bienvenu O.J., III, Liang K.Y., Hoehn-Saric R., Cullen B., Grados M., Beaty T.H., Shugart Y.Y. (2000a) Complex segregation analysis provides compelling evidence for a major gene underlying obsessive-compulsive disorder and for heterogeneity by sex. Am J Hum Genet 67:1611–1616

Nestadt G., Samuels J., Riddle M., Bienvenu O.J., Liang K.Y., LaBuda M., Walkup J., Grados M., Hoehn-Saric R. (2000b) A family study of obsessive-compulsive disorder. Arch Gen Psychiatry 57:358–363

Newman S.C. Bland R.C. (2006) A population-based family study of DSM-III generalized anxiety disorder. Psychol Med 36:1275–1281

Noyes R., Jr., Clarkson C., Crowe R.R., Yates W.R., McChesney C.M. (1987) A family study of generalized anxiety disorder. Am J Psychiatry 144:1019–1024

Ozaki N., Goldman D., Kaye W.H., Plotnicov K., Greenberg B.D., Lappalainen J., Rudnick G., Murphy D.L. (2003) Serotonin transporter missense mutation associated with a complex neuropsychiatric phenotype. Mol Psychiatry 8:933–936

Pauls D.L., Crowe R.R., Noyes R., Jr. (1979a) Distribution of ancestral secondary cases in anxiety neurosis (panic disorder). J Affect Disord 1:387–390

Pauls D.L., Noyes R., Jr., Crowe R.R. (1979b) The familial prevalence in second-degree relatives of patients with anxiety neurosis (panic disorder). J Affect Disord 1:279–285

Pauls D.L., Bucher K.D., Crowe R.R., Noyes R., Jr. (1980) A genetic study of panic disorder pedigrees. Am J Hum Genet 32:639–644

Pauls D.L., Towbin K.E., Leckman J.F., Zahner G.E., Cohen D.J. (1986) Gilles de la Tourette's syndrome and obsessive-compulsive disorder. Evidence supporting a genetic relationship. Arch Gen Psychiatry 43:1180–1182

Pauls D.L., Alsobrook J.P., Goodman W., Rasmussen S., Leckman J.F. (1995) A family study of obsessive-compulsive disorder. Am J Psychiatry 152:76–84

Perna G., Cocchi S., Bertani A., Arancio C., Bellodi L. (1995) Sensitivity to 35% CO2 in healthy first-degree relatives of patients with panic disorder. Am J Psychiatry 152:623–625

Perna G., Bertani A., Caldirola D., Bellodi L. (1996) Family history of panic disorder and hypersensitivity to CO2 in patients with panic disorder. Am J Psychiatry 153:1060–1064

Perna G., Caldirola D., Arancio C., Bellodi L. (1997) Panic attacks: a twin study. Psychiatry Res 66:69–71

Phillips K., Fulker D.W., Rose R.J. (1987) Path analysis of seven fear factors in adult twin and sibling pairs and their parents. Genet Epidemiol 4:345–355

Pine D.S., Klein R.G., Roberson-Nay R., Mannuzza S., Moulton J.L., III, Woldehawariat G., Guardino M. (2005) Response to 5% carbon dioxide in children and adolescents: relationship to panic disorder in parents and anxiety disorders in subjects. Arch Gen Psychiatry 62:73–80

Pooley E.C., Fineberg N., Harrison P.J. (2007) The met158 allele of catechol-O-methyltransferase (COMT) is associated with obsessive-compulsive disorder in men: case-control study and meta-analysis. Mol Psychiatry 12:556–561

Rasmussen S.A. Eisen J.L. (1992) The epidemiology and clinical features of obsessive compulsive disorder. Psychiatry Clin North Am 15:743–758

Reschke A.H., Mannuzza S., Chapman T.F., Lipsitz J.D., Liebowitz M.R., Gorman J.M., Klein D.F., Fyer A.J. (1995) Sodium lactate response and familial risk for panic disorder. Am J Psychiatry 152:277–279

Rettew D.C., Vink J.M., Willemsen G., Doyle A., Hudziak J.J., Boomsma D.I. (2006) The genetic architecture of neuroticism in 3301 Dutch adolescent twins as a function of age and sex: a study from the Dutch twin register. Twin Res Hum Genet 9:24–29

Robins L.N., Helzer J.E., Weissman M.M., Orvaschel H., Gruenberg E., Burke J.D., Jr., Regier D.A. (1984) Lifetime prevalence of specific psychiatric disorders in three sites. Arch Gen Psychiatry 41:949–958

Rothe C., Koszycki D., Bradwejn J., King N., Deluca V., Tharmalingam S., Macciardi F., Deckert J., Kennedy J.L. (2006) Association of the Val158Met catechol O-methyltransferase genetic polymorphism with panic disorder. Neuropsychopharmacology 31:2237–2242

Roy M.A., Neale M.C., Pedersen N.L., Mathe A.A., Kendler K.S. (1995) A twin study of generalized anxiety disorder and major depression. Psychol Med 25:1037–1049

The Genetics of Anxiety Disorders | 183

Ruscio A.M., Chiu W.T., Roy-Byrne P., Stang P.E., Stein D.J., Wittchen H.U., Kessler R.C. (2007) Broadening the definition of generalized anxiety disorder: Effects on prevalence and associations with other disorders in the National Comorbidity Survey Replication. J Anxiety Disord 21:662–676

Samochowiec J., Hajduk A., Samochowiec A., Horodnicki J., Stepien G., Grzywacz A., Kucharska-Mazur J. (2004) Association studies of MAO-A, COMT, and 5-HTT genes polymorphisms in patients with anxiety disorders of the phobic spectrum. Psychiatry Res 128:21–26

Samuels J., Shugart Y.Y., Grados M.A., Willour V.L., Bienvenu O.J., Greenberg B.D., Knowles J.A., McCracken J.T., Rauch S.L., Murphy D.L., Wang Y., Pinto A., Fyer A.J., Piacentini J., Pauls D.L., Cullen B., Rasmussen S.A., Hoehn-Saric R., Valle D., Liang K.Y., Riddle M.A., Nestadt G. (2007) Significant linkage to compulsive hoarding on chromosome 14 in families with obsessive-compulsive disorder: results from the OCD collaborative genetics study. Am J Psychiatry 164:493–499

Scherrer J.F., True W.R., Xian H., Lyons M.J., Eisen S.A., Goldberg J., Lin N., Tsuang M.T. (2000) Evidence for genetic influences common and specific to symptoms of generalized anxiety and panic. J Affect Disord 57:25–35

Scott L.J., Mohlke K.L., Bonnycastle L.L., Willer C.J., Li Y., Duren W.L., Erdos M.R., Stringham H.M., Chines P.S., Jackson A.U., Prokunina-Olsson L., Ding C.J., Swift A.J., Narisu N., Hu T., Pruim R., Xiao R., Li X.Y., Conneely K.N., Riebow N.L., Sprau A.G., Tong M., White P.P., Hetrick K.N., Barnhart M.W., Bark C.W., Goldstein J.L., Watkins L., Xiang F., Saramies J., Buchanan T.A., Watanabe R.M., Valle T.T., Kinnunen L., Abecasis G.R., Pugh E.W., Doheny K.F., Bergman R.N., Tuomilehto J., Collins F.S., Boehnke M. (2007) A genome-wide association study of type 2 diabetes in finns detects multiple susceptibility variants. Science 316:1341–1345

Sen S., Burmeister M., Ghosh D. (2004) Meta-analysis of the association between a serotonin transporter promoter polymorphism (5-HTTLPR) and anxiety-related personality traits. Am J Med Genet B Neuropsychiatry Genet 127:85–89

Shekhar A., McCann U.D., Meaney M.J., Blanchard D.C., Davis M., Frey K.A., Liberzon I., Overall K.L., Shear M.K., Tecott L.H., Winsky L. (2001) Summary of a National Institute of Mental Health workshop: developing animal models of anxiety disorders. Psychopharmacology (Berl) 157:327–339

Shifman S., Bhomra A., Smiley S., Wray N.R., James M.R., Martin N.G., Hettema J.M., An S.S., Neale M.C., van den Oord E.J., Kendler K.S., Chen X., Boomsma D.I., Middeldorp C.M., Hottenga J.J., Slagboom P.E., Flint J. (2008) A whole genome association study of neuroticism using DNA pooling. Mol Psychiatry 13:302–312

Shugart Y.Y., Samuels J., Willour V.L., Grados M.A., Greenberg B.D., Knowles J.A., McCracken J.T., Rauch S.L., Murphy D.L., Wang Y., Pinto A., Fyer A.J., Piacentini J., Pauls D.L., Cullen B., Page J., Rasmussen S.A., Bienvenu O.J., Hoehn-Saric R., Valle D., Liang K.Y., Riddle M.A., Nestadt G. (2006) Genomewide linkage scan for obsessive-compulsive disorder: evidence for susceptibility loci on chromosomes 3q, 7p, 1q, 15q, and 6q. Mol Psychiatry 11:763–770

Skre I., Onstad S., Torgersen S., Lygren S., Kringlen E. (1993) A twin study of DSM-III-R anxiety disorders. Acta Psychiatry Scand 88:85–92

Slater E. Shields J. (1969) Genetical aspects of anxiety. In Lader M.H. (ed) Studies of anxiety. Headly Brothers, Ashford, Kent, pp. 62–71

Sobin C. Karayiorgou M. (2000) The genetic basis and neurobiological characteristics of obsessive-compulsive disorder. In Pfaff D.W, Berrettini W.H., Joh T.H., Maxson S.C. (eds) Genetic influences on neural and behavioral functions. CRC Press Boca Raton, pp. 83–104

Stein B., Chartier J., Kozak M.V., King N., Kennedy J.L. (1998a) Genetic linkage to the serotonin transporter protein and 5HT2A receptor genes excluded in generalized social phobia. Psychiatry Res 81:283–291

Stein M.B., Chartier M.J., Hazen A.L., Kozak M.V., Tancer M.E., Lander S., Furer P., Chubaty D., Walker J.R. (1998b) A direct-interview family study of generalized social phobia. Am J Psychiatry 155:90–97

Sundet J.M., Skre I., Okkenhaug J.J., Tambs K. (2003) Genetic and environmental causes of the interrelationships between self-reported fears. A study of a non-clinical sample of Norwegian identical twins and their families. Scand J Psychol 44:97–106

Swedo S.E., Rapoport J.L., Cheslow D.L., Leonard H.L., Ayoub E.M., Hosier D.M., Wald E.R. (1989) High prevalence of obsessive-compulsive symptoms in patients with Sydenham's chorea. Am J Psychiatry 146:246–249

Swedo S.E., Leonard H.L., Mittleman B.B., Allen A.J., Rapoport J.L., Dow S.P., Kanter M.E., Chapman F., Zabriskie J. (1997) Identification of children with pediatric autoimmune neuropsychiatric disorders associated with streptococcal infections by a marker associated with rheumatic fever [see comments]. Am J Psychiatry 154:110–112

Swedo S.E., Leonard H.L., Garvey M., Mittleman B., Allen A.J., Perlmutter S., Lougee L., Dow S., Zamkoff J., Dubbert B.K. (1998) Pediatric autoimmune neuropsychiatric disorders associated with streptococcal infections: clinical description of the first 50 cases [published erratum appears in Am J Psychiatry 1998 Apr;155(4):578]. Am J Psychiatry 155:264–271

Tadic A., Rujescu D., Szegedi A., Giegling I., Singer P., Moller H.J., Dahmen N. (2003) Association of a MAOA gene variant with generalized anxiety disorder, but not with panic disorder or major depression. Am J Med Genet B Neuropsychiatry Genet 117:1–6

Talati A., Ponniah K., Strug L.J., Hodge S.E., Fyer A.J., Weissman M.M. (2008) Panic disorder, social anxiety disorder, and a possible medical syndrome previously linked to chromosome 13. Biol Psychiatry 63:594–601

Thorgeirsson T.E., Oskarsson H., Desnica N., Kostic J.P., Stefansson J.G., Kolbeinsson H., Lindal E., Gagunashvili N., Frigge M.L., Kong A., Stefansson K., Gulcher J.R. (2003) Anxiety with panic disorder linked to chromosome 9q in Iceland. Am J Hum Genet 72:1221–1230

Torgersen S. (1983) Genetic factors in anxiety disorders. Arch Gen Psychiatry 40:1085–1089

van Grootheest D.S., Cath D.C., Beekman A.T., Boomsma D.I. (2005) Twin studies on obsessive-compulsive disorder: a review. Twin Res Hum Genet 8:450–458

van Grootheest D.S., Cath D.C., Beekman A.T., Boomsma D.I. (2007) Genetic and environmental influences on obsessive-compulsive symptoms in adults: a population-based twin-family study. Psychol Med 1–10

van B.N. Griez E. (2000) Reactivity to a 35% CO2 challenge in healthy first-degree relatives of patients with panic disorder. Biol Psychiatry 47:830–835

Veenstra-VanderWeele J., Kim S.J., Gonen D., Hanna G.L., Leventhal B.L., Cook E.H., Jr. (2001) Genomic organization of the SLC1A1/EAAC1 gene and mutation screening in early-onset obsessive-compulsive disorder. Mol Psychiatry 6:160–167

Vieland V.J., Hodge S.E., Lish J.D., Adams P., Weissman M.M. (1993) Segregation analysis of panic disorder. Psychiatry Genet 3:63–71

Vieland V.J., Goodman D.W., Chapman T., Fyer A.J. (1996) New segregation analysis of panic disorder. Am J Med Genet 67:147–153

Weissman M.M., Bland R.C., Canino G.J., Greenwald S., Hwu H.G., Lee C.K., Newman S.C., Oakley-Browne M.A., Rubio-Stipec M., Wickramaratne P.J. (1994) The cross national epidemiology of obsessive compulsive disorder. The Cross National Collaborative Group. J Clin. Psychiatry 55 Suppl:5–10

Weissman M.M., Bland R.C., Canino G.J., Faravelli C., Greenwald S., Hwu H.G., Joyce P.R., Karam E.G., Lee C.K., Lellouch J., Lépine J.P., Newman S.C., Oakley-Browne M.A., Rubio-Stipec M., Wells J.E., Wickramaratne P.J., Wittchen H.U., Yeh E.K. (1997) The cross-national epidemiology of panic disorder. Arch Gen Psychiatry 54:305–309

Weissman M.M., Fyer A.J., Haghighi F., Heiman G.A., Deng Z., Hen R., Hodge S.E., Knowles J.A. (2000) Potential panic disorder syndrome: clinical and genetic linkage evidence. Am J Med Genet (Neuropsychiatry Genet) 96:24–35

Weissman M.M., Gross R., Fyer A., Heiman G.A., Gameroff M.J., Hodge S.E., Kaufman D., Kaplan S.A., Wickramaratne P.J. (2004) Interstitial cystitis and panic disorder: a potential genetic syndrome. Arch Gen Psychiatry 61:273–279

Wellcome Trust Case Control Consortium (2007) Genome-wide association study of 14, 000 cases of seven common diseases and 3, 000 shared controls. Nature 447:661–678

The Genetics of Anxiety Disorders

Wendland J.R., Kruse M.R., Cromer K.C., Murphy D.L. (2007) A large case-control study of common functional SLC6A4 and BDNF variants in obsessive-compulsive disorder. Neuropsychopharmacology 32:2543–2551

Willis-Owen S.A.G., Turri M.G., Munafo M.R., Surtees P.G., Wainwright N.W.J., Brixey R.D., Flint J. (2005) The serotonin transporter length polymorphism, neuroticism, and depression: a comprehensive assessment of association. Biol Psychiatry 58:451–456

Willour V.L., Yao S.Y., Samuels J., Grados M., Cullen B., Bienvenu O.J., III, Wang Y., Liang K. Y., Valle D., Hoehn-Saric R., Riddle M., Nestadt G. (2004) Replication study supports evidence for linkage to 9p24 in obsessive-compulsive disorder. Am J Hum Genet 75:508–513

Woo J.M., Yoon K.S., Yu B.H. (2002) Catechol O-Methyltransferase genetic polymorphism in panic disorder. Am J Psychiatry 159:1785–1787

Woo J.M., Yoon K.S., Choi Y.H., Oh K.S., Lee Y.S., Yu B.H. (2004) The association between panic disorder and the L/L genotype of catechol-O-methyltransferase. J Psychiatric Res 38:365–370

Yamada K., Hattori E., Shimizu M., Sugaya A., Shibuya H., Yoshikawa T. (2001) Association studies of the cholecystokinin B receptor and A2a adenosine receptor genes in panic disorder. J Neural Transm 108:837–848

Molecular Biology of Addiction and Substance Dependence

Sibylle G Schwab, Adrian Scott, and Dieter B Wildenauer

Contents

1 Introduction 188
2 Heritability of Substance Dependence 188
3 Biology of Drug-Induced Changes in the Brain 191
4 Molecular Genetic Evidence of Substance Dependence 193
    4.1 Heroin 194
    4.2 Nicotine 196
    4.3 Alcohol 198
5 Conclusion 199
References 199

Abstract Addictions are chronic, devastating psychiatric disorders that are characterized by a compulsive pattern of drug taking or behavior. The development of addictions relies on life style choices but also on genetic, heritable factors that put some individuals at an increased risk for these disorders. Genetic factors for addictive disorders were established by multiple epidemiological studies, showing heritability between 40 and 60%. In an effort to dissect addictive disorders, genetic research was focused on the direct interaction of a variety of drugs with the respective biological systems. Another focus was the identification of long-lasting changes in the brain reward mechanism, using animal models, and more recently brain imaging studies in humans.

In the following, we are briefly summarizing the evidence obtained so far for an inheritable trait in addiction as well as what is currently known about the underlying biology.

S.G. Schwab (✉)
Western Australian Institute for Medical Research and Centre for Medical Research,
Sir Charles Gairdner Hospital, B-Block Ground Floor, University of Western Australia,
Nedlands, WA 6009, Australia
sschwab@cyllene.uwa.edu.au

D.B. Wildenauer (ed.), *Molecular Biology of Neuropsychiatric Disorders*,
Nucleic Acids and Molecular Biology, © Springer-Verlag Berlin Heidelberg 2009

1 Introduction

It is generally thought that the development of addictions involves a transition state from a casual drug use to a compulsive pattern of drug use. However, the majority of those who try substances with addictive potential do not become addicted. For example, for heroin, the transition rate is estimated to be around 23%. This compares to an estimated transition rate for tobacco and alcohol of approximately 32 and 15%, respectively (Hiroi and Agatsuma 2005)

The term addiction describes a behavioral pattern that is characterized by compulsion, loss of control and continued repetition of a behavior or activity, even when adverse consequences are associated with this behavior.

According to DSM-IV (APA 1994), a diagnosis of drug dependence is met if at least three of the following criteria apply:

- Tolerance, as defined by increased amount of drug intake
- Characteristic withdrawal symptoms after interruption of drug intake
- Increased or prolonged use
- Unsuccessful attempts to cease or control use
- Significant time spent in activities to acquire and use the drug
- Important social, occupational, or recreational activities sacrificed because of substance abuse
- Persistent use despite clear evidence of physical and psychological harm related to substance use

Addictions belong to the group of chronic psychiatric disorders. Individuals suffering from addiction are exposed to a major threat to their health and their social standing, both in the short term and equally important in the long term. According to the World Health Organization (WHO), 76.3 million people worldwide abuse or are dependent on alcohol and at least 15.3 million persons have drug use disorders (WHO 2008). Alcohol contributes to traumatic outcomes that kill or disable at a relatively young age and therefore contribute to far more years of life lost to death and disability compared to tobacco or illegal drugs. The burden of drug abuse disorders is not equally distributed among countries. The disease burden measured in disability adjusted life years (DALYs) is significantly higher in Europe and the Western Pacific than in Africa and the Eastern Mediterranean (WHO 2004). In addition, restrictive drug policies and religious influences are known to have an impact on specific drug abuse patterns. These might even include behaviors like compulsive gambling or binge eating, which are also thought to rely on the same reward mechanisms as dependence on legal and illegal substances.

2 Heritability of Substance Dependence

It is well known that drug disorders accumulate in families, suggesting the presence of both genetic and environmental influences. This has been established in a limited number of twin and adoption studies. Most of these studies have focused on alcohol,

Molecular Biology of Addiction and Substance Dependence
189

nicotine and illicit drugs such as marijuana, sedatives, stimulants (including amphetamine and cocaine) heroin and other opiates, and psychodelics (LSD or mescaline).

Simultaneous abuse and dependence on multiple drugs is observed quite commonly as is comorbidity with other psychiatric and personality disorders. Epidemiological studies have therefore also focused on dissecting general addictive predisposition versus substance specific risks, as well as the influence of the environment on addiction.

In 1996, Tsuang and colleagues (Tsuang et al. 1996) studied a sample of 3,372 male twin pairs drawn from the Vietnam Era Twin Registry. The study focused on abuse/dependence of five different categories of drugs, specifically marijuana, sedatives, stimulants, heroin/opiates and PCP/psychodelics. Of this male twin sample, 10.1% had abused or been dependent on at least one illicit drug from these categories. It can be reasonably assumed that a proportion of this magnitude is likely being dependent on drug abuse. Warner et al. (1995) identified a similar number of dependent subjects in the National Comorbidity Survey which focused on males aged between 35–44 years. In the study of (Tsuang et al. 1996), monozygotic (MZ) twins revealed a statistically higher concordance rate for drug abuse or dependence (26.2%) compared to dizygotic twins (16.5%), indicating that genetic factors play a role. This was essentially true for marijuana and stimulant use, whereas for opiates, due to the relatively low prevalence of less than 2%, the study had probably not enough power to unequivocally show that specific genetic factors are in place. In addition, biometric modeling using this data set suggested that genetic factors, the environment shared by the twins, and the non-shared environment had a significant influence of similar magnitude (34, 28 and 38% of the variance, respectively).

In 1998, a series of three papers with a focus on the co-occurrence of drugs of abuse were published in the Archives of General Psychiatry. Tsuang and colleagues further investigated the data obtained from the Vietnam Era Twin Registry (Tsuang et al. 1998). They concluded that each category of drug, except the psychodelics, had genetic influences to itself, with heroin revealing the largest drug-specific genetic influence (38% of the variance). They also reported on evidence for a shared common risk factor that underlies the abuse of marijuana, sedatives, heroin or opiates, and psychodelics.

Merikangas and colleagues performed a controlled family study including 231 probands with dependence on opioids, cocaine, cannabis, and/or alcohol, 61 control probands, and their 1,267 adult first-degree relatives (Merikangas et al. 1998). An 8-fold increased risk of drug disorders was observed among relatives of probands with drug disorders, suggesting genetic and environmental factors contributing to this increase in risk. In addition, they found evidence of specificity of familial aggregation of the predominant drug of abuse. According to their studies, alcoholism and drug disorders are independent, and comorbidity with other drug abuse is not an indicator of severity of substance disorder in general. A similar conclusion was drawn by Bierut et al., who studied the familial transmission of alcohol, marijuana, and cocaine dependence and habitual smoking in the Collaborative Study on the Genetics of Alcoholism (Bierut et al. 1998). For siblings of alcohol dependent probands, about 50% of brothers and 24% of sisters were alcohol dependent, but this elevated risk was not further increased by comorbid substance dependence in probands. In addition, they showed an effect of birth cohort and sex on the development of substance dependence.

Based on the Virginia Twin Registry, Kendler et al. investigated the specificity of genetic and environmental risk factors for use and abuse/dependence of cannabis, cocaine, hallucinogens, sedatives, stimulants, and opiates in 1,196 male twin pairs (Kendler et al. 2003). They addressed the central question, to what extent the risk factors for use or misuse of a particular class of psychoactive substances are specific to that class using multivariate twin modeling. In this study sample, they found high levels of comorbidity involving the different substance categories both for use and abuse/dependence with one genetic factor strongly influencing the risk for illicit use and abuse/dependence. In 2007, they extended this study including 4,865 members of male – male and female–female twin pairs drawn form the Virginia Adult Twin Study of Psychiatric and Substance Disorder (Kendler et al. 2007). The centre of this study was on symptoms of abuse and dependence of cannabis, cocaine, alcohol, caffeine, and nicotine. In general, the pattern of genetic and environmental risk factors for psychoactive substance dependence was similar in male and females. A best fit exploratory model was established, containing two genetic factors and one individual environmental factor contributing to all analyzed substance dependences. The first genetic factor had a strong load on cocaine and cannabis dependence (illicit drugs) compared to the second, which had a more specific load on alcohol and nicotine dependence (licit drugs). In addition, they report that genetic influences on nicotine and in particular caffeine dependence appeared to be specific to these drugs.

Multiple studies published in the early 1980s suggested two types of inheritance of alcoholism, with "type II" alcoholics showing increased antisocial behavior (Cloninger et al. 1981; Bohman et al. 1982; von Knorring et al. 1987). Cadoret et al. (1995) studied the influence of antisocial behavior on alcoholism further, and his group was able to show a strong correlation of drug abuse with antisocial personality disorder, suggesting that one pathway to drug abuse had its origin in antisocial parents. Using a sample of 95 male adoptees, Cadoret et al. was able to confirm a model of two independent genetic factors which were involved in drug abuse/dependence. In addition, disturbed adoptive parents were associated with adoptee drug abuse (Cadoret et al. 1995). A similar result was received when studying 102 women who had been adopted at birth. Using log-linear analyses a major pathway of genetic etiology was revealed, that started with a biological parent with antisocial personality and led to an adoptee with conduct disorder and later drug abuse/dependence (Cadoret et al. 1996). It was concluded that one element of familial factors is genetic and that, in addition, the family environment directly affects behavior (i.e., aggressivity) that leads to drug abuse/dependency. To further study the role of conduct disorder in explaining the comorbidity between alcohol and illicit drug dependence in adolescence, Button and colleagues studied 645 monozygotic and 702 dizygotic twin pairs (drawn from the Colorado Longitudinal Twin Study and the Colorado Twin Registry), 429 biological sibling pairs, and 96 adoptive sibling pairs (drawn from the Colorado Adoption Project and from the Adolescent Substance Abuse Family Study). They showed heritability estimates for conduct disorder, alcohol dependence, and illicit drug dependence of 58, 66 and 36%, respectively. The genetic correlation between alcohol dependence and illicit drug dependence could be partially explained by the genetic risk they both share with conduct disorder (Button et al. 2007).

Since tobacco is the substance causing the maximum health damage globally, epidemiological studies into nicotine dependence have been performed independently of other substances of abuse. Sullivan and Kendler summarized the genetic epidemiology of smoking, revisiting data from family, adoption, and twin studies (Sullivan and Kendler 1999). According to this summary, liability to initiate smoking resulted from genetic influences (approximately 60%) and from environmental influences shared by a member of a twin pair (about 20%), with another 20% specific to an individual. A subset of those who initiated smoking progressed to nicotine dependence. In this transition process, genetic factors appeared to be more prominent (about 70%) and the influence of the shared environment seemed to be negligible.

In summary, vulnerability to substance dependence reveals a strong genetic component that can explain 40–60% of the overall vulnerability. In addition, there seems to be some evidence for genetic factors predisposing to a specific substance of abuse, in addition to a general genetic "addictive" factor. Environmental factors seem to be equally important with some evidence for comorbid personality disorders playing a crucial role.

3 Biology of Drug-Induced Changes in the Brain

The majority of individuals experimenting with potentially addictive substances do not become addicted. This is one of the puzzling observations inherent to substance dependence. Even for potent addictive drugs like cocaine, the transition rate to addiction within 10 years of abuse is estimated to be only around 15–16% (Wagner and Anthony 2002).

The key question therefore remains: how does the transition from a casual use to a compulsive pattern of drug use occur?

Drugs of abuse are chemically quite diverse and target a whole variety of different primary activator systems, i.e., receptors, ion channels, transportes. Most of the primary activation sites of the diverse drugs have been identified and will be described in more detail below. After both acute and chronic exposure, they cause common effects which converge on a common circuitry in the brains limbic system. These brain reward circuits include, but are not restricted to, dopamine projections from the ventral tegmental area and substantia nigra to the nucleus accumbens and striatum, as well as glutamate inputs from the prefrontal cortex, amygdala, and hippocampus. This circuit is normally involved in pleasure, incentive motivation, and learning, and drug-induced adaptations have been found in this system (Robinson and Berridge 2003). A specific focus has been on associative learning processes which are thought to involve emotional memories important to addictive behavior (Ramsay and Woods 1997; Everitt et al. 2003; Wise 2004).

It is now well established that all addictive drugs increase levels of synaptic dopamine within the nucleus accumbens (Wise 1998; Hyman et al. 2006). This elevation of dopamine levels can occur directly or indirectly. Even though, under normal circumstances, dopamine seems to play a central role in reward-related learning, it appears to be not necessarily required for hedonic responses. This conclusion has been drawn from experiments using genetically engineered dopamine-deficient mice that essentially do not produce dopamine. However, dopamine was necessary for mice to seek rewards during goal-directed behavior (Robinson et al. 2005).

A brain region that has been continuously involved in working memory processes, control of executive function, attention, and behavioral inhibition is the prefrontal cortex (PFC). Similar to the nucleus accumbens, the PFC receives dopamine innervation from the ventral tegmental areas (Montague et al. 1996). Excessive dopamine signaling and activation of areas within the PFC can be hypothesized to underlie the learning of drug-related cues, as suggested in reinforcement learning models. By engaging dopamine-mediated reinforcement learning signals, drugs might generate a feedback loop that ultimately could result in a vicious cycle of action and learning (Montague et al. 2004). Imaging studies in humans revealing a reduced baseline activity in several regions of the PFC support an important role of this brain region in drug addiction processes (Volkow et al. 2004). It should be also mentioned in this context that not only dopaminergic innervations from the ventral tegmental areas occur, but also glutamatergic projections from the PFC to the nucleus accumbens and ventral tegmental areas. Therefore, a complex system of regulation can be hypothesized. The chronic drug-treated state is associated with reduced basal activity of cortical pyramidal neurons and a reduced sensitivity of the neurons to activation by natural rewards (Nestler 2005).

Another system that has been involved in circuitry adaptation is the central corticotropin releasing factor (CRF) system. After chronic administration of drugs, activation of CRF containing neurons in the amygdala is observed for numerous drugs after abrupt withdrawal (Heinrichs and Koob 2004). It is thought that this hyperfunctional CRF system plays a crucial role during withdrawal and partly mediates the negative emotional symptoms including somatic features that occur upon withdrawal. It has also been shown by Sillaber et al. (2002) that, in mice lacking a functional corticotrophin releasing hormone-1 receptor (CRH-1), stress led to an enhanced and progressively increasing intake of alcohol. The effect of the repeated alcohol drinking behaviors was in addition associated with an upregulation of the N-methyl-d-aspartate (NMDA) receptor subunit NR2B (Sillaber et al. 2002). Other lines of evidence support the idea that an increase in glucocorticoid hormones could increase vulnerability to drug abuse. This is thought to occur through an action of glucocorticoids on mesolimbic dopaminergic neurons. During chronic stress, repeated increase in glucocorticoids and dopamine might result in sensitization of the reward system and trigger the development of addiction (Marinelli and Piazza 2002).

Molecular Biology of Addiction and Substance Dependence   193

Some of the basic mechanisms involved in addiction must be associated with synaptic strength and synaptic plasticity. Synaptic plasticity plays a key role in learning and memory processes and involves long-term potentiation and long-term depression. From experiments using NMDA receptor antagonists, which prevented the development of sensitization to psychostimulants (Wolf 1998; Hyman et al. 2006), it was concluded that these receptors might be important to addiction. Since NMDA receptors are known to be involved in the generation of long term potentiation and depression, it was suggested that addictive drugs might cause changes to synaptic plasticity in the ventral tegmental areas (Hyman et al. 2006). In addition, the AMPA (α-Amino-3-hydroxy-5-Methylisoxazole-4-Proprionic Acid) versus NMDA receptor ratio seems to be crucial for synaptic strength (Ungless et al. 2001). It can be concluded that synaptic plasticity in the ventral tegmental area is relevant to behavior and specifically to addiction. It was proposed that long-term potentiation induced in dopaminergic neurons by addictive drugs or stress might play an important role in enhancing the rewarding properties of addictive drugs (Hyman et al. 2006).

Finally, we would like to touch on some of the biochemical changes observed in drug addiction. One of the best studied transcription factors that should be mentioned in this context is ΔFosB. This transcription factor is induced by a number of potentially addictive drugs in the nucleus accumbens and dorsal striatum after chronic administration (Hope et al. 1994; Nestler 2001). This transcription factor displays a relatively stable protein, which is quite a unique feature. Therefore, ΔFosB persists in the relevant brain areas for a long time period (up to a couple of months), suggesting that it contributes to the long-term effects observed in addiction (Nestler 2001). Accumulation of ΔFosB within neurons of the nucleus accumbens is thought to provide evidence that it contributes to a state of sensitization (Harris and Aston-Jones 1994; Harris and Aston-Jones 2003).

CREB (cAMP response element-binding) is another transcription factor thought to be involved in adaptation processes in the nucleus accumbens. Activation of CREB in the nucleus accumbens has been shown to decrease behavioral responses to cocaine, opiates, and alcohol. In contrast, decreased CREB activity increases these responses (Nestler 2005).

In summary, research into the biology of addiction has been focusing on the changes in the rewarding brain circuitry system. A primary role for dopamine and projections of dopaminergic neurons from the ventral tegmental area and substantia nigra to the nucleus accumbens has been established. An interaction with glutamatergic signaling from the PFC appears to be crucial. The CRF seems to play a particular role in stress induced situations in the context of addiction.

4 Molecular Genetic Evidence of Substance Dependence

The basis for molecular genetics studies on substance dependence is (1) the evidence for heritability for substance dependence, and (2) evidence from biological studies into the mechanisms involved in reward.

Once a drug is administered, it is absorbed and distributed to its sites of action where it interacts with targets (i.e., receptors, enzymes etc.), undergoes metabolism and is then excreted. The interaction of a drug with respective receptors as well as downstream effects resulting of this interaction is described as *pharmacodynamic action*, whereas absorption, distribution and metabolism of a drug represent the *pharmacokinetic aspects* of drug interaction. Proteins involved in pharmacokinetics as well as in pharmacodynamics are known to be variable in their effects and this variability can be under genetic control and consequently be determined in the DNA sequence coding for the protein. *Pharmacogenetics* describes the genetic variability which has an impact on drug effects in the human body and associates this variability to the different outcome of drug–body interaction.

Below, we would like to focus on three different addictive drugs which do have a major health impact in western societies: heroin, nicotine, and alcohol.

4.1 Heroin

In many countries, the majority of illicit drug users seeking treatment are primarily addicted to opiates, preferentially to heroin. Since heroin is commonly used by injecting, the health risk, in particular by transmission of HIV and hepatitis, is substantial and adds tremendously to the costs implicated in treatment. A high rate of comorbidity with other common psychiatric disorders, i.e., current major depression (23%), antisocial personality disorder (75%) or borderline personality disorder (51%) is also reported (Mills et al. 2004)

The World Health Organization (WHO) (Ali and Godwing 2002) estimated that approximately 185 million individuals are using illicit drugs. According to the WHO, addiction contributed in the year 2000 to 12.4% of deaths worldwide.

Even though animal models might be powerful in detecting genetic susceptibility loci for heroin dependence, there is so far only scarce literature available. It should also be kept in mind that animal studies can provide only a proxy for the condition observed in humans. A study on a QTL for opioid dependence by Berrettini et al. (1994) has identified a locus on mouse chromosome 10. This locus harbors the mouse μ-opioid receptor gene, which is assumed to be primarily involved in opioid dependence. In 2005, the same group was able to replicate this finding (Ferraro et al. 2005). Follow-up of this finding revealed functional relevant promoter variants in the mouse gene (Doyle et al. 2006) as well as putatively relevant splice variants (Doyle et al. 2007).

The main focus of molecular genetic research into heroin dependence has been on studying candidate genes involved in opioidergic pathways or genes, related to reward circuits like the dopaminergic system. Recently, two genome-wide linkage scans for heroin addiction have become available (Gelernter et al. 2006; Glatt et al. 2006). Using a mixed population of European Americans and African Americans (Gelernter et al. 2006), and a population of Han Chinese families (Glatt et al. 2006), both studies were able to identify loci within a 25 cM interval on chromosome 17,

Molecular Biology of Addiction and Substance Dependence

where the gene for the serotonin transporter is located. In the following, we would like to summarize briefly the pathways that have been targeted in candidate gene studies in heroin dependence.

4.1.1 Opioid Related Genes

There are three types of opioid receptors classified according to their specific ligands: the μ-opioid receptor (OPRM1, main endogenous ligand β-endorphin and the different varieties of enkephalins), the κ-opioid receptor (OPRK1, main endogenous ligand dynorphin A and B, and α-neoendorphin), and the δ-opioid receptor (OPRD1, main endogenous ligands are the different varieties of enkephalins). The endogenous ligands for these receptors are derived from three precursor proteins by proteolytic processing. The precursor proteins for the different varieties of enkephalins are the proenkephalin (PENK), β-endorphin, which is processed from proopiomelanocortin (POMC), and the dynorphins and α-neoendorphin, which are retrieved from prodynorphin (PDYN). The opioid receptors mediate both the analgesic and rewarding properties of opioid compounds as well as opioid effects on the hypothalamic-pituitary-adrenal (HPA) stress-responsive axis, respiratory and pulmonary function, gastrointestinal motility, immune responses, and other functions. The μ-opioid receptor has been acknowledged to be mainly involved in analgesic properties together with rewarding responses to morphine (Berrettini et al. 1994), and has therefore been the main focus in genetic studies of heroin addiction.

A polymorphism (G118A) located in exon 1 of the OPRM1 gene results in an Asp to Asn change at amino acid position 40. Receptors with 40Asp (encoded by the 118G allele), have been shown to bind β-endorphin with greater affinity (Kreek et al. 2005) and to have a more intense activation of the HPA axis stress-mediated response as measured by plasma cortisol levels (Wand et al. 2002). However, association studies have not been conclusive, with eight studies showing evidence for association and four studies rejecting association with heroin addiction (Table 1). Like for most of the association studies targeting heroin addiction, sample sizes were either small or ethnically heterogeneous, both factors well known to adversely affect association studies. In addition, only a few polymorphisms have been tested within each candidate gene. These polymorphisms might not present the complete variety of the genes and their related proteins.

Other opioid-related genes have also been targeted in association studies. However, no conclusive results regarding association of the studied DNA variant with heroin addiction have been obtained (Table 1).

4.1.2 Monoaminergic Pathways Related Genes

There is a large body of evidence indicating interactions between the dopaminergic, serotonergic and opioidergic neurotransmitter systems in reward, drug dependence, and withdrawal (Kreek et al. 2005). Relevant genes analyzed previously include

Table 1 Association studies with heroin addiction targeting opioid related genes

Reference	No. heroin cases/ethnicity	Gene symbol	No of DNA variants analyzed	Association
Zhang et al. 2006	382 EA	OPRM1	13	Yes
Bart et al. 2004	139 SW	OPRM1	2	Yes
Crowley et al. 2003	89 AA + 124 EA	OPRM1	5	No
Luo et al. 2003	318 EA + 124 AA	OPRM1	10	Yes
Tan et al. 2003	20 IN, 25 MA, 52 CH	OPRM1	1	Yes
Shi et al. 2002	139 CH	OPRM1	2	Yes
Szeto et al. 2001	200 CH	OPRM1	2	Yes
Franke et al. 2001	287 GE + 111 GE trios	OPRM1	1	No
Hoehe et al. 2000	172 EA, AA	OPRM1	42	Yes
Li et al. 2000a	282 CH	OPRM1	2	No
Bond et al. 1998	113 EA, AA, HI	OPRM1	1	Ambiguous
Berrettini et al. 1997	55 EA, AA	OPRM1	1	Ambiguous
Yuferov et al. 2004	145 EA, AA, HI	OPRK1	1	Ambiguous
Xu et al. 2002	450 CH	OPRD1	2	No
Franke et al. 1999	233 GE + 90 GE trios	OPRD1	1	No
Mayer et al. 1997	103 GE	OPRD1	1	Yes
Comings et al. 1999	31 CA	PENK	1	Yes
Kreek et al. 2005	not identified	PENK	1	Yes
Zimprich et al. 2000	118 GE	PDYN	1	No

EA European American, *AA* African American, *SW* Swedish, *IN* Indian, *MA* Malay, *CH* Chinese, *GE* German, *HI* Hispanic, *CA* Caucasian

genes for dopamine receptors (DRD2, DRD3, DRD4 and DRD5), dopamine transporter (SLC6A3), serotonin transporter (SLC6A4), and the catechol-O-methyltransferase (COMT) (Table 2).

Similarly to the association studies targeting the opioidergic genes, inconsistent results have been obtained in different studies. Again, sample sizes and study design (i.e., ethnicity, phenotype characterization) might be held responsible for obtaining contradictory results (Table 2).

4.2 Nicotine

Nicotine is the psychoactive substance from the tobacco plant *Nicotiana tobacum*. Tobacco continues to be the drug causing the maximum health damage globally. It is estimated that about one-third of the global population aged over 15 smokes (WHO 2008), resulting in lung cancer, cardiovascular disease, and other respiratory disorders which finally result in premature death.

Nicotine acts on neuronal nicotinic acetylcholine receptors in the brain to release dopamine and other neurotransmitters that sustain addiction.

Molecular Biology of Addiction and Substance Dependence 197

Table 2 Association studies with heroin addiction targeting monoaminergic related genes

Reference	Sample	Candidate gene no. of variants analyzed)	Association
Szilagyi et al. 2005	53 Hungarian cases	DRD4 (1), 5-HTT (1)	Yes (for DRD4)
Li et al. 2000c	405 Chinese cases	DRD4 (2)	No
Franke et al. 2000	285 German cases + 111 German trios	DRD4 (1)	No
Kotler et al. 1997	141 Israeli cases	DRD4	Yes
Li et al. 2006	420 Chinese cases	DRD2 (1), DRD5(1), SLC6A3 (1)	Yes (for DRD2)
Xu et al. 2004	486 Chinese + 471 German cases	DRD2 (10)	Yes
Li et al. 2002	121 + 344 Chinese cases	DRD2 (3), DRD3 (1), 5-HT2A (2), 5-HTT (2), GABRG2 (1)	Yes (DRD2)
Kotler et al. 1999	193 Non-Ashkenazi cases	DRD3 (1), 5-HTT (1)	No
Tan et al. 1999	63 Chinese cases	5-HTT (1)	Yes
Proudnikov et al. 2006	235 cases	HTR1B	Yes
Gerra et al. 2005	104 Italian cases	SLC6A3 (=DAT, 1)	Ambiguous
Cheng et al. 2005	200 Taiwanese cases	BDNF (1)	Yes
Horowitz et al. 2000	38 Israeli trios + 101 Israeli cases	COMT (1)	Yes
Li et al. 2000b	375 Chinese cases	CNR1	No

Multiple association studies targeting specific variants within the nicotine addiction relevant genes have become available. A number of different loci have been suggested, as summarized in recent reviews (Ho and Tyndale 2007; Lessov-Schlaggar et al. 2008). Identified genes are comprised of but not restricted to genes coding for the various nicotine receptor subtypes, dopamine receptors or transporters, and GABA receptors. In addition, studies included subphenotypes of nicotine addiction, such as quantities of cigarettes smoked per day, as an estimator for the severity of addiction. Comorbidity with alcohol was also used in genetic association studies (Sullivan et al. 2008).

Converging evidence on involvement of subunits of the nicotine receptors in nicotine addiction has only recently been established. It is a unique feature of more recent studies that study samples are impressive in size with studying up to, or more than, 1,000 cases and controls, thus having enough power to reveal reliable effects. Saccone et al. (2007) analyzed, in 1,050 cases and 879 controls, 3,713 DNA sequence variants (SNPs). The cases were evaluated using the Fagerstrom test for nicotine dependence (Heatherton et al. 1991). A score of four on this scale was required for inclusion in this study. Some evidence for an involvement of DNA variation in the GABRA4 gene and KCNJ6 gene was detected, but strongest evidence for association was observed for a single DNA variation located in the gene encoding the β3-subunit of the nicotinic receptor (CHRNB3). In addition, strong evidence for association was observed for a DNA variant located in the gene encoding the α5 subunit of the nicotinic receptor (CHRNA5). Interestingly, two reports have become available, one supporting the genetic association of nicotine dependence with the α5 subunit (Berrettini et al. 2008) and one supporting the evidence for association with the β 3 subunit (Zeiger et al. 2008) of the nicotine receptor gene. Berrettini et al. (2008) studied three samples drawn from European populations totalling about 15,000 individuals. Two samples with about 7,500 individuals each, who were all assessed for the number of cigarettes smoked per day, revealed association with this measurement of severity of nicotine dependence with a common haplotype in the CHRNA3-CHRNA5 nicotinic receptor subunit gene cluster. This association was confirmed in a third set of 7,500 European individuals, suggesting an important role of these genes in predisposition to nicotine dependence. Zeiger et al. (2008) analyzed DNA variation in CHRNB3 and CHRNA6 genes with tobacco and alcohol phenotypes in 1,056 ethnically diverse individuals ascertained from clinical and community settings. Two single nucleotide polymorphisms were found to be significantly associated with three subjective response factors to initial tobacco use. They were able to replicate this finding in 1,524 families from the National Longitudinal Study of Adolescent Health (Harris et al. 2006).

Not only genes targeting the primary receptors involved in nicotine dependence have been a recent focus. Significant association has also been shown for the neurexin gene (NRXN1), the protein product of which is thought to be involved in synaptogenesis and synaptic maintenance, as well as calcium channel and NMDA receptor recruitment (Saccone et al. 2007). Nussbaum and colleagues followed up the association for NRXN1 and analyzed DNA sequence variation in this gene in 2,037 individuals from 602 nuclear families of African American or European American origin (Nussbaum et al. 2008). The study individuals were all characterized for smoking quantity, heaviness of smoking, and the Fagerstrom Test for nicotine dependence (Heatherton et al. 1991). Converging evidence from single nucleotide polymorphism analysis and haplotype analysis was obtained for all three nicotine dependence measures, with one SNP mainly driving the results. Therefore, it might be concluded the neurexin gene could play an important role in the etiology of nicotine dependence. Using the same set of families, (Huang et al. 2008) was able to show association of SNPs located in the dopamine D1 receptor gene. Interestingly, the SNP revealing the main association also showed some evidence for regulation of transcription of this gene.

4.3 Alcohol

Alcohol is one of the most widely abused drugs next to nicotine. According to WHO, the global health damage related to alcohol abuse/dependence results in even more lost lifetime than nicotine, due to fatal accidents, especially motor vehicle-related accidents, and violence occurring after excessive alcohol consumption in young people (WHO 2004). In addition, alcohol dependence reveals a major negative impact on social networks like families and communities.

As recently reviewed by (Kohnke 2008) and (Ducci and Goldman 2008), numerous association studies are available with a focus on the dopaminergic system, including the dopamine metabolizing enzymes, the GABAergic and glutamatergic system. Genes from the serotonergic system, cholinergic system, opioid-related genes, and neuropeptide Y have also attracted the attention of researchers. A number of studies have also taken subphenotypes into account. These subphenotypes include, but are not restricted to, a classification of Jellinek (1960), or more widely used, the classification system according to Cloninger (1987). The latter describes two main subforms of alcoholism with type I alcoholism revealing a later onset in life and comorbid neurotic features in contrast to type II alcoholism, which starts early in life and is accompanied by antisocial behavior, elevated novelty seeking, and reduced harm avoidance scores. In addition, heritability seems to be specifically high for type II alcoholism, making it an ideal subphenotype for genetic studies.

Another interesting point that should be mentioned is the high co-occurrence of alcohol and nicotine dependence (True et al. 1999). Analysis of genes involved in nicotine dependence might therefore be of primary interest to research into alcohol dependence as well. One recent study by (Wang et al. 2008) showed association of variants in the CHRNA5 gene with alcohol dependence in a set of 262 families, comprising 2,309 individuals. The authors of this study, however, conclude that their finding might be unique to alcohol dependence since no overlap of the involved genomic regions was observed. Research using a family-based association design in alcohol and nicotine dependence performed by our own group indicates that the synaptic vesicular amine transporter gene (SLC18A2) might contribute to the development of both, nicotine and alcohol dependence (Schwab et al. 2005).

5 Conclusion

Studies targeting genetic variation and association for substance dependence have been a major focus for the dissection of substance dependence. However, only recently, due to major advances in knowledge and technology related to the human genome, have large-scale studies become feasible. Recent emerging findings are promising, but need to be validated in order to provide translatable results for addiction therapy.

References

Ali R, Godwing L (2002) Pharmacotherapy of opioid depdenence in South-East Asia and Western Pacific Regions: key informant country reports, Australia. WHO, Geneva

APA (1994) Diagnostic and statistical manual of mental disorders, 4th edn. American Psychiatric Press, Washington, DC

Bart G, Heilig M, LaForge KS, Pollak L, Leal SM, Ott J, Kreek MJ (2004) Substantial attributable risk related to a functional mu-opioid receptor gene polymorphism in association with heroin addiction in central Sweden. Mol Psychiatry 9:547–549

Berrettini W, Yuan X, Tozzi F, Song K, Francks C, Chilcoat H, Waterworth D, Muglia P, Mooser V (2008) Alpha-5/alpha-3 nicotinic receptor subunit alleles increase risk for heavy smoking. Mol Psychiatry 13:368–373

Berrettini WH, Ferraro TN, Alexander RC, Buchberg AM, Vogel WH (1994) Quantitative trait loci mapping of three loci controlling morphine preference using inbred mouse strains. Nat Genet 7:54–58

Berrettini WH, Hoehe MR, Ferrada TN, Gottheil E (1997) Human mu-opioid receptor gene polymorphism and vulnerability to substance dependence. Addict Biol 2:303–308

Bierut LJ, Dinwiddie SH, Begleiter H, Crowe RR, Hesselbrock V, Nurnberger JI, Jr., Porjesz B, Schuckit MA, Reich T (1998) Familial transmission of substance dependence: alcohol, marijuana, cocaine, and habitual smoking: a report from the collaborative study on the genetics of alcoholism. Arch Gen Psychiatry 55:982–988

Bohman M, Cloninger CR, Sigvardsson S, von Knorring AL (1982) Predisposition to petty criminality in Swedish adoptees. I. Genetic and environmental heterogeneity. Arch Gen Psychiatry 39:1233–1241

Bond C, LaForge KS, Tian M, Melia D, Zhang S, Borg L, Gong J, Schluger J, Strong JA, Leal SM, Tischfield JA, Kreek MJ, Yu L (1998) Single-nucleotide polymorphism in the human mu opioid receptor gene alters beta-endorphin binding and activity: possible implications for opiate addiction. Proc Natl Acad Sci USA 95:9608–9613

Button TM, Rhee SH, Hewitt JK, Young SE, Corley RP, Stallings MC (2007) The role of conduct disorder in explaining the comorbidity between alcohol and illicit drug dependence in adolescence. Drug Alcohol Depend 87:46–53

Cadoret RJ, Yates WR, Troughton E, Woodworth G, Stewart MA (1995) Adoption study demonstrating two genetic pathways to drug abuse. Arch Gen Psychiatry 52:42–52

Cadoret RJ, Yates WR, Troughton E, Woodworth G, Stewart MA (1996) An adoption study of drug abuse/dependency in females. Compr Psychiatry 37:88–94

Cheng CY, Hong CJ, Yu YW, Chen TJ, Wu HC, Tsai SJ (2005) Brain-derived neurotrophic factor (Val66Met) genetic polymorphism is associated with substance abuse in males. Brain Res Mol Brain Res 140:86–90

Cheung KH, Osier MV, Kidd JR, Pakstis AJ, Miller PL, Kidd KK (2000) ALFRED: an allele frequency database for diverse populations and DNA polymorphisms. Nucleic Acids Res 28:361–363

Cloninger CR (1987) Neurogenetic adaptive mechanisms in alcoholism. Science 236:410–416

Cloninger CR, Bohman M, Sigvardsson S (1981) Inheritance of alcohol abuse. Cross-fostering analysis of adopted men. Arch Gen Psychiatry 38:861–868

Comings DE, Blake H, Dietz G, Gade-Andavolu R, Legro RS, Saucier G, Johnson P, Verde R, MacMurray JP (1999) The proenkephalin gene (PENK) and opioid dependence. Neuroreport 10:1133–1135

Crowley JJ, Oslin DW, Patkar AA, Gottheil E, DeMaria PA, Jr., O'Brien CP, Berrettini WH, Grice DE (2003) A genetic association study of the mu opioid receptor and severe opioid dependence. Psychiatr Genet 13:169–173

Doyle GA, Sheng XR, Schwebel CL, Ferraro TN, Berrettini WH, Buono RJ (2006) Identification and functional significance of polymorphisms in the mu-opioid receptor gene (Oprm) promoter of C57BL/6 and DBA/2 mice. Neurosci Res 55:244–254

Doyle GA, Rebecca Sheng X, Lin SS, Press DM, Grice DE, Buono RJ, Ferraro TN, Berrettini WH (2007) Identification of three mouse mu-opioid receptor (MOR) gene (Oprm1) splice variants containing a newly identified alternatively spliced exon. Gene 388:135–147

Ducci F, Goldman D (2008) Genetic approaches to addiction: genes and alcohol. Addiction 103: 1414–1428

Edenberg HJ (2007) The genetics of alcohol metabolism: role of alcohol dehydrogenase and aldehyde dehydrogenase variants. Alcohol Res Health 30:5–13

Everitt BJ, Cardinal RN, Parkinson JA, Robbins TW (2003) Appetitive behavior: impact of amygdala-dependent mechanisms of emotional learning. Ann N Y Acad Sci 985:233–250

Ferraro TN, Golden GT, Smith GG, Martin JF, Schwebel CL, Doyle GA, Buono RJ, Berrettini WH (2005) Confirmation of a major QTL influencing oral morphine intake in C57 and DBA mice using reciprocal congenic strains. Neuropsychopharmacology 30:742–746

Franke P, Nothen MM, Wang T, Neidt H, Knapp M, Lichtermann D, Weiffenbach O, Mayer P, Hollt V, Propping P, Maier W (1999) Human delta-opioid receptor gene and susceptibility to heroin and alcohol dependence. Am J Med Genet 88:462–464

Franke P, Nothen MM, Wang T, Knapp M, Lichtermann D, Neidt H, Sander T, Propping P, Maier W (2000) DRD4 exon III VNTR polymorphism-susceptibility factor for heroin dependence? Results of a case-control and a family-based association approach. Mol Psychiatry 5:101–104

Franke P, Wang T, Nothen MM, Knapp M, Neidt H, Albrecht S, Jahnes E, Propping P, Maier W (2001) Nonreplication of association between mu-opioid-receptor gene (OPRM1) A118G polymorphism and substance dependence. Am J Med Genet 105:114–119

Gelernter J, Panhuysen C, Wilcox M, Hesselbrock V, Rounsaville B, Poling J, Weiss R, Sonne S, Zhao H, Farrer L, Kranzler HR (2006) Genomewide linkage scan for opioid dependence and related traits. Am J Hum Genet 78:759–769

Gerra G, Garofano L, Pellegrini C, Bosari S, Zaimovic A, Moi G, Avanzini P, Talarico E, Gardini F, Donnini C (2005) Allelic association of a dopamine transporter gene polymorphism with antisocial behaviour in heroin-dependent patients. Addict Biol 10:275–281

Glatt SJ, Su JA, Zhu SC, Zhang R, Zhang B, Li J, Yuan X, Li J, Lyons MJ, Faraone SV, Tsuang MT (2006) Genome-wide linkage analysis of heroin dependence in Han Chinese: results from wave one of a multi-stage study. Am J Med Genet B Neuropsychiatr Genet 141B:648–652

Harris GC, Aston-Jones G (1994) Involvement of D2 dopamine receptors in the nucleus accumbens in the opiate withdrawal syndrome. Nature 371:155–157

Harris GC, Aston-Jones G (2003) Altered motivation and learning following opiate withdrawal: evidence for prolonged dysregulation of reward processing. Neuropsychopharmacology 28:865–871

Harris KM, Halpern CT, Smolen A, Haberstick BC (2006) The National Longitudinal Study of Adolescent Health (Add Health) twin data. Twin Res Hum Genet 9:988–997

Heatherton TF, Kozlowski LT, Frecker RC, Fagerstrom KO (1991) The Fagerstrom Test for Nicotine Dependence: a revision of the Fagerstrom Tolerance Questionnaire. Br J Addict 86:1119–1127

Heinrichs SC, Koob GF (2004) Corticotropin-releasing factor in brain: a role in activation, arousal, and affect regulation. J Pharmacol Exp Ther 311:427–440

Hiroi N, Agatsuma S (2005) Genetic susceptibility to substance dependence. Mol Psychiatry 10:336–344

Ho MK, Tyndale RF (2007) Overview of the pharmacogenomics of cigarette smoking. Pharmacogenomics J 7:81–98

Hoehe MR, Kopke K, Wendel B, Rohde K, Flachmeier C, Kidd KK, Berrettini WH, Church GM (2000) Sequence variability and candidate gene analysis in complex disease: association of mu opioid receptor gene variation with substance dependence. Hum Mol Genet 9:2895–2908

Hope BT, Nye HE, Kelz MB, Self DW, Iadarola MJ, Nakabeppu Y, Duman RS, Nestler EJ (1994) Induction of a long-lasting AP-1 complex composed of altered Fos-like proteins in brain by chronic cocaine and other chronic treatments. Neuron 13:1235–1244

Horowitz R, Kotler M, Shufman E, Aharoni S, Kremer I, Cohen H, Ebstein RP (2000) Confirmation of an excess of the high enzyme activity COMT val allele in heroin addicts in a family-based haplotype relative risk study. Am J Med Genet 96:599–603

Huang W, Ma JZ, Payne TJ, Beuten J, Dupont RT, Li MD (2008) Significant association of DRD1 with nicotine dependence. Hum Genet 123:133–140

Hyman SE, Malenka RC, Nestler EJ (2006) Neural mechanisms of addiction: the role of reward-related learning and memory. Annu Rev Neurosci 29:565–598

Jellinek EM (1960) Alcoholism, a genus and some of its species. Can Med Assoc J 83:1341–1345

Kendler KS, Jacobson KC, Prescott CA, Neale MC (2003) Specificity of genetic and environmental risk factors for use and abuse/dependence of cannabis, cocaine, hallucinogens, sedatives, stimulants, and opiates in male twins. Am J Psychiatry 160:687–695

Kendler KS, Myers J, Prescott CA (2007) Specificity of genetic and environmental risk factors for symptoms of cannabis, cocaine, alcohol, caffeine, and nicotine dependence. Arch Gen Psychiatry 64:1313–1320

Kohnke MD (2008) Approach to the genetics of alcoholism: a review based on pathophysiology. Biochem Pharmacol 75:160–177

Kotler M, Cohen H, Segman R, Gritsenko I, Nemanov L, Lerer B, Kramer I, Zer-Zion M, Kletz I, Ebstein RP (1997) Excess dopamine D4 receptor (D4DR) exon III seven repeat allele in opioid-dependent subjects. Mol Psychiatry 2:251–254

Kotler M, Cohen H, Kremer I, Mel H, Horowitz R, Ohel N, Gritsenko I, Nemanov L, Katz M, Ebstein R (1999) No association between the serotonin transporter promoter region (5-HTTLPR) and the dopamine D3 receptor (Ball D3DR) polymorphisms and heroin addiction. Mol Psychiatry 4:313–314

Kreek MJ, Bart G, Lilly C, LaForge KS, Nielsen DA (2005) Pharmacogenetics and human molecular genetics of opiate and cocaine addictions and their treatments. Pharmacol Rev 57:1–26

Lessov-Schlaggar CN, Pergadia ML, Khroyan TV, Swan GE (2008) Genetics of nicotine dependence and pharmacotherapy. Biochem Pharmacol 75:178–195

Li T, Liu X, Zhu ZH, Hu X, Sham PC, Collier DA (2000a) Association analysis of polymorphisms in the mu-opioid gene and heroin abuse in Chinese subjects. Addict Biol 5:181–186

Li T, Liu X, Zhu ZH, Zhao J, Hu X, Ball DM, Sham PC, Collier DA (2000b) No association between (AAT)n repeats in the cannabinoid receptor gene (CNR1) and heroin abuse in a Chinese population. Mol Psychiatry 5:128–130

Li T, Zhu ZH, Liu X, Hu X, Zhao J, Sham PC, Collier DA (2000c) Association analysis of polymorphisms in the DRD4 gene and heroin abuse in Chinese subjects. Am J Med Genet 96:616–621

Li T, Liu X, Zhao J, Hu X, Ball DM, Loh el W, Sham PC, Collier DA (2002) Allelic association analysis of the dopamine D2, D3, 5-HT2A, and GABA(A)gamma2 receptors and serotonin transporter genes with heroin abuse in Chinese subjects. Am J Med Genet 114:329–335

Li Y, Shao C, Zhang D, Zhao M, Lin L, Yan P, Xie Y, Jiang K, Jin L (2006) The effect of dopamine D2, D5 receptor and transporter (SLC6A3) polymorphisms on the cue-elicited heroin craving in Chinese. Am J Med Genet B Neuropsychiatr Genet 141B:269–273

Luo X, Kranzler HR, Zhao H, Gelernter J (2003) Haplotypes at the OPRM1 locus are associated with susceptibility to substance dependence in European-Americans. Am J Med Genet B Neuropsychiatr Genet 120B:97–108

Marinelli M, Piazza PV (2002) Interaction between glucocorticoid hormones, stress and psychostimulant drugs. Eur J Neurosci 16:387–394

Mayer P, Rochlitz H, Rauch E, Rommelspacher H, Hasse HE, Schmidt S, Hollt V (1997) Association between a delta opioid receptor gene polymorphism and heroin dependence in man. Neuroreport 8:2547–2550

Merikangas KR, Stolar M, Stevens DE, Goulet J, Preisig MA, Fenton B, Zhang H, O'Malley SS, Rounsaville BJ (1998) Familial transmission of substance use disorders. Arch Gen Psychiatry 55:973–979

Mills KL, Teesson M, Darke S, Ross J, Lynskey M (2004) Young people with heroin dependence: findings from the Australian Treatment Outcome Study (ATOS). J Subst Abuse Treat 27:67–73

Montague PR, Dayan P, Sejnowski TJ (1996) A framework for mesencephalic dopamine systems based on predictive Hebbian learning. J Neurosci 16:1936–1947

Montague PR, Hyman SE, Cohen JD (2004) Computational roles for dopamine in behavioural control. Nature 431:760–767

Nestler EJ (2001) Molecular basis of long-term plasticity underlying addiction. Nat Rev Neurosci 2:119–128

Nestler EJ (2005) Is there a common molecular pathway for addiction? Nat Neurosci 8:1445–1449

Nussbaum J, Xu Q, Payne TJ, Ma JZ, Huang W, Gelernter J, Li MD (2008) Significant association of the neurexin-1 gene (NRXN1) with nicotine dependence in European- and African-American smokers. Hum Mol Genet 17:1569–1577

Osier MV, Cheung KH, Kidd JR, Pakstis AJ, Miller PL, Kidd KK (2002) ALFRED: an allele frequency database for anthropology. Am J Phys Anthropol 119:77–83

Molecular Biology of Addiction and Substance Dependence 203

Proudnikov D, LaForge KS, Hofflich H, Levenstien M, Gordon D, Barral S, Ott J, Kreek MJ (2006) Association analysis of polymorphisms in serotonin 1B receptor (HTR1B) gene with heroin addiction: a comparison of molecular and statistically estimated haplotypes. Pharmacogenet Genomics 16:25–36

Ramsay DS, Woods SC (1997) Biological consequences of drug administration: implications for acute and chronic tolerance. Psychol Rev 104:170–193

Robinson S, Sandstrom SM, Denenberg VH, Palmiter RD (2005) Distinguishing whether dopamine regulates liking, wanting, and/or learning about rewards. Behav Neurosci 119:5–15

Robinson TE, Berridge KC (2003) Addiction. Annu Rev Psychol 54:25–53

Saccone SF, Hinrichs AL, Saccone NL, Chase GA, Konvicka K, Madden PA, Breslau N, Johnson EO, Hatsukami D, Pomerleau O, Swan GE, Goate AM, Rutter J, Bertelsen S, Fox L, Fugman D, Martin NG, Montgomery GW, Wang JC, Ballinger DG, Rice JP, Bierut LJ (2007) Cholinergic nicotinic receptor genes implicated in a nicotine dependence association study targeting 348 candidate genes with 3713 SNPs. Hum Mol Genet 16:36–49

Schwab SG, Franke PE, Hoefgen B, Guttenthaler V, Lichtermann D, Trixler M, Knapp M, Maier W, Wildenauer DB (2005) Association of DNA polymorphisms in the synaptic vesicular amine transporter gene (SLC18A2) with alcohol and nicotine dependence. Neuropsychopharmacology 30:2263–2268

Shi J, Hui L, Xu Y, Wang F, Huang W, Hu G (2002) Sequence variations in the mu-opioid receptor gene (OPRM1) associated with human addiction to heroin. Hum Mutat 19:459–460

Sillaber I, Rammes G, Zimmermann S, Mahal B, Zieglgansberger W, Wurst W, Holsboer F, Spanagel R (2002) Enhanced and delayed stress-induced alcohol drinking in mice lacking functional CRH1 receptors. Science 296:931–933

Sullivan PF, Kendler KS (1999) The genetic epidemiology of smoking. Nicotine Tob Res 1 Suppl 2:S51–57; discussion S69–S70

Sullivan PF, Kuo PH, Webb BT, Neale MC, Vittum J, Furberg H, Walsh D, Patterson DG, Riley B, Prescott CA, Kendler KS (2008) Genomewide linkage survey of nicotine dependence phenotypes. Drug Alcohol Depend 93:210–216

Szeto CY, Tang NL, Lee DT, Stadlin A (2001) Association between mu opioid receptor gene polymorphisms and Chinese heroin addicts. Neuroreport 12:1103–1106

Szilagyi A, Boor K, Szekely A, Gaszner P, Kalasz H, Sasvari-Szekely M, Barta C (2005) Combined effect of promoter polymorphisms in the dopamine D4 receptor and the serotonin transporter genes in heroin dependence. Neuropsychopharmacol Hung 7:28–33

Tan EC, Yeo BK, Ho BK, Tay AH, Tan CH (1999) Evidence for an association between heroin dependence and a VNTR polymorphism at the serotonin transporter locus. Mol Psychiatry 4:215–217

Tan EC, Chong SA, Mahendran R, Tan CH, Teo YY (2003) Mu opioid receptor gene polymorphism and neuroleptic-induced tardive dyskinesia in patients with schizophrenia. Schizophr Res 65:61–63

True WR, Xian H, Scherrer JF, Madden PA, Bucholz KK, Heath AC, Eisen SA, Lyons MJ, Goldberg J, Tsuang M (1999) Common genetic vulnerability for nicotine and alcohol dependence in men. Arch Gen Psychiatry 56:655–661

Tsuang MT, Lyons MJ, Eisen SA, Goldberg J, True W, Lin N, Meyer JM, Toomey R, Faraone SV, Eaves L (1996) Genetic influences on DSM-III-R drug abuse and dependence: a study of 3,372 twin pairs. Am J Med Genet 67:473–477

Tsuang MT, Lyons MJ, Meyer JM, Doyle T, Eisen SA, Goldberg J, True W, Lin N, Toomey R, Eaves L (1998) Co-occurrence of abuse of different drugs in men: the role of drug-specific and shared vulnerabilities. Arch Gen Psychiatry 55:967–972

Ungless MA, Whistler JL, Malenka RC, Bonci A (2001) Single cocaine exposure in vivo induces long-term potentiation in dopamine neurons. Nature 411:583–587

Volkow ND, Fowler JS, Wang GJ, Swanson JM (2004) Dopamine in drug abuse and addiction: results from imaging studies and treatment implications. Mol Psychiatry 9:557–569

Von Knorring L, von Knorring AL, Smigan L, Lindberg U, Edholm M (1987) Personality traits in subtypes of alcoholics. J Stud Alcohol 48:523–527

Wagner FA, Anthony JC (2002) From first drug use to drug dependence; developmental periods of risk for dependence upon marijuana, cocaine, and alcohol. Neuropsychopharmacology 26:479–488

Wand GS, McCaul M, Yang X, Reynolds J, Gotjen D, Lee S, Ali A (2002) The mu-opioid receptor gene polymorphism (A118G) alters HPA axis activation induced by opioid receptor blockade. Neuropsychopharmacology 26:106–114

Wang JC, Grucza R, Cruchaga C, Hinrichs AL, Bertelsen S, Budde JP, Fox L, Goldstein E, Reyes O, Saccone N, Saccone S, Xuei X, Bucholz K, Kuperman S, Nurnberger J, Jr., Rice JP, Schuckit M, Tischfield J, Hesselbrock V, Porjesz B, Edenberg HJ, Bierut LJ, Goate AM (2008) Genetic variation in the CHRNA5 gene affects mRNA levels and is associated with risk for alcohol dependence. Mol Psychiatry, Apr 15 [Epub ahead of print]

Warner LA, Kessler RC, Hughes M, Anthony JC, Nelson CB (1995) Prevalence and correlates of drug use and dependence in the United States. Results from the National Comorbidity Survey. Arch Gen Psychiatry 52:219–229

WHO (2004) Global Status Report on Alcohol. http://wwwwhoint/substance_abuse/publications/global_status_report_2004_overviewpdf

WHO (2008) http://www.who.int/substance_abuse/facts/en/.

Wise RA (1998) Drug-activation of brain reward pathways. Drug Alcohol Depend 51:13–22

Wise RA (2004) Dopamine, learning and motivation. Nat Rev Neurosci 5:483–494

Wolf ME (1998) The role of excitatory amino acids in behavioral sensitization to psychomotor stimulants. Prog Neurobiol 54:679–720

Xu K, Liu XH, Nagarajan S, Gu XY, Goldman D (2002) Relationship of the delta-opioid receptor gene to heroin abuse in a large Chinese case/control sample. Am J Med Genet 110:45–50

Xu K, Lichtermann D, Lipsky RH, Franke P, Liu X, Hu Y, Cao L, Schwab SG, Wildenauer DB, Bau CH, Ferro E, Astor W, Finch T, Terry J, Taubman J, Maier W, Goldman D (2004) Association of specific haplotypes of D2 dopamine receptor gene with vulnerability to heroin dependence in 2 distinct populations. Arch Gen Psychiatry 61:597–606

Yuferov V, Fussell D, LaForge KS, Nielsen DA, Gordon D, Ho A, Leal SM, Ott J, Kreek MJ (2004) Redefinition of the human kappa opioid receptor gene (OPRK1) structure and association of haplotypes with opiate addiction. Pharmacogenetics 14:793–804

Zeiger JS, Haberstick BC, Schlaepfer I, Collins AC, Corley RP, Crowley TJ, Hewitt JK, Hopfer CJ, Lessem J, McQueen MB, Rhee SH, Ehringer MA (2008) The neuronal nicotinic receptor subunit genes (CHRNA6 and CHRNB3) are associated with subjective responses to tobacco. Hum Mol Genet 17:724–734

Zhang H, Luo X, Kranzler HR, Lappalainen J, Yang BZ, Krupitsky E, Zvartau E, Gelernter J (2006) Association between two mu-opioid receptor gene (OPRM1) haplotype blocks and drug or alcohol dependence. Hum Mol Genet 15:807–819

Zimprich A, Kraus J, Woltje M, Mayer P, Rauch E, Hollt V (2000) An allelic variation in the human prodynorphin gene promoter alters stimulus-induced expression. J Neurochem 74:472–477

Neurobiology of Suicide

Brigitta Bondy and Peter Zill

Contents

1 Introduction.. 206
2 The Clinical Phenotype... 206
   2.1 The Link Between Mental Morbidity and Suicide 207
3 Pathophysiological Mechanisms... 208
   3.1 The Neurochemical Basis of Suicidality.. 208
   3.2 Genetic Contribution to Suicidal Behavior.. 213
   3.3 Suicide as Interplay of Genes and Environment.................................... 219
4 Conclusions... 220
References.. 221

Abstract The concept that neurobiological systems and mechanisms are contributing to the complex trait of suicidal behaviour has stimulated much work by several direct and indirect methods, as post-mortem analyses, studies in body fluids and blood platelets, or imaging techniques.

These approaches yielded interesting data concerning the involvement of the serotonergic, as well as noradrenergic and other neurotransmitter systems. Newer aspects comprise alterations in signal transduction mechanisms, the hypothalamic-pituitary-adrenal axis (HPA) and an excess activity of the noradrenergic system. The latter are both involved in the response to stressful life events and in the pathophysiology of depression and might thus have impact on suicide risk. However, multiple systems play a role in regulating the risk of suicide, and under this perspective new candidates have to be discovered.

On the basis of the neurobiological findings genetic studies have been carried out since more than a decade, which are now converging on several key areas. These studies focused mainly on the serotonergic pathway as the intent to die and the lethality of suicide acts were related to the serotonergic system. Two genes, the tryptophan hydroxylase 1 (TPH1) and the serotonin transporter gene

---

B. Bondy (✉)
Department of Psychiatry, Section of Psychiatric Genetics and Neurochemistry,
Ludwig-Maximilians-University Munich, Nussbaumstrasse 7, Munich, Germany
Brigitta.Bondy@med.uni-muenchen.de

D.B. Wildenauer (ed.), *Molecular Biology of Neuropsychiatric Disorders*,
Nucleic Acids and Molecular Biology, © Springer-Verlag Berlin Heidelberg 2009

(5-HTTLPR) have so far emerged as being involved in the vulnerability for suicidality. Predominantly negative were the findings with any type of candidate genes of the underlying neurobiological pathways.

This chapter reviews the status of current knowledge in the area of neurobiology and genetics of suicidality, points to the weakness of the investigations and presents new approaches beyond the serotonergic system.

1 Introduction

Suicide is a significant public health issue and a major cause of death throughout the world. Although the epidemiological data vary from country to country with the highest annual rates in a group of Eastern European countries which share similar historical and sociocultural characteristics, such as Estonia, Latvia, Lithuania, Finland and Hungary, and, to a lesser extent, the Russian federation, the WHO estimates that suicide accounts for almost 2% of the world's deaths (World Health Organization 2000). There is a relatively constant predominance of completed suicide rates in males over those in females with the exception of China, where suicide rates are higher in females (Phillips et al. 2002; Bertolote and Fleischmann 2005). Attempted suicide is more frequent than completed suicide with a life time prevalence of about 3.5%, and it is usually estimated that up to 10% of suicide attempters will commit suicide within 10 years (Suominen et al. 2004). Although for both men and women there is a clear tendency towards an increase in suicide rates with age, numerically more suicides are committed by younger people, and recent evidence suggests that suicide rates of young people are increasing in many geographic areas (Fleischmann et al. 2005). Several explanations have been considered for national and regional variations, including climate, religion, social, and political systems, but a more likely scenario is that the genetic contributions to suicide will be represented by small size effects of many gene variants associated with processes involved in suicidal behavior and by interaction of these genetic factors with environmental ones (Bertolote and Fleischmann 2005; Fleischmann et al. 2005).

2 The Clinical Phenotype

Suicidal behavior includes a wide spectrum and refers to the occurrence of suicide attempts which range from fatal acts (completed suicide) through highly lethal, but failed suicide attempts (where high intention and planning are evident and survival is fortuitous) to low lethality, usually impulsive attempts triggered by a social crisis which contain a strong element of an appeal for help. Suicidal ideation, which comprises suicidal thoughts or threats but without action, is more common than suicide attempts and completed suicide, and its prevalence varies widely, being almost twice in females compared to males. It was further discovered that suicidal ideations are almost always associated with a psychiatric disorder (Mann 2003).

Although the exact clinical definition of suicidal behavior remains unsatisfactory and is a source of confusion, the phenomenon of suicidality is often viewed as occurring on a continuum of increasing severity from ideation through attempts to completed suicide (Courtet et al. 2005), and may be classified according to the intent to die, method and lethality (violent or non-violent), cognitive impairments (impulsivity, aggressiveness), or mitigating circumstances (Leboyer et al. 2005). The method of suicide is not randomly distributed. For example, violent methods, assessed with a higher level of lifetime aggression and a higher level of impulsivity, are more often applied by males than by females and often associated with lifetime substance abuse or dependence (Dumais et al. 2005).

2.1 The Link Between Mental Morbidity and Suicide

Several arguments suggest that suicidal behavior is a disorder on its own, although the presence of a psychiatric disorder and particularly major depression is a well-established risk factor for suicidality across all groups. Approximately 90% of all completed suicides have a diagnosable psychiatric disorder according to the Diagnostic and Statistical Manual of Mental Disorder (Mann 2003). The majority of all suicides occur in relation to mood disorders, but also other psychiatric disturbances such as schizophrenia, personality disorders, alcoholism, and drug abuse are similarly related to suicide (Fleischmann et al. 2005) (for review, see Mann 2003). Other clinical features that increase the liability for suicidal behavior include hopelessness, a history of physical or sexual abuse during childhood, a history of head injury or neurological disorder, and cigarette smoking (Mann 2003; Baud 2005). However, although the presence of a psychopathology is a strong predictor for suicide, only a minority of people with these diagnoses commit suicide. Thus, a psychiatric diagnosis might be a necessary, but insufficient, risk factor for suicide, indicating a predisposition that is independent of the main psychiatric disorders (Turecki 2005; Mann 2003).

Personality disorders are now widely recognized as increasing the risk for suicide, independently from the presence of any axis I disorder. Especially borderline and antisocial personality disorders are the ones most frequently associated with attempted and completed suicide. Comorbidity with major depression or alcohol dependency is further increasing the risk (Baud 2005). In alcohol-dependent patients suicidal behavior is a common and important problem, as up to 40% of alcohol-dependent patients attempt suicide at some time, and up to 7% end up their lives by committing suicide (Harris and Barraclough 1998). Although the intricate relationships and the causal links between alcohol abuse or dependence are not clearly understood, some personality traits are probably key common predisposing factors. Impulsive–aggressive traits, neuroticism and anxiety as well as anger-related traits were proposed as intermediary phenotypes and risk factors for suicidal behavior (Baud 2005). These traits seem to be independent from the role of associated axis I disorders, particularly major depression, and may be part of a developmental cascade that increases suicide risk among a subset of patients (Turecki 2005).

Further, as personality traits themselves are partly under genetic control, it was suggested that they may contribute to the familial loading of suicide attempts and completions (Baud 2005; Turecki 2005). On the other hand, there is no doubt that impulsive–aggressive personality disorders and alcohol abuse/dependence are two independent predictors of suicide in major depression (Dumais et al. 2005).

3 Pathophysiological Mechanisms

Suicidal behavior is complex in terms of symptomatology and multifactorial in aetiopathogenesis. Although the specific causes have not been elucidated in detail, there is overwhelming evidence that suicidality is caused by a multiple interaction of genetic risk factors, environmental and neurobiological factors (Fig. 1) (Wasserman et al. 2007). Cross-influences can be seen between all these pathways, and genes contribute to all of them (Kendler et al. 2006).

3.1 The Neurochemical Basis of Suicidality

Neurobiological systems and mechanisms, which are potentially related to suicidal behavior, have been investigated by several direct and indirect methods, as postmortem analyses, studies in body fluids and blood platelets, or imaging techniques. These approaches yielded interesting data concerning the involvement of the serotonergic and other neurotransmitter systems. Newer aspects comprise alterations in

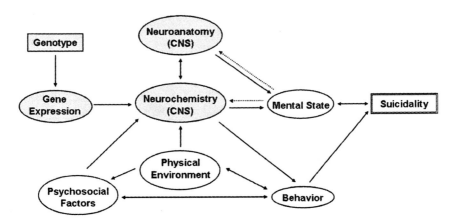

Fig. 1 The complex network of gene–environment interactions in a schematic model of suicidality. Relatively stable and constant parameters are *boxed*; dynamic and acute parameters are *encircled*. *Solid arrows* show well-established interactions, whereas *dotted arrows* show hypothesized interactions. (Adapted from Wasserman et al. 2007)

the hypothalamic–pituitary–adrenal axis (HPA) and an excess activity of the noradrenergic system, which are both involved in the response to stressful life events and in the pathophysiology of depression, and might thus have impact on suicide risk (Mann 2003). However, multiple systems play a role in regulating the risk of suicide, and under this perspective new candidates have to be discovered.

3.1.1 The Serotonergic System

Serotonergic alterations which are involved in a variety of psychopathological dimensions such as anxiety, depressed mood, impulsivity and aggression, were especially the focus of many investigations. The initial, seminal finding from Asberg et al. (1976), that a low cerebrospinal fluid (CSF) 5-hydroxyindolacetic acid (5-HIAA) concentration could be related to the incidence of violent suicidal acts, was repeatedly replicated in more than 20, mostly retrospective, studies in patients suffering from several mood disorders, schizophrenia, and personality disorders (Asberg 1997). Additionally, prospective studies provide estimates of the predictive utilities of these biological measures (for review, see Mann and Currier 2007). All these studies indicated a negative correlation between CSF 5-HIAA and lethality of the suicide attempt and the switch from attempt to completed suicide (Nordstrom et al. 1994).

Today, there is no doubt that suicidal acts show a clear relation to aggressive and impulsive behavior, traits that are also associated with alterations of the serotonergic neurotransmission, being demonstrated by the low brain serotonin (5-HT) turnover rate in impulsive violent offenders, especially when intoxicated (Virkkunen et al. 1995). Even determinations in blood samples as peripheral model of serotonergic activity revealed lower 5-HIAA plasma levels in impulsive suicide attempters but not in nonimpulsive attempters and controls, thus supporting the hypothesis (Spreux-Varoquaux et al. 2001). Although the relationship between aggression/impulsivity, suicidal behavior and the serotonergic system are complex, these findings suggest that impulsivity may serve as a link between low 5-HIAA concentrations and suicidal behavior.

Postmortem studies using brains of suicide victims revealed evidence for reduced serotonin transporter (5-HTT) sites in the prefrontal cortex, hypothalamus, occipital cortex, and brainstem (Mann et al. 1996). Some, but not all studies, have observed alterations on the receptor level, as postsynaptic 5-HT1A and 5-HT2A receptors were found to be upregulated in prefrontal cortex and this increase was suggested as being a compensatory mechanism to the low activity of the serotonergic neurones (Mann 2003). It is interesting to note that the 5-HT1A upregulation seems to be localized to the ventral prefrontal cortex, a region that is involved in behavioral and cognitive inhibition (Arango et al. 1995), and low serotonergic input might contribute to impaired inhibition, creating a greater propensity to act on suicidal or aggressive feelings. The quantitative alterations of receptor density seem to be associated with suicidal behavior, independently from psychiatric diagnoses (Mann 2003).

Using in vivo functional imaging with SPECT (single photon emission computerized tomography), a reduction in 5-HT2A binding index was demonstrated in anxious and depressed suicide attempters but an increase in 5-HT2A binding in impulsive suicide attempters (Audenaert et al. 2006). This finding is consistent with the postmortem 5-HT2A binding studies in humans and the impulsive animal research.

The findings from postmortem studies are underlined by challenge investigation with the 5-HT2B agonist fenfluramine. With this neuroendocrine probe, the degree of serotonergic stimulation can be measured via the determination of prolactin. Results have generally shown a decreased prolactin response in suicide attempters versus healthy controls or depressed patients (Mann and Malone 1997). As already described for the 5-HIAA studies, lethality and impulsivity has been considered as modulators of this relationship. High lethality attempters show decreased prolactin response as compared to low lethality attempters (Malone et al. 1996). A blunted prolactin response has also been seen in impulsive and aggressive individuals as compared to controls (Sher et al. 2003). Although some of the reports are not consistent and/or discussed as being more related to major depression than to suicide (Meyer et al. 2003), these findings support the hypothesis of a decreased serotonergic activity in suicidal behavior.

With respect to the many well-known pitfalls of postmortem investigations, peripheral markers for serotonergic activity were focus of investigations for decades, and blood platelets were considered as ideal peripheral model. Reduced 5-HT uptake, fewer serotonin transporter binding sites, and an elevated density of 5-HT2A receptors have been observed in suicidal patients, including a positive correlation between the 5-HT2A receptor upregulation and lethality of the suicidal act (Pandey et al. 1995). Although these data seem to support the postmortem findings, it is currently unclear whether platelets reflect the neuronal serotonergic mechanisms in detail, as a critical review claimed that methodological flaws limited the validity of platelets as a biological model (Muller-Oerlinghausen et al. 2004).

3.1.2 The Dopaminergic System

The hypothesis of a possible role of dopaminergic function in the control of suicidal behavior was less extensively studied. Few postmortem studies covered alterations of the dopaminergic system. A recent in vivo investigation of homovanillinic acid (HVA), the main dopamine metabolite, in CSF of depressed suicide attempters demonstrated reduced HVA levels in attempters, but not in depressed nonattempters (Sher et al. 2006), thus suggesting a relation of dopamine to suicide but not to depression. However, these findings are still not convincingly confirmed concerning either suicidal behavior or aggression or impulsivity (for review, see Mann and Currier 2007). Interestingly, some studies have found that the ratio between the metabolites shows stronger associations with suicidal behavior, although the underlying mechanisms of this observation are unclear. Lower HVA/5-HIAA and HVA/3-methoxy-4-hydroxyphenylglycol (MHPG), a major metabolite of noradrenalin, have been reported in suicide attempters (Engstrom et al. 1999), as well as a significant correlation between CSF HVA/5-HIAA ratio and aggression and violence, suggesting a dysfunction in both investigated systems (Soderstrom et al.

Neurobiology of Suicide   211

2001). Additionally, no alterations were found for mRNA levels of the D1 and D2 receptors in the caudate nuclei of suicide victim (Hurd et al. 1997) and similarly not for D4 receptor binding (Sumiyoshi et al. 1995). The results with tyrosine hydroxylase (TH), the rate-limiting enzyme for noradrenalin (NA) and dopamine (DA) synthesis were also divergent, as both increased (Ordway et al. 1994a) and decreased immunoreactivity were observed (Biegon and Fieldust 1992).

In conclusion, although the dopaminergic system is involved in the pathophysiology of depression (Kapur and Mann 1992), its implication in suicidal behavior is still not established.

3.1.3 The Noradrenergic System

Only a limited number of postmortem studies of the noradrenergic system in suicidal behavior exist. The main findings were decreased noradrenalin (NA) levels in brainstem due to decreased noradrenergic neurons in the locus coeruleus, increased alpha2-adrenergic receptor densities, suggested as being upregulated due to the NA deficit and a cortical noradrenergic overactivity (lower alpha and beta1-adrenergic receptor binding) (Arango et al. 1993, 1996; Ordway et al. 1994b). However, increased TH and alpha2-adrenergic receptor densities could be indicative for noradrenergic depletion compensatory to increased NA release. This hypothesis is important with regard to the relation between the noradrenergic system and stress response, as severe anxiety or agitation are associated with noradrenergic over activity, higher suicide risk and over activity of the hypothalamic–pituitary–adrenal (HPA) axis (Mann 2003). Studies about the predictive value of 3-methoxy-4-hydroxyphenylglycol (MHPG), a major metabolite of noradrenalin, as biomarker for suicidal behavior are still without conclusive results (Mann and Currier 2007).

3.1.4 The Hypothalamic–Pituitary–Adrenal Axis and the Response to Stress

There is growing evidence that other brain systems besides the classical neurotransmitter ones, such as the hypothalamic–pituitary–adrenal (HPA) axis and the response to stress are involved in suicidal behavior. According to the stress-vulnerability model early traumatic life experiences, chronic illness, chronic alcohol, and substance abuse together with external, environmental factors are part of the vulnerability for psychiatric disturbances, including suicidal behavior (Wasserman et al. 2007; Sher 2007). Exposure to stress activates the HPA axis in order to sustain the internal environment via increase in cortisol secretion and to adapt to stress situation. However, exposure to continuous or repeated stress leads to a dysregulation of the HPA axis with enhanced release of corticotrophin-releasing hormone (CRH), adrenocorticotropin (ACTH) and cortisol, to negative feedback mechanisms together with altered glucocorticoid and CRH receptor function (Smith and Vale 2006). The failure to suppress cortisol in the combined dexamethasone/CRH test (Dex/CRH) is among the most robust neurobiological findings in depression, a disorder highly related to stressful life events (Holsboer 2001). Simultaneously to the HPA axis, stress also activates the noradrenergic activity in the locus coeruleus neurons and the autonomous nervous system with lots of reciprocal interactions between them at different levels (Mastorakos et al. 2006). Moreover, the serotonergic system is also involved, as it seems to exert a protective effect in the hippocampus and attenuates the behavioral consequences of stress (Joca et al. 2007). As glucocorticoids modulate the activity of the raphe-hippocampal system, an abnormal interaction between the HPA mechanisms and the serotonergic system may substantially contribute to suicidal behavior.

Several studies were carried out to compare serum cortisol levels, derived from the Dex/CRH test between suicidal and non-suicidal individuals, to identify the role of the HPA axis. Although several groups could confirm the association between HPA axis hyperactivity and suicidality, that nonsuppression of cortisol in response to dexamethason is predictive of later death by suicide (Westrin and Nimeus 2003), others did not find any relation (Pitchot et al. 1995) or even obtained opposite effects. In the study of Pfennig et al. (2005), suicidal behavior, including past and recent suicide attempts, was associated with a lower cortisol response in the Dex/CRH test, with lowest cortisol levels in patients with a recent suicide attempt. However, in summarizing the different findings in a meta-analysis, evidence emerged that nonsuppression of cortisol confers a four-fold risk for a suicide attempt (Coryell et al. 2006).

The impact of a dysbalanced HPA axis was at least partly underlined by several results from post-mortem studies. CRH and arginin-vasopressin (ACV) are the main activators of the HPA system and an increase of these neuropeptides was shown in brains from depressed suicide victims (Arato et al. 1989; Nemeroff et al. 1984; Raadsheer et al. 1994). Although not all studies were able to find elevated CRH or ACV concentrations in the brains of suicide victims (Merali et al. 2004, 2006), the HPA axis and the response to stress with the involvement of multiple interacting systems is one of the key areas in suicide research.

3.1.5 Other Systems

There is growing evidence during the last years that, in addition to quantitative receptor alterations, the intracellular signaling pathways through which these receptors mediate their physiological response are involved in the development and progression of suicidal behavior (Mann 2003). Postmortem studies implicate abnormalities in both the phosphoinoside (PI) and the adenylyl cyclase cAMP (cAMP) systems in the pathophysiology of suicide. It has been show that the activity of protein kinase C (PKC), which is a part of the 5-HT2A receptor mediated PI signal cascade, as well as protein kinase A (PKA), part of the cAMP pathway, is diminished in the frontal cortex of suicide victims (Pandey et al. 2005; Dwivedi et al. 2004). Furthermore, additional components of these signaling ways seem to be altered, as Dwivedi et al. reported decreased activities of some mitogen activated protein kinases (MAP) in the brain of suicide victims and depressed suicide attempters (Dwivedi et al. 2006). Morever, the levels of the cyclic-AMP-responsive element (CREB), a transcription factor which is activated by the cAMP cascade, are

Neurobiology of Suicide

decreased in the prefrontal cortex of suicide victims (Pandey et al. 2006). One target gene of CREB is the neurotrophic factor BDNF (brain-derived neurotrophic factor). Several studies have demonstrated that BDNF levels are also reduced in the prefrontal cortex of suicide victims (Dwivedi et al. 2003) and in the plasma of depressed patients with suicidal behavior (Kim et al. 2007). These findings clearly demonstrate that disturbances in signal transduction components and mechanisms, starting behind the receptor level up to modulations of gene and protein expression on the genomic level, may play a hitherto underestimated role in the development and treatment of suicidal behavior.

Recently, there have been indications for increased suicide risk with decreasing cholesterol levels. Although the underlying mechanisms are poorly understood at the moment, previous investigations of platelets and plasma of suicide attempters revealed a significant relation between low cholesterol concentrations and suicidal behavior (Diaz-Sastre et al. 2007; Marcinko et al. 2007). These findings need further analysis, because other groups did not find any relation (Fiedorowicz and Coryell 2007).

In conclusion, neurobiological investigations to date have clearly provided evidence for a major role of the serotonergic system in the pathophysiology of suicidal behavior. It is important to mention that, although the findings concerning other neurotransmitter systems or signal transduction mechanism are not convincingly confirmed, interactions between the different systems should also be considered. Suicide research has also implicated disturbances in the human stress response system. Findings of an altered HPA system, due to an inadequate response to stress in suicidal behavior, are partially inconclusive and still poorly understood, but further investigations will provide a better insight in the pathological mechanisms and will probable also allow a better differentiation between depression and suicidality.

3.2 Genetic Contribution to Suicidal Behavior

3.2.1 Epidemiological Studies

It is long known that suicidal behavior runs within families and that this familial transmission cannot solely be explained by the transmission of psychiatric disorders, as the highest suicide rates were not observed in the biological relatives of patients with affective disorders, which themselves have a strong genetically driven vulnerability (Brent and Mann 2005). Although the causes of suicide are complex and no simple explanation of the phenomenon exists, various genetic, biological, psychological, sociological, and economic factors are involved (Wasserman et al. 2007). Evidence for a genetic liability has been convincing for many years. Twin studies have shown that monozygotic twins have a significantly higher concordance rate for completed and attempted suicide than dizygotic twins (Roy 1993). This finding was replicated in a large Australian twin study, in which the risk of a serious suicide attempt by a monozygotic twin was 17-fold increased if the co-twin made a serious suicide attempt (Statham et al. 1998). Also, adoption studies supported the genetic component, as among the biological relatives of an adoptee who committed suicide, the rate of suicide is elevated about six times compared to the biological relatives of non-suicidal adopted persons (for review, see Brent and Mann 2005). From these data, it was estimated that 43% of the variability in suicidal behavior may be explained by genetics, with the remaining 57% by environmental factors (Roy 1993; Roy et al. 1995; McGuffin et al. 2001).

The heritability of suicide and suicidal behavior seems to be determined through at least two components, the heritable liability to psychiatric disorders and the heritable liability to impulsive-aggression or other personality traits. And thus the concordance of both liability factors results in the highest risk for suicidal behavior (Brent and Mann 2005).

3.2.2 Genetic Studies

On the basis of the neurobiological findings, genetic studies have been carried out for more than a decade, and which are now converging on several key areas. As there is convincing evidence that a serotonergic dysfunction is involved in the biological susceptibility to suicide, especially in high-lethality suicide, the majority of the studies were performed with candidate genes of the serotonergic pathway. Newer aspects, deviant from the classical paradigms, will also be reviewed here.

Tryptophan Hydroxylase

Tryptophan hydroxylase (TPH), the rate-limiting enzyme in the biosynthesis of serotonin (5-HT), is a critical component for the amount of 5-HT available in the synaptic cleft, and the *TPH* gene was among the first candidate genes for association studies of suicidality. In the meantime, two different TPH isoforms have been identified, referred to as TPH1 and TPH2, the genes for both being located on different chromosomes (Walther et al. 2003). It is further known that TPH2 is expressed in brain but not in peripheral tissues in men (Zill et al. 2004a). Although the exact biological differentiations of both isoforms are not completely clarified, TPH2 is discussed as the brain-specific enzyme while the TPH1 might be responsible for peripheral serotonin generation (Zill et al. 2007).

The data concerning TPH as a pre-synaptic marker in postmortem brains of suicide victims were unequivocal for a long time in immunoreactivity studies (Boldrini et al. 2005). Investigating the TPH2 mRNA either by in situ hybridization or quantitative real-time PCR, discrepant results were also obtained, with either higher levels in the raphe nuclei of drug free suicide victims (Bach-Mizrachi et al. 2006) or no alterations between suicides and controls in prefrontal cortex being reported (De Luca et al. 2005). In evaluating these discrepant results, one has to acknowledge that different areas were investigated in the different studies, and that the studies that have been carried out before the identification of TPH2 did not distinguish between the two isoforms of the enzyme. All together, the discovery of TPH2 was supposed to explain all the previously puzzling data in the past 30 years about divergent protein/mRNA ratios and biochemical characteristics of TPH from peripheral sources and from CNS.

The *TPH1* gene is located on chromosome 11p15.3-p14 and has two common intronic polymorphisms consisting of an A to C substitution at nucleotides 779 (A779C), being originally classified as U and L (upper and lower band) and at 218 (A218C). Nielsen et al. were the first who reported an association between the UU genotype and higher CSF 5-HIAA concentration in a group of violent alcoholic offenders (Nielsen et al. 1994). These findings were repeatedly replicated and extended by the same group with the final conclusion that the A779C variant of the *TPH1* gene might predispose to suicidality, but mainly among impulsive offenders, as they are phenotypically more extreme (Nielsen et al. 1998; Virkkunen et al. 1995; Roy et al. 2001).

Many studies followed these first reports and most of them were carried out in the frame of association studies with depressed, bipolar, schizophrenic, or alcoholic patients. Although the numbers of patients within the diagnostic categories seemed to be sufficient, those with suicidal attempts were small in most studies. Despite some positive reports, the list of negative findings is long (for review, see Bondy et al. 2006). Due to the discrepancy of the results together with the small numbers of patients, the diagnostic heterogeneity with either committed suicide as clear-cut act or a history of suicidal attempts, and finally due to the use of the different markers, the impact of the *TPH1* gene on suicidal behavior still remains ambiguous. Almost all studies had a lack in statistical power, ethnic heterogeneity, and variations in the sampling strategy, in particular for controls. Some meta-analyses have combined the data of all investigated polymorphisms and again come to discrepant results (Rujescu et al. 2003; Bellivier et al. 2004). Nevertheless, a recent meta-analysis finally confirmed a weak association between suicidal behavior and the A779C/A218C polymorphisms (Li and He 2006).

Concerning the relation between personality traits and the *TPH1* gene, it is noteworthy that two independent studies observed a relation between the A779 allele and aggressive disposition in healthy individuals without psychopathology. In both studies, investigating probands of American (Manuck et al. 1999) and German descent (Rujescu et al. 2002), the A-allele scored significantly higher on measures of aggression, irritability, and anger-related traits.

The identification of the neuronal *TPH2* gene (Walther and Bader 2003), being located on chromosome 12q15, promised a step forward in investigating the genetic contribution to suicidality, as this isoform apparently plays a more important role in the synthesis of brain serotonin and might thus be a better candidate. Although the number of studies using *TPH2* as candidate gene is rare, there is increasing recent supportive evidence for various polymorphisms and haplotypes contributing to the risk for suicidality and several suicide-related personality traits (Gutknecht et al. 2006;, Zhou et al. 2005; Ke et al. 2006; Lopez et al. 2007; Zill et al. 2004b). However, negative findings also exist suggesting that the observed association might be based on the association with affective psychoses (Harvey et al. 2007). The functional consequences of all the investigated TPH2 polymorphisms are still unknown and the data are still limited, nevertheless, the *TPH2* gene deserves further evaluation as candidate gene for suicidal behavior.

The Serotonin Transporter

The majority of association studies were carried out with the serotonin transporter (*5-HTT*) gene, as the 5-HTT is terminating serotonin action via its uptake from the synaptic cleft. A common polymorphism in the 5 -regulatory region with a 44 base pair deletion, resulting in either the S (short) or L (long) allele (5-HTTLPR) with altered transcriptional activity, was associated with anxiety-related personality traits and with several clinical features in depression (Heils et al. 1996; Lesch et al. 1996). Based on these findings, it was hypothesized that the short form of the *5-HTT* gene might be associated with impulsive aggression and suicidal behavior. Much work was carried out to reliably associate the 5-HTTLPR with suicidal behavior, but the results were not consistent (for review, see Bondy et al. 2006). Recent meta-analyses could not verify an association between one of the alleles and suicidality per se, but clearly gave evidence for an association between the S allele (with lower transcriptional activity) and violent suicide, independently from clinical diagnosis, demographic, or socio-cultural parameters (Li and He 2007; Bondy et al. 2006).

The relation of the S allele of the 5-HTTLPR and some personality traits was not only shown in suicidal behavior but also in alcohol- or heroin-dependent persons. Dissocial alcoholics carrying the SS genotype exhibited significantly lower scores of harm avoidance and higher novelty-seeking scores (Sander et al. 1998). Similarly, the SS genotype was more frequent among a group of violent heroin-dependent persons compared with addicted individuals without aggressive behavior (Gerra et al. 2004) and was further associated with an increased availability to experiment illegal drugs, particularly in subjects with more aggressiveness (Gerra et al. 2005).

Gender-specific associations of the 5-HTTLPR and suicide were also proposed, as the S-allele was related to both female (Baca-Garcia et al. 2002) or male suicide attempters (Limosin et al. 2005). An explanation for this discrepant data and the gender specific effect might be that intermediate links between alcohol dependence and suicidal behavior may involve different genetic factors in men and women. A recent study investigating the 5-HTTLPR in the continuum between compulsivity and impulsivity in females observed that the frequency of S individuals (either homo- or heterozygotic) was low in patients with obsessive compulsive disorder (OCD), intermediate in nonimpulsive controls and higher in impulsive suicide attempters (Baca-Garcia et al. 2005), and thus it was suggested that genetic variants may be more likely related to behavioral dimensions instead of to specific psychiatric disorders.

Serotonin Receptors

There is considerable evidence that the density of the 5-HT2A receptor is upregulated in parietal cortical regions of depressed suicide victims (for review, see Mann 2003), and it was suggested that this increase may be at least partly regulated genetically. The *5-HT2A* receptor gene is located on chromosome 13q14-q21 and, despite the numerous known SNPs spanning the gene, only a small number of them have been investigated, most notably the T102C and A-1438G variants. However, the genetic findings of associations between *5-HT2A* receptor gene variants and suicide have been controversial, and there is no convincing evidence for an association between any of these polymorphisms and suicide in Caucasian, Jewish, or Brazilian populations (for review, see Bondy et al. 2006).

Other polymorphisms in genes coding for the different types of the 5-HT receptors have been investigated, such as the 5-HT1A autoreceptors or the 5-HT1B receptors. However, the constantly negative findings and a recent meta-analysis revealed no evidence for association with suicidal behavior (Kia-Keating et al. 2007).

Genes Involved in Transmitter Synthesis and Degradation

Central monoamines are important modulators of mood and behavior, and a deficit in both serotonin and noradrenalin plays an important role in the pathophysiology of stress and the response to stress reactions (Charney 1998).

Monoamine oxydase A (MAOA) is a mitochondrial membrane enzyme which is involved in human behavior due to its key role in the metabolism of biological amines (Ma et al. 2004). The general idea underlying the investigation of MAOA activity in relation to violent behavior is that low MAO activity results in elevated levels of serotonin, noradrenalin, and dopamine in the brain, manifesting as mood disorder and/or aggressive behavior (Mann 2003). Several findings with peripheral tissues underlined its importance, as low platelet MAO activity seemed to predispose for violent behavior and increased risk for antisocial and criminal acts (Skondras et al. 2004).

A number of polymorphisms have been described for the *MAOA* gene. A 30 bp repeat (VNTR) has been shown to affect the transcriptional activity of the gene. Alleles with 3.5 or 4 copies of this repeat are transcribed 2–10 times more efficiently than those with 3 or 5 copies (Sabol et al. 1998). Manuk et al. reported that healthy men carrying the high activity alleles (3.5 and 4 copies) of the VNTR expressed a lower CNS serotonergic responsiveness in the fenfluramine challenge test and more impulsive aggression (Manuck et al. 2000). Concerning suicide, the results are not convincing and most results were negative (Bondy et al. 2006). Although at present there seems to be no relation between the *MAOA* gene and the vulnerability for suicidality, the association between the MAOA VNTR polymorphism and aggression (Eisenberger et al. 2007) strengthens the hypothesis that an excess of high activity *MAOA* gene promoter alleles may influence the methods used in suicide attempts.

Catechol-O-methyltransferase (COMT), the major catecholamine degrading enzyme has a functional polymorphism at position 158 (Val/Met) which is responsible for substantial variability in COMT enzymatic activity. The Val allele is associated with high (H allele) and the Met allele with low activity (L allele). For different ethnic groups it was shown that the Met allele has higher propensity for violence, and despite some negative findings a recent meta-analysis gave evidence for an weak association between the 158Met allele and suicidal behavior in males which could be related to the increased lethality (Kia-Keating et al. 2007).

Genes of the Dopaminergic System

The interest in the dopamine system, including DA receptors and the DA transporter, derived from the involvement of this system in several psychiatric disorders, especially in alcoholism and its relationship to the reward syndrome which is mediated via the dopamine D2 receptors (Franken et al. 2005). A deficiency or absence of dopamine D2 receptors then predisposes individuals to a higher risk for multiple addictive, impulsive or compulsive behaviors (Bowirrat and Oscar-Berman 2005). A polymorphism in the D2 receptor gene (*TaqI*) was repeatedly investigated as risk for addictive behavior, but with inconsistent results (for review, see Noble 2003). A recent study investigated an insertion/deletion polymorphism within the promoter region of the dopamine D2 receptor at position −141 (−141C Ins/Del), which shows a lower transcription activity (Arinami et al. 1997). In a large sample of more than 1,000 chronic alcoholic patients of German descent by Putzhammer et al., a significant excess of the −141C Del allele in alcoholics with positive family history and in alcoholics with suicidality was observed (Johann et al. 2005). From these results, it might be concluded that the dopamine D2 receptor might confer a risk to suicidality in patients with high familial loading for alcoholism.

Further, there was an extended discussion about an association between the risk-taking behavior and the length of the dopamin-D4 receptor (*DRD4*) gene exon III repeat alleles (for review, see Kluger 2003). However, most studies revealed negative results, either with Israeli suicidal patients (Zalsman et al. 2004) or with Swedish suicide attempters (Persson et al. 1999).

Search for New Candidate Genes and Miscellaneous Results

Despite the studies which followed the original concepts and hypotheses of serotonergic dysfunction, several recent studies expanded the research to identify new mechanisms and candidate genes. Thus, as the hypothlamaic–pituitry–adrenal (HPA) axis plays a critical role in the reaction to stress, and as alterations in this system may be linked to aggression and impulsivity, polymorphisms in regulating genes, such as the corticotrophin-releasing hormone (CRH), are gaining importance (De Luca et al. 2007). Further, the effects of cannabinoids on anxiety-related responses, which involve endocannabinoid receptors (CB1), γ-aminobutyric acid (GABA) as main inhibitory transmitter, and the neuropeptide cholecystokinin (CCK) (Viveros et al. 2005), were in the focus of recent investigations. However, these studies are scarce and just beginning (for review, see Bondy et al. 2006).

Neurotrophins might play a role in the aetiology of mood disorders and suicidal behavior, as a significant reduction of BDNF mRNA levels was shown in prefrontal cortex and hippocampus of suicide subjects (Dwivedi et al. 2003). The common Val66Met polymorphism of the BDNF (brain-derived neurotrophic factor) gene was investigated in Chinese unipolar and bipolar patients, but no relation between the disorder or a history of suicide attempts was observed (Hong et al. 2003). A further study comprising a missense polymorphism (S205L) of the low-affinity neurotrophin receptor gene (*p75NTR*) found a positive result in Japanese suicide attempters (Kunugi et al. 2004). The possible involvement of *14–3–3 epsilon* gene, which is related to neurogenesis, was identified as possible candidate gene, as it was upregulated using DNA microarrays with brains of suicide victims (Yanagi et al. 2005). These authors investigated several SNPs within this gene and found a haplotype associated with completed suicide, thus suggesting that a dysregulation of neurogenesis might be involved in suicide (Yanagi et al. 2005).

This latter study clearly elucidated that microarray analysis provides the opportunity to study thousands of genes at once and may thus give a "snapshot" about the brain gene activity before completing suicide (Bunney et al. 2003). Thus, microarray analysis is especially useful to identify new candidate genes and to gain new insight into the biological mechanism of suicide beyond the serotonergic system. Despite the fact that there are several methodological pitfalls, such as postmortem delay and the quality of the mRNA preparation, microarray analysis has become a standard tool in many areas of biomedical research (Gwadry et al. 2005).

3.3 Suicide as Interplay of Genes and Environment

Gene-environment interactions are highly relevant in complex disorders in which an overlapping of several contributing factors is crucial in pathogenetic mechanisms. This approach assumes that environmental pathogens cause the disorder and that genes or an interaction of different genes influence the susceptibility to the pathogens. These environmental causes are considered only as contributors because exposure to them does not always generate the disorder (Caspi and Moffitt 2006). Findings from molecular genetic studies and from the behavioral genetic literature suggest that gene–environment interactions are mediated by personality and behavioral characteristics (Jaffee and Price 2007).

These studies are new in the field of suicide research, but some of the initial findings are intriguing. Based on the first study by Caspi et al. (2003) demonstrating that the L-allele of the 5-HTTLPR polymorphism was moderating the influence of stressful life events on later development of depression and increased suicidality, several investigations followed.

It is well known that early life trauma is a risk factor for the development of suicidal behavior (Roy and Janal 2006), and investigating the recently identified triallelic 5-HTTLPR in substance-dependent patients with and without childhood trauma and suicide, it could clearly be demonstrated that subjects with the low expression allele had higher risk than subjects without that allele (Roy et al. 2007). The genetic impact in the response to stress could be replicated by Stein et al. (2007), who observed a significant interaction between levels of childhood emotional and physical maltreatment and the 5-HTTLPR polymorphism. Subjects with the SS genotype had significantly higher levels of anxiety sensitivity. Taken together, there is growing literature documenting gene ×environment interactions involving the 5-HTTLPR variant, and it was hypothesized that this variant regulates some aspect of brain function being relevant for stress sensitivity. Although extensive studies have to follow confirming these previous results, this is a promising step forward in identifying vulnerability genes.

4 Conclusions

Suicidal behavior is a complex disorder and thus the predisposition towards suicide consists of numerous neurobiological mechanisms, genetic factors, as well as environmental influences which manifest themselves as suicidal behavior only when a certain threshold of predisposition is crossed (Balazic and Marusic 2005). There is growing evidence that at least three neurobiological systems play a major role in the pathogenesis of suicidal behavior. First, a large number of studies using postmortem samples, CSF, platelets, and imaging techniques have repeatedly and convincingly associated a deficient serotonergic system with suicidal behavior. Second, diminished cortisol suppression after Dex/CRH test and postmortem studies suggest a hyperactivity of the HPA system in suicidality. Third, an elevated release of noradrenalin and subsequent alterations in the noradrenergic neurotransmission have been observed in suicide victims and attempters. The involvement of the noradrenergic and HPA system in the pathophysiology of suicidality suggests an inadequate response to stress (e.g., stressful life events) as a major risk factor in the development of suicidal behavior. Moreover, a contribution of the dopaminergic system, of signal transduction mechanisms, and of the cholesterol metabolism has been indicated in suicidal behavior, but further studies have to confirm these observations.

Considering neurobiological findings that the serotonergic transmission plays a pivotal role in individual differences in mood, impulsiveness, and aggression, personality traits which are related to suicidal behavior, it is no surprise that molecular genetic studies in suicide research have focused during the last decade particularly on serotonergic genes. A growing number of molecular genetic studies have been carried out to identify candidate genes that may be involved in the pathophysiological mechanisms of suicidal behavior. Despite tremendous effort and a plenty of investigations, only two genes, one coding for the tryptophan hydroxylase 1 (TPH1 A218C) and the other for the serotonin transporter (5-HTTLPR), were reliably suggestive to be involved in the vulnerability for suicidal behavior. In summarizing all findings with TPH1, it seems that the intronic polymorphisms A779C and/or A218C are associated with suicidality via serotonergic dysfunction, reflected by the low 5-HIAA levels. The serotonin transporter polymorphism (5-HTTLPR) was further shown to be involved in violent behavior in subjects with type II alcoholism (Sander et al. 1998) and aggressive behavior in heroin addicts (Gerra et al. 2005). These data, together with the results of the association studies in suicidal behavior, favor the assumption that the serotonin transporter gene is not involved in suicidal behavior in general, but in violent and repeated suicide attempts. The findings with any type of the serotonin receptors were predominantly negative and not convincing

Neurobiology of Suicide

for studies with catecholamine synthesizing and metabolizing enzymes nor with the dopaminergic receptors.

One has to concede that the majority of these studies were case-control association studies which are hampered by many pitfalls, such as limited sample size, ethnic stratification between cases and controls, and the use of different phenotypes (e.g., suicidal ideation, suicide attempts, completed suicide).

Given the multi-causal and genetically complex nature of suicidal behavior it cannot be expected that one neurobiological system, respectively a single gene, is responsible for the development of suicidal behavior. There is no doubt that the interaction of several neurobiological systems with different effect sizes and several candidate genes from different biological systems, including gene–gene interactions of these numerous susceptibility genes, will finally modulate and regulate the predisposition for suicidal behavior. For example, a meta-analysis of prospective studies about CSF 5-HIAA levels and dexamethasone suppression in the HPA axis in suicide yielded odds ratios for prediction of suicidal behavior of 4.48 and 4.65, respectively (Mann et al. 2006). With regard to the known relation between the serotonergic and HPA systems (Porter et al. 2004), a combined analysis of both systems should thus improve the prediction of the suicidal risk.

Furthermore, genes interact not only among themselves but also with environmental factors, but so far little attention has been paid to this possible interplay. A positive example in this direction is the before-mentioned study of Caspi et al. who found that the functional 5-HTTLPR moderate the influence of stressful life events on depression (Caspi et al. 2003). This relation is of great importance in suicide research, due to the repeatedly described alterations in the stress response-regulating HPA system in suicidal behavior.

Taken together, there is no doubt that disturbances in neurobiological systems including mutations/polymorphisms in susceptibility genes are two of the multiple factors implicated in the phenomenon of suicide and represent serious risk factors, in particular when an individual is confronted with stress, such as negative life events and somatic disorders. Thus, further studies are needed to identify more robustly the predisposing factors. Some biological factors may be related to aggressiveness and impulsivity, which have their effects independently of, or additively to, a mental disorder (Balazic and Marusic 2005).

The neurobiological research will finally improve prevention and treatment of suicidal behavior.

References

Arango V, Ernsberger P, Sved AF, Mann JJ (1993) Quantitative autoradiography of alpha 1- and alpha 2-adrenergic receptors in the cerebral cortex of controls and suicide victims. Brain Res 630:271–282

Arango V, Underwood MD, Gubbi AV, Mann JJ (1995) Localized alterations in pre- and postsynaptic serotonin binding sites in the ventrolateral prefrontal cortex of suicide victims. Brain Res 688:121–133

Arango V, Underwood MD, Mann JJ (1996) Fewer pigmented locus coeruleus neurons in suicide victims: preliminary results. Biol Psychiatry 39:112–120

Arato M, Banki CM, Bissette G, Nemeroff CB (1989) Elevated CSF CRF in suicide victims. Biol Psychiatry 25:355–359

Arinami T, Gao M, Hamaguchi H, Toru M (1997) A functional polymorphism in the promoter region of the dopamine D2 receptor gene is associated with schizophrenia. Hum Mol Genet 6:577–582

Asberg M (1997) Neurotransmitters and suicidal behavior. The evidence from cerebrospinal fluid studies. Ann NY Acad Sci 836:158–81.

Asberg M, Traskman L, Thoren P (1976) 5-HIAA in the cerebrospinal fluid. A biochemical suicide predictor? Arch Gen Psychiatry 33:1193–1197

Audenaert K, Peremans K, Goethals I, van HC (2006) Functional imaging, serotonin and the suicidal brain. Acta Neurol Belg 106:125–131

Baca-Garcia E, Salgado BR, Segal HD, Lorenzo CV, Acosta MN, Romero MA, Hernandez MD, Saiz-Ruiz J, Fernandez PJ, de LJ (2005) A pilot genetic study of the continuum between compulsivity and impulsivity in females: the serotonin transporter promoter polymorphism. Prog Neuropsychopharmacol Biol Psychiatry 29:713–717

Baca-Garcia E, Vaquero C, az-Sastre C, Saiz-Ruiz J, Fernandez-Piqueras J, de LJ (2002) A gender-specific association between the serotonin transporter gene and suicide attempts. Neuropsychopharmacology 26:692–695

Bach-Mizrachi H, Underwood MD, Kassir SA, Bakalian MJ, Sibille E, Tamir H, Mann JJ, Arango V (2006) Neuronal tryptophan hydroxylase mRNA expression in the human dorsal and median raphe nuclei: major depression and suicide. Neuropsychopharmacology 31:814–824

Balazic J, Marusic A (2005) The completed suicide as interplay of genes and environment. Forensic Sci Int 147 Suppl:S1–S3

Baud P (2005) Personality traits as intermediary phenotypes in suicidal behavior: genetic issues. Am J Med Genet C Semin Med Genet 133:34–42

Bellivier F, Chaste P, Malafosse A (2004) Association between the *TPH* gene A218C polymorphism and suicidal behavior: a meta-analysis. Am J Med Genet 124B:87–91

Bertolote JM, Fleischmann A (2005) Suicidal behavior prevention: WHO perspectives on research. Am J Med Genet C Semin Med Genet 133:8–12

Biegon A, Fieldust S (1992) Reduced tyrosine hydroxylase immunoreactivity in locus coeruleus of suicide victims. Synapse 10:79–82

Boldrini M, Underwood MD, Mann JJ, Arango V (2005) More tryptophan hydroxylase in the brainstem dorsal raphe nucleus in depressed suicides. Brain Res 1041:19–28

Bondy B, Buettner A, Zill P (2006) Genetics of suicide. Mol Psychiatry 11:336–351

Bowirrat A, Oscar-Berman M (2005) Relationship between dopaminergic neurotransmission, alcoholism, and Reward Deficiency syndrome. Am J Med Genet B Neuropsychiatr Genet 132:29–37

Brent DA, Mann JJ (2005) Family genetic studies, suicide, and suicidal behavior. Am J Med Genet C Semin Med Genet 133:13–24

Bunney WE, Bunney BG, Vawter MP, Tomita H, Li J, Evans SJ, Choudary PV, Myers RM, Jones EG, Watson SJ, Akil H (2003) Microarray technology: a review of new strategies to discover candidate vulnerability genes in psychiatric disorders. Am J Psychiatry 160:657–666

Caspi A, Moffitt TE (2006) Gene-environment interactions in psychiatry: joining forces with neuroscience. Nat Rev Neurosci 7:583–590

Caspi A, Sugden K, Moffitt TE, Taylor A, Craig IW, Harrington H, McClay J, Mill J, Martin J, Braithwaite A, Poulton R (2003) Influence of life stress on depression: moderation by a polymorphism in the *5-HTT* gene. Science 301:386–389

Charney DS (1998) Monoamine dysfunction and the pathophysiology and treatment of depression. J Clin Psychiatry 59 Suppl 14:11–14

Coryell W, Young E, Carroll B (2006) Hyperactivity of the hypothalamic-pituitary-adrenal axis and mortality in major depressive disorder. Psychiatr Res 142:99–104

Courtet P, Jollant F, Castelnau D, Buresi C, Malafosse A (2005) Suicidal behavior: relationship between phenotype and serotonergic genotype. Am J Med Genet C Semin Med Genet 133:25–33

De Luca V, Tharmalingam S, Kennedy JL (2007) Association study between the corticotropin-releasing hormone receptor 2 gene and suicidality in bipolar disorder. Eur Psychiatry 22(5):282–287

Neurobiology of Suicide

De Luca V, Likhodi O, Van Tol HH, Kennedy JL, Wong AH (2005) Tryptophan hydroxylase 2 gene expression and promoter polymorphisms in bipolar disorder and schizophrenia. Psychopharmacology (Berl) 183(3):378–382

Diaz-Sastre C, Baca-Garcia E, Perez-Rodriguez MM, Garcia-Resa E, Ceverino A, Saiz-Ruiz J, Oquendo MA, de LJ (2007) Low plasma cholesterol levels in suicidal males: a gender- and body mass index-matched case-control study of suicide attempters and nonattempters. Prog Neuropsychopharmacol Biol Psychiatry 31:901–905

Dumais A, Lesage AD, Alda M, Rouleau G, Dumont M, Chawky N, Roy M, Mann JJ, Benkelfat C, Turecki G (2005) Risk factors for suicide completion in major depression: a case-control study of impulsive and aggressive behaviors in men. Am J Psychiatry 162:2116–2124

Dwivedi Y, Rizavi HS, Conley RR, Pandey GN (2006) ERK MAP kinase signaling in post-mortem brain of suicide subjects: differential regulation of upstream Raf kinases Raf-1 and B-Raf. Mol Psychiatry 11:86–98

Dwivedi Y, Rizavi HS, Conley RR, Roberts RC, Tamminga CA, Pandey GN (2003) Altered gene expression of brain-derived neurotrophic factor and receptor tyrosine kinase B in postmortem brain of suicide subjects. Arch Gen Psychiatry 60:804–815

Dwivedi Y, Rizavi HS, Shukla PK, Lyons J, Faludi G, Palkovits M, Sarosi A, Conley RR, Roberts RC, Tamminga CA, Pandey GN (2004) Protein kinase A in postmortem brain of depressed suicide victims: altered expression of specific regulatory and catalytic subunits. Biol Psychiatry 55:234–243

Eisenberger NI, Way BM, Taylor SE, Welch WT, Lieberman MD (2007) Understanding genetic risk for aggression: clues from the brain's response to social exclusion. Biol Psychiatry 61:1100–1108

Engstrom G, Alling C, Blennow K, Regnell G, Traskman-Bendz L (1999) Reduced cerebrospinal HVA concentrations and HVA/5-HIAA ratios in suicide attempters. Monoamine metabolites in 120 suicide attempters and 47 controls. Eur Neuropsychopharmacol 9:399–405

Fiedorowicz JG, Coryell WH (2007) Cholesterol and suicide attempts: a prospective study of depressed inpatients. Psychiatry Res 152:11–20

Fleischmann A, Bertolote JM, Belfer M, Beautrais A (2005) Completed suicide and psychiatric diagnoses in young people: a critical examination of the evidence. Am J Orthopsychiatry 75:676–683

Franken IH, Booij J, van den BW (2005) The role of dopamine in human addiction: from reward to motivated attention. Eur J Pharmacol 526:199–206

Gerra G, Garofano L, Castaldini L, Rovetto F, Zaimovic A, Moi G, Bussandri M, Branchi B, Brambilla F, Friso G, Donnini C (2005) Serotonin transporter promoter polymorphism genotype is associated with temperament, personality traits and illegal drugs use among adolescents. J Neural Transm 112:1397–1410

Gerra G, Garofano L, Santoro G, Bosari S, Pellegrini C, Zaimovic A, Moi G, Bussandri M, Moi A, Brambilla F, Donnini C (2004) Association between low-activity serotonin transporter genotype and heroin dependence: behavioral and personality correlates. Am J Med Genet B Neuropsychiatr Genet 126:37–42

Gutknecht L, Jacob C, Strobel A, Kriegebaum C, Muller J, Zeng Y, Markert C, Escher A, Wendland J, Reif A, Mossner R, Gross C, Brocke B, Lesch KP (2006) Tryptophan hydroxylase-2 gene variation influences personality traits and disorders related to emotional dysregulation. Int J Neuropsychopharmacol 10(3):309–320

Gwadry FG, Sequeira A, Hoke G, Ffrench-Mullen JM, Turecki G (2005) Molecular characterization of suicide by microarray analysis. Am J Med Genet C Semin Med Genet 133:48–56

Harris EC, Barraclough B (1998) Excess mortality of mental disorder. Br J Psychiatry 173:11–53

Harvey M, Gagne B, Labbe M, Barden N (2007) Polymorphisms in the neuronal isoform of tryptophan hydroxylase 2 are associated with bipolar disorder in French Canadian pedigrees. Psychiatr Genet 17:17–22

Heils A, Teufel A, Petri S, Stober G, Riederer P, Bengel D, Lesch KP (1996) Allelic variation of human serotonin transporter gene expression. J Neurochem 66:2621–2624

Holsboer F (2001) Stress, hypercortisolism and corticosteroid receptors in depression: implications for therapy. J Affect Disord 62:77–91

Hong CJ, Huo SJ, Yen FC, Tung CL, Pan GM, Tsai SJ (2003) Association study of a brain-derived neurotrophic-factor genetic polymorphism and mood disorders, age of onset and suicidal behavior. Neuropsychobiology 48:186–189

Hurd YL, Herman MM, Hyde TM, Bigelow LB, Weinberger DR, Kleinman JE (1997) Prodynorphin mRNA expression is increased in the patch versus matrix compartment of the caudate nucleus in suicide subjects. Mol Psychiatry 2:495–500

Jaffee SR, Price TS (2007) Gene-environment correlations: a review of the evidence and implications for prevention of mental illness. Mol Psychiatry 12:432–442

Joca SR, Ferreira FR, Guimaraes FS (2007) Modulation of stress consequences by hippocampal monoaminergic, glutamatergic and nitrergic neurotransmitter systems. Stress 10:227–249

Johann M, Putzhammer A, Eichhammer P, Wodarz N (2005) Association of the -141C Del variant of the dopamine D2 receptor (DRD2) with positive family history and suicidality in German alcoholics. Am J Med Genet B Neuropsychiatr Genet 132:46–49

Kapur S, Mann JJ (1992) Role of the dopaminergic system in depression. Biol Psychiatry 32:1–17

Ke L, Qi ZY, Ping Y, Ren CY (2006) Effect of SNP at position 40237 in exon 7 of the *TPH2* gene on susceptibility to suicide. Brain Res 1122:24–26

Kendler KS, Gardner CO, Prescott CA (2006) Toward a comprehensive developmental model for major depression in men. Am J Psychiatry 163:115–124

Kia-Keating BM, Glatt SJ, Tsuang MT (2007) Meta-analyses suggest association between COMT, but not HTR1B, alleles, and suicidal behavior. Am J Med Genet B Neuropsychiatr Genet 144B(8):1048–1053

Kim YK, Lee HP, Won SD, Park EY, Lee HY, Lee BH, Lee SW, Yoon D, Han C, Kim DJ, Choi SH (2007) Low plasma BDNF is associated with suicidal behavior in major depression. Prog Neuropsychopharmacol Biol Psychiatry 31:78–85

Kluger J (2003) Medicating young minds. Time Magazine 162(14):48–56, November 3, 2003

Kunugi H, Hashimoto R, Yoshida M, Tatsumi M, Kamijima K (2004) A missense polymorphism (S205L) of the low-affinity neurotrophin receptor *p75NTR* gene is associated with depressive disorder and attempted suicide. Am J Med Genet B Neuropsychiatr Genet 129:44–46

Leboyer M, Slama F, Siever L, Bellivier F (2005) Suicidal disorders: a nosological entity per se? Am J Med Genet C Semin Med Genet 133C:3–7

Lesch KP, Bengel D, Heils A, Sabol SZ, Greenberg BD, Petri S, Benjamin J, Muller CR, Hamer DH, Murphy DL (1996) Association of anxiety-related traits with a polymorphism in the serotonin transporter gene regulatory region. Science 274:1527–1531

Li D, He L (2007) Meta-analysis supports association between serotonin transporter (5-HTT) and suicidal behavior. Mol Psychiatry 12:47–54

Li D, He L (2006) Further clarification of the contribution of the tryptophan hydroxylase (*TPH*) gene to suicidal behavior using systematic allelic and genotypic meta-analyses. Hum Genet 119:233–240

Limosin F, Loze JY, Boni C, Hamon M, Ades J, Rouillon F, Gorwood P (2005) Male-specific association between the 5-HTTLPR S allele and suicide attempts in alcohol-dependent subjects. J Psychiatr Res 39:179–182

Lopez VA, tera-Wadleigh S, Cardona I, Kassem L, McMahon FJ (2007) Nested association between genetic variation in tryptophan hydroxylase II, bipolar affective disorder, and suicide attempts. Biol Psychiatry 61:181–186

Ma J, Yoshimura M, Yamashita E, Nakagawa A, Ito A, Tsukihara T (2004) Structure of rat monoamine oxidase A and its specific recognitions for substrates and inhibitors. J Mol Biol 338:103–114

Malone KM, Corbitt EM, Li S, Mann JJ (1996) Prolactin response to fenfluramine and suicide attempt lethality in major depression. Br J Psychiatry 168:324–329

Mann JJ (2003) Neurobiology of suicidal behaviour. Nat Rev Neurosci 4:819–828

Mann JJ, Arango V, Henteleff RA, Lagattuta TF, Wong DT (1996) Serotonin 5-HT3 receptor binding kinetics in the cortex of suicide victims are normal. J Neural Transm 103:165–171

Mann JJ, Currier D (2007) A review of prospective studies of biologic predictors of suicidal behavior in mood disorders. Arch Suicide Res 11:3–16

Mann JJ, Currier D, Stanley B, Oquendo MA, Amsel LV, Ellis SP (2006) Can biological tests assist prediction of suicide in mood disorders? Int J Neuropsychopharmacol 9:465–474

Mann JJ, Malone KM (1997) Cerebrospinal fluid amines and higher-lethality suicide attempts in depressed inpatients. Biol Psychiatry 41:162–171

Manuck SB, Flory JD, Ferrell RE, Dent KM, Mann JJ, Muldoon MF (1999) Aggression and anger-related traits associated with a polymorphism of the tryptophan hydroxylase gene. Biol Psychiatry 45:603–614

Manuck SB, Flory JD, Ferrell RE, Mann JJ, Muldoon MF (2000) A regulatory polymorphism of the monoamine oxidase-A gene may be associated with variability in aggression, impulsivity, and central nervous system serotonergic responsivity. Psychiatry Res 95:9–23

Marcinko D, Pivac N, Martinac M, Jakovljevic M, Mihaljevic-Peles A, Muck-Seler D (2007) Platelet serotonin and serum cholesterol concentrations in suicidal and non-suicidal male patients with a first episode of psychosis. Psychiatry Res 150:105–108

Mastorakos G, Pavlatou MG, Mizamtsidi M (2006) The hypothalamic-pituitary-adrenal and the hypothalamic- pituitary-gonadal axes interplay. Pediatr Endocrinol Rev 3 Suppl 1:172–81

McGuffin P, Marusic A, Farmer A (2001) What can psychiatric genetics offer suicidology? Crisis 22:61–65

Merali Z, Du L, Hrdina P, Palkovits M, Faludi G, Poulter MO, Anisman H (2004) Dysregulation in the suicide brain: mRNA expression of corticotropin-releasing hormone receptors and GABA(A) receptor subunits in frontal cortical brain region. J Neurosci 24:1478–1485

Merali Z, Kent P, Du L, Hrdina P, Palkovits M, Faludi G, Poulter MO, Bedard T, Anisman H (2006) Corticotropin-releasing hormone, arginine vasopressin, gastrin-releasing peptide, and neuromedin B alterations in stress-relevant brain regions of suicides and control subjects. Biol Psychiatry 59:594–602

Meyer JH, McMain S, Kennedy SH, Korman L, Brown GM, DaSilva JN, Wilson AA, Blak T, Eynan-Harvey R, Goulding VS, Houle S, Links P (2003) Dysfunctional attitudes and 5-HT2 receptors during depression and self-harm. Am J Psychiatry 160:90–99

Muller-Oerlinghausen B, Roggenbach J, Franke L (2004) Serotonergic platelet markers of suicidal behavior – do they really exist? J Affect Disord 79:13–24

Nemeroff CB, Widerlov E, Bissette G, Walleus H, Karlsson I, Eklund K, Kilts CD, Loosen PT, Vale W (1984) Elevated concentrations of CSF corticotropin-releasing factor-like immunore-activity in depressed patients. Science 226:1342–1344

Nielsen DA, Goldman D, Virkkunen M, Tokola R, Rawlings R, Linnoila M (1994) Suicidality and 5-hydroxyindoleacetic acid concentration associated with a tryptophan hydroxylase polymor-phism. Arch Gen Psychiatry 51:34–38

Nielsen DA, Virkkunen M, Lappalainen J, Eggert M, Brown GL, Long JC, Goldman D, Linnoila M (1998) A tryptophan hydroxylase gene marker for suicidality and alcoholism. Arch Gen Psychiatry 55:593–602

Noble EP (2003) D2 dopamine receptor gene in psychiatric and neurologic disorders and its phe-notypes. Am J Med Genet B Neuropsychiatr Genet 116:103–125

Nordstrom P, Samuelsson M, Asberg M, Traskman-Bendz L, berg-Wistedt A, Nordin C, Bertilsson L (1994) CSF 5-HIAA predicts suicide risk after attempted suicide. Suicide Life Threat Behav 24:1–9

Ordway GA, Smith KS, Haycock JW (1994a) Elevated tyrosine hydroxylase in the locus coeru-leus of suicide victims. J Neurochem 62:680–685

Ordway GA, Widdowson PS, Smith KS, Halaris A (1994b) Agonist binding to alpha 2-adrenocep-tors is elevated in the locus coeruleus from victims of suicide. J Neurochem 63:617–624

Pandey GN, Dwivedi Y, Ren X, Rizavi HS, Mondal AC, Shukla PK, Conley RR (2005) Brain region specific alterations in the protein and mRNA levels of protein kinase A subunits in the post-mortem brain of teenage suicide victims. Neuropsychopharmacology 30:1548–1556

Pandey GN, Dwivedi Y, Ren X, Rizavi HS, Roberts RC, Conley RR (2006) Cyclic AMP response element-binding protein in post-mortem brain of teenage suicide victims: specific decrease in the prefrontal cortex but not the hippocampus. Int J Neuropsychopharmacol:1–9.:1–9

Pandey GN, Pandey SC, Dwivedi Y, Sharma RP, Janicak PG, Davis JM (1995) Platelet serotonin-2A receptors: a potential biological marker for suicidal behavior. Am J Psychiatry 152:850–855

Persson ML, Geijer T, Wasserman D, Rockah R, Frisch A, Michaelovsky E, Jonsson EG, Apter A, Weizman A (1999) Lack of association between suicide attempt and a polymorphism at the dopamine receptor D4 locus. Psychiatr Genet 9:97–100

Pfennig A, Kunzel HE, Kern N, Ising M, Majer M, Fuchs B, Ernst G, Holsboer F, Binder EB (2005) Hypothalamus-pituitary-adrenal system regulation and suicidal behavior in depression. Biol Psychiatry 57:336–342

Phillips MR, Li X, Zhang Y (2002) Suicide rates in China, 1995–99. Lancet 359:835–840

Pitchot W, Hansenne M, Gonzalez MA, Ansseau M (1995) The dexamethasone suppression test in violent suicide attempters with major depression. Biol Psychiatry 37:273–274

Porter RJ, Gallagher P, Watson S, Young AH (2004) Corticosteroid-serotonin interactions in depression: a review of the human evidence. Psychopharmacology (Berl) 173:1–17

Raadsheer FC, Hoogendijk WJ, Stam FC, Tilders FJ, Swaab DF (1994) Increased numbers of corticotropin-releasing hormone expressing neurons in the hypothalamic paraventricular nucleus of depressed patients. Neuroendocrinology 60:436–444

Roy A (1993) Genetic and biologic risk factors for suicide in depressive disorders. Psychiatr Q 64:345–358

Roy A, Hu XZ, Janal MN, Goldman D (2007) Interaction between childhood trauma and serotonin transporter gene variation in Suicide. Neuropsychopharmacology 32(9):2046–2052

Roy A, Janal M (2006) Gender in suicide attempt rates and childhood sexual abuse rates: is there an interaction? Suicide Life Threat Behav 36:329–335

Roy A, Rylander G, Forslund K, Asberg M, Mazzanti CM, Goldman D, Nielsen DA (2001) Excess tryptophan hydroxylase 17 779C allele in surviving cotwins of monozygotic twin suicide victims. Neuropsychobiology 43:233–236

Roy A, Segal NL, Sarchiapone M (1995) Attempted suicide among living co-twins of twin suicide victims. Am J Psychiatry 152:1075–1076

Rujescu D, Giegling I, Bondy B, Gietl A, Zill P, Moller HJ (2002) Association of anger-related traits with SNPs in the *TPH* gene. Mol Psychiatry 7:1023–1029

Rujescu D, Giegling I, Sato T, Hartmann AM, Moller HJ (2003) Genetic variations in tryptophan hydroxylase in suicidal behavior: analysis and meta-analysis. Biol Psychiatry 54:465–473

Sabol SZ, Hu S, Hamer D (1998) A functional polymorphism in the monoamine oxidase A gene promoter. Hum Genet 103:273–279

Sander T, Harms H, Dufeu P, Kuhn S, Hoehe M, Lesch KP, Rommelspacher H, Schmidt LG (1998) Serotonin transporter gene variants in alcohol-dependent subjects with dissocial personality disorder. Biol Psychiatry 43:908–912

Sher L (2007) The role of the hypothalamic-pituitary-adrenal axis dysfunction in the pathophysiology of alcohol misuse and suicidal behavior in adolescents. Int J Adolesc Med Health 19:3–9

Sher L, Mann JJ, Traskman-Bendz L, Winchel R, Huang YY, Fertuck E, Stanley BH (2006) Lower cerebrospinal fluid homovanillic acid levels in depressed suicide attempters. J Affect Disord 90:83–89

Sher L, Oquendo MA, Li S, Ellis S, Brodsky BS, Malone KM, Cooper TB, Mann JJ (2003) Prolactin response to fenfluramine administration in patients with unipolar and bipolar depression and healthy controls. Psychoneuroendocrinology 28:559–573

Skondras M, Markianos M, Botsis A, Bistolaki E, Christodoulou G (2004) Platelet monoamine oxidase activity and psychometric correlates in male violent offenders imprisoned for homicide or other violent acts. Eur Arch Psychiatry Clin Neurosci 254:380–386

Smith SM, Vale WW (2006) The role of the hypothalamic-pituitary-adrenal axis in neuroendocrine responses to stress. Dialogues Clin Neurosci 8:383–395

Soderstrom H, Blennow K, Manhem A, Forsman A (2001) CSF studies in violent offenders. I. 5-HIAA as a negative and HVA as a positive predictor of psychopathy. J Neural Transm 108:869–878

Neurobiology of Suicide

Spreux-Varoquaux O, Alvarez JC, Berlin I, Batista G, Despierre PG, Gilton A, Cremniter D (2001) Differential abnormalities in plasma 5-HIAA and platelet serotonin concentrations in violent suicide attempters: relationships with impulsivity and depression. Life Sci 69:647–657

Statham DJ, Heath AC, Madden PA, Bucholz KK, Bierut L, Dinwiddie SH, Slutske WS, Dunne MP, Martin NG (1998) Suicidal behaviour: an epidemiological and genetic study. Psychol Med 28:839–855

Stein MB, Schork NJ, Gelernter J (2007) Gene-by-environment (serotonin transporter and childhood maltreatment) interaction for anxiety sensitivity, an intermediate phenotype for anxiety disorders. Neuropsychopharmacology [Epub ahead of print]

Sumiyoshi T, Stockmeier CA, Overholser JC, Thompson PA, Meltzer HY (1995) Dopamine D4 receptors and effects of guanine nucleotides on [3H]raclopride binding in postmortem caudate nucleus of subjects with schizophrenia or major depression. Brain Res 681:109–116

Suominen K, Isometsa E, Suokas J, Haukka J, Achte K, Lonnqvist J (2004) Completed suicide after a suicide attempt: a 37-year follow-up study. Am J Psychiatry 161:562–563

Turecki G (2005) Dissecting the suicide phenotype: the role of impulsive-aggressive behaviours. J Psychiatry Neurosci 30:398–408

Virkkunen M, Goldman D, Nielsen DA, Linnoila M (1995) Low brain serotonin turnover rate (low CSF 5-HIAA) and impulsive violence. J Psychiatry Neurosci 20:271–275

Viveros MP, Marco EM, File SE (2005) Endocannabinoid system and stress and anxiety responses. Pharmacol Biochem Behav 81:331–342

Walther DJ, Bader M (2003) A unique central tryptophan hydroxylase isoform. Biochem Pharmacol 66:1673–1680

Walther DJ, Peter JU, Bashammakh S, Hortnagl H, Voits M, Fink H, Bader M (2003) Synthesis of serotonin by a second tryptophan hydroxylase isoform. Science 299:76

Wasserman D, Geijer T, Sokolowski M, Rozanov V, Wasserman J (2007) Nature and nurture in suicidal behavior, the role of genetics: some novel findings concerning personality traits and neural conduction. Physiol Behav 92(1-2):245–249

Westrin A, Nimeus A (2003) The dexamethasone suppression test and CSF-5-HIAA in relation to suicidality and depression in suicide attempters. Eur Psychiatry 18:166–171

World Health Organization (2000) World Health Organization. The World Health Report 2000-health systems: improving performance. Geneve WHO 2000:

Yanagi M, Shirakawa O, Kitamura N, Okamura K, Sakurai K, Nishiguchi N, Hashimoto T, Nushida H, Ueno Y, Kanbe D, Kawamura M, Araki K, Nawa H, Maeda K (2005) Association of *14–3–3 epsilon* gene haplotype with completed suicide in Japanese. J Hum Genet 50:210–216

Zalsman G, Frisch A, Lewis R, Michaelovsky E, Hermesh H, Sher L, Nahshoni E, Wolovik L, Tyano S, Apter A, Weizman R, Weizman A (2004) DRD4 receptor gene exon III polymorphism in inpatient suicidal adolescents. J Neural Transm 111:1593–1603

Zhou Z, Roy A, Lipsky R, Kuchipudi K, Zhu G, Taubman J, Enoch MA, Virkkunen M, Goldman D (2005) Haplotype-based linkage of tryptophan hydroxylase 2 to suicide attempt, major depression, and cerebrospinal fluid 5-hydroxyindoleacetic acid in 4 populations. Arch Gen Psychiatry 62:1109–1118

Zill P, Buttner A, Eisenmenger W, Bondy B, Ackenheil M (2004a) Regional mRNA expression of a second tryptophan hydroxylase isoform in postmortem tissue samples of two human brains. Eur Neuropsychopharmacol 14:282–284

Zill P, Buttner A, Eisenmenger W, Moller HJ, Ackenheil M, Bondy B (2007) Analysis of tryptophan hydroxylase I and II mRNA expression in the human brain: a post-mortem study. J Psychiatr Res 41:168–173

Zill P, Buttner A, Eisenmenger W, Moller HJ, Bondy B, Ackenheil M (2004b) Single nucleotide polymorphism and haplotype analysis of a novel tryptophan hydroxylase isoform (*TPH2*) gene in suicide victims. Biol Psychiatry 56:581–586v

Molecular Genetics of Alzheimer's Disease

Giuseppe Verdile and Ralph N. Martin

Contents

1 Introduction 230
2 Familial Alzheimer's Disease 230
   2.1 The Amyloid Precursor Protein (APP) Gene 231
   2.2 The Presenilin Genes 239
3 Apolipoprotein E (APOE ε4): Strongest Genetic Risk Factor for Sporadic AD 246
   3.1 The Apolipoprotein ε4 Allele 246
   3.2 Apolipoprotein E (ApoE) and Its Role in AD 250
   3.3 Mechanisms by Which ApoE May Contribute to AD 252
4 Other Genetic Risk Factors 254
5 Concluding Remarks 257
References 257

Abstract Alzheimer's disease (AD) is the most common form of dementia, with currently ~24 million affected worldwide, a number that is expected to increase ~4-fold in the next 40 years. Since the first descriptions of the neuropathological hallmarks over 100 years ago, it is now apparent that the disease is a genetically complex, heterogenous, progressive neurodegenerative disorder. The identification of genetic mutations that account for the early onset forms of the disease and genes associated with the increased risk of developing the disease have played important roles in AD aetiology and the underlying mechanisms that contribute to the pathogenesis of the disease. In particular, they have provided insight into the production, accumulation and clearance of beta amyloid (Aβ), a molecule that plays a key role in the pathogenesis of AD. This review focuses on the genes in which mutations account for the majority of early onset familial cases (*APP*, *PSEN1*, and *PSEN2*), and those that are the major genetic risk factors for the disease (i.e., *APOE* ε4) and outlines their contributions to disease pathogenesis.

G. Verdile and R.N. Martin (✉)
Centre of Excellence for Alzheimer's Disease Research and Care, School of Exercise, Biomedical and Health Sciences, Edith Cowan University, Joondalup, WA 6027, Australia

D.B. Wildenauer (ed.), *Molecular Biology of Neuropsychiatric Disorders*,
Nucleic Acids and Molecular Biology, © Springer-Verlag Berlin Heidelberg 2009

1 Introduction

The histopathological characteristics of Alzheimer's disease (AD) was first described over 100 years ago by Dr. Alois Alzheimer, a German physician who published a case report, detailing the pathological changes in the cerebral cortex of a 55-year-old woman with progressive dementia. The woman, known by the name of Auguste, was presented to Dr Alzheimer with behavioral and psychiatric symptoms, including paranoia, delusions, hallucinations, and impaired memory, that progressively worsened over 5 years, until she died of a short illness in 1906. On postmortem examination of her brain, Dr. Alzheimer wrote the following in his journal "in the centre of an almost normal cell there stands out one or several fibers due to their characteristic thickness and peculiar impregnability. Numerous small miliary foci are found in the superior layers. They are determined by the storage of a peculiar substance in the cerebral cortex. All in all we have to face a peculiar disease process" (English translation; Alzheimer et al. 1995). The "fibers" and "small miliary foci" referred to in his descriptions are neurofibrillary tangles and neuritic amyloid plaques, respectively, that characterize the disease and are used for definite diagnosis upon postmortem examination of the brain.

Today, AD is classified as a progressive, neurodegenerative disorder that is the most common form of dementia with 50–70% of all cases histopathologically confirmed as AD at post-mortem (Burns 2002). Globally, the incidence of dementia was estimated in 2001 to be 24.3 million, with ~4.6 million new cases annually (one new case every 7 s) (Ferri et al. 2005). This number is estimated to increase to 81.1 million by 2040 (Ferri et al. 2005). The ageing population, improved medical health, and lifestyle risk factors are thought to contribute to the significant increase in the number of cases of AD.

Extensive research, particularly in the past 20 years, has focused on understanding mechanisms that underlie the disease process. This research has led to the identification of a number of genetic, environmental, and lifestyle factors that significantly contribute to the increased risk of developing AD and play major roles in the disease pathogenesis. This review will focus on genetic mutations, which account for a small number of familial AD cases and the genetic factors that increase the risk of developing AD.

There are a number of genetic mutations, gene polymorphisms, and gene promoter polymorphisms that predispose or increase the risk of an individual of developing AD. Of these, genetic mutations in the amyloid precursor protein (*APP*) gene and presenilins 1 and 2 (*PSEN1* and *PSEN2*) genes, which account for the majority of early onset familial AD cases, and allelic variations and polymorphisms in the *APOE* gene (the major genetic risk factor for AD) are widely accepted as the major genes involved in AD though others remain to be identified.

2 Familial Alzheimer's Disease

Familial forms of AD (FAD) account for a small percentage (5–10%) of all clinically diagnosed cases and can be classified as early (age of onset <65 years) or late onset (age of onset >65 years) AD (Thomas and Fenech 2007). Three genes are implicated in the pathogenesis of early onset forms of AD, the amyloid precursor protein

(*APP*) gene located on chromosome 21q21 and the presenilin (*PSEN1* and *PSEN2*) genes located on chromosomes 14q24.3 and 1q31-q42, respectively.

2.1 The Amyloid Precursor Protein (APP) Gene

Senile plaques and neurofibrillary tangles, which are characteristic of AD, are also found in Down syndrome (DS) patients over 35 years old (Mann 1989; Cork 1990; Tanzi 1991). Since DS individuals have an additional full-length copy or fragment of chromosome 21, this chromosome became one of the first targets of genetic linkage studies aimed at localizing the familial AD gene defect (St George-Hyslop et al. 1987; Tanzi 1991). St. George-Hyslop and colleagues (1987) found DNA markers, located on the proximal arm (q arm) of chromosome 21, which displayed genetic linkage with AD. This study was performed on four large familial AD pedigrees exhibiting an early age of onset, and thus the locus was referred to as the familial Alzheimer's disease (FAD) gene.

The APP gene encodes a group of glycosylated, transmembrane proteins containing different isoforms generated by alternate mRNA/ exon splicing of exons 7, 8, and 15 (Kang et al. 1987; Tanzi et al. 1988; Kitaguchi et al. 1988, 1990; Yoshikai et al. 1990). The isoforms can be broadly divided into two groups based on the presence or absence of the kunitz-type serine protease inhibitor (KPI) domain encoded by exon 7 (Kitaguchi et al. 1988; Ponte et al. 1988). This domain is of particular interest since it might have an autoregulatory function, modulating APP cleavage by controlling proteolytic events near the cell membrane (Li et al. 1995). Neuronal cells predominately express isoforms lacking the KPI domain (APP365, APP695, and APP714). Nevertheless, all APP isoforms are precursors of the beta amyloid (Aβ) protein.

The APP molecule is proteolytically processed by two competing pathways, the nonamyloidogenic and amyloidogenic (Aβ forming) pathways (review in Verdile et al. 2004a, 2007). These competing pathways are regulated by many factors including diet, hormonal status, and genetic mutations that determine whether or not Aβ is generated (review in Gandy and Petanceska 2001; Verdile et al. 2004a). Three major secretases are postulated to be involved in the proteolytic cleavage of APP. These include α-secretase (of which the metalloproteases ADAM17/ TACE and ADAM 10 are likely candidates), beta APP cleaving enzyme (BACE, formally known as β-secretase), and the γ-secretase. The α-secretase enzyme cleaves within the Aβ domain of APP, thus precluding the formation Aβ and generating non-amyloidogenic fragments and a secreted form of APP (α-APPs). In the amyloidogenic pathway, BACE cleaves near the N-terminus of the Aβ domain on the APP molecule, liberating another soluble form of APP, β-APPs, and a C-terminal fragment (C99) containing the whole Aβ domain. The final step in the amyloidogenic pathway is the intramembranous cleavage of the C99 fragment by γ-secretase, to liberate the Aβ peptide ([review in Verdile et al. 2004a, 2007; Fig. 1).

The dysregulated proteolytic processing of APP is thought to be the primary event in the accumulation of Aβ, particularly the longer, more neurotoxic, Aβ42 species. This process is central to the "amyloid hypothesis," which provides insight

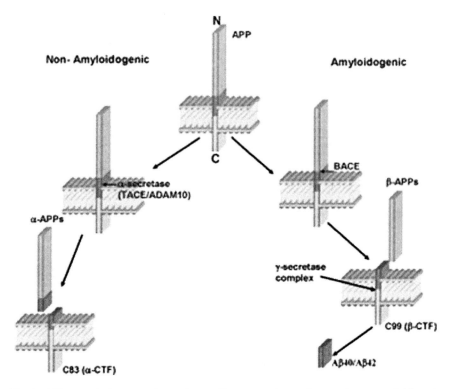

Fig. 1 APP proteolytic processing pathways. Two competing pathways, the nonamyloidogenic and amyloidogenic pathways, proteolytically cleave the majority of APP within the cell. In the nonamyloidogenic pathway, APP is cleaved with in the Aβ domain by α-secretase to liberate a secreted form of APP (α-APPs). A C-terminal fragment (C83/αCTF) is left embedded in the membrane for further cleavage into nonamyloidogenic fragments. In the amyloidogenic pathway, APP is first cleaved by BACE to liberate β-apps. The C-terminal fragment (C99/β-CTF) left embedded in the transmembrane (TM) is cleaved by the γ-secretase enzyme, which consists of four proteins (PS1, Nicastrin, APH-1, and PEN-2) that interact with each other in a high molecular weight complex. This cleavage event liberates the Aβ40/Aβ42 peptides

into mechanisms underlying the disease pathogenesis. In this hypothesis, the overproduction, accumulation, and aggregation of Aβ42, plays a pivotal role in neuronal loss or dysfunction that includes the generation of free radicals, mitochondrial damage, and inflammatory processes (review in Verdile et al. 2004a; Reddy 2006). Many factors alter APP processing towards the accumulation of Aβ42 which include the functional effects of mutations in the APP, PSEN1, and PSEN2 genes.

2.1.1 APP Mutations

While some early genetic linkage studies failed to find associations between APP mutations and AD (Tanzi et al. 1987; Van Broeckhoven et al. 1987), the screening of more families that had good linkage to chromosome 21 found several APP mutations (Murrell et al. 1991; Goate et al. 1991; see Table 1 for more information). To

Molecular Genetics of Alzheimer's Disease

Table 1 APP mutations and their effect on Aβ production

Mutation	Phenotype	Average age of onset	Effects on Aβ production	References
E665D	Not segregating with AD (not pathogenic)		No functional data available.	Peacock et al. 1993, 1994
K670M/N671L Swedish mutation	AD	55	Transfected cells: ~6 fold increase in $A\beta_{40}$ and ~5-fold increase in $A\beta_{42}$. $A\beta_{42}/A\beta_{40}$ ratio remains unchanged	Mullan et al. 1992; Citron et al. 1992
				Cai et al. 1993; Felsenstein et al. 1994
A673T	Not pathogenic (detected in 1 patient with myocardial infarction)		No functional data available.	Peacock et al. 1993
H677R	Not segregating with AD (not pathogenic)	55	No functional data available.	Janssen et al. 2003
D678N Tottori mutation	AD	61	No functional data available	Wakutani et al. 2004
A692G Flemish mutation	AD/ Cerebral Hemorrhage	46	Transfected cells: 1.5–2 fold increase in $A\beta_{40}$; 2.0–2.5-fold increase in $A\beta_{42}$; 1–1.5-fold increase in $A\beta_{42}/40$ ratio.	Hendriks et al. 1992; Haass et al. 1994; Roks et al. 2000
E693Q Dutch mutation	Hereditary cerebral hemorrhage with amyloidosis- Dutch type (HCHWA-D)	57	No change	Levy et al. 1990; Van Broeckhoven et al. 1990; Fernandez-Madrid et al. 1991; Felsenstein et al. 1994; Maruyama et al. 1991; Sahasrabudhe et al. 1992; Nilsberth et al. 2001; De Jonghe et al. 1998
			Decrease in $A\beta_{40}$ and $A\beta_{42}$ levels in media from transfected cells. Decreased $A\beta_{42}/40$ ratio in media from transfected cells. Accelerated Aβ aggregation	
E693K Italian mutation	CAA		Decrease $A\beta_{40}$ and $A\beta_{42}$ in media from transfected cells	Tagliavini et al. 1999; Nilsberth et al. 2001

(continued)

Table 1 (continued)

Mutation	Phenotype	Average age of onset	Effects on Aβ production	References
E693G Arctic mutation	AD	60	Decreased $A\beta_{40}$ and $A\beta_{42}$ in plasma and conditioned media from transfected cells. Accelerated Aβ aggregation	Kamino et al. 1992; Nilsberth et al. 2001
D694N Iowa mutation	AD/CAA/ Cerebral hemorrhage	62	Extensive distribution of $A\beta_{40}$ in plaques. No functional data available.	Grabowski et al. 2001
L705V	CAA	64	No functional data available	Obici et al. 2005
A713T	No phenotype (detected in 1 AD patient and in 5 unaffected, aged relatives). Pathogenic nature unclear		No functional data available	Carter et al. 1992
A713T	AD/ CAA	54	No functional data available	Rossi et al. 2004; Armstrong et al. 2004
A713V	No phenotype (detected in 1 patient with schizophrenia and in 1 aged unaffected individual).		No functional data available	Jones et al. 1992; Forsell and Lannfelt 1995
T714A Iranian mutation	AD	50	No functional data available	Pasalar et al. 2002; Zekanowski et al. 2003
T714I Austrian mutation	AD	36	Transfected cells: 1.5-fold increase in Aβ40; 3.5-fold increase in Aβ42; 8.5 fold increase in Aβ42/40 ratio. Mouse primary neurons: 5-fold increase in Aβ40; 1.5-fold increase in Aβ42; 8-fold increase in Aβ42/Ab40 ratio. Plasma: ~ 2-fold increase in Aβ42/Aβ40 ratio.	De Jonghe et al. 2001; Edwards-Lee et al. 2005; Raux et al. 2005
V715M French mutation	AD	51	Reduces $A\beta_{40}$ production without affecting $A\beta_{42}$ production. Increased insoluble intracellular Aβ42.	Ancolio et al. 1999; Campion et al. 1999

Molecular Genetics of Alzheimer's Disease

Mutation	Phenotype		Functional effect	References
V715A German Mutation	AD	45	Increased Aβ42/40 in HEK293 cells. Reduced Aβ40 and increased Aβ42 in mouse primary neuronal cells.	De Jonghe et al. 2001; Janssen et al. 2003; Cruts et al. 2003; Zekanowski et al. 2003
I716V Florida Mutation	AD	53	Increases Aβ42 production, Aβ40 levels remain unchanged.	Eckman et al. 1997
I716T	AD	36	No functional data available	Terreni et al. 2002
V717I London mutation	AD	52	No Change. Increased Aβ42 from stably transfected M17 cells. Increased Aβ42 in brain homogenates from transgenic mice. Aβ42 predominant in senile plaques of the cortices from two FAD individuals. Transfected cells: 1 to 5-fold reduction in Aβ40 levels, ~1.5-fold increase in Aβ42 levels and ~2-fold increase in Aβ42/Aβ40 ratio. Mouse primary neuronal cells: ~1.5-fold reduction in Aβ40 levels and 2-fold increase in Aβ42. Plasma: 1.5-fold increase in Aβ42 levels.	Goate et al. 1991; Naruse et al. 1991; Yoshioka et al. 1991; Maruyama et al. 1991; Yanagisawa et al. 1992; Cai et al. 1993; Suzuki et al. 1994; Felsenstein et al. 1994; Iwatsubo et al. 1994; Tamaoka et al. 1994; Sorbi et al. 1995; Brooks et al. 1995; Matsumura et al. 1996; Campion et al. 1996; Moechars et al. 1999; Finckh et al. 2000
V717L	AD	45	Mouse primary neurons: 1.5-fold reduction in Aβ40 and ~3-fold increase in Aβ42 levels.	Murrell et al. 2000; De Jonghe et al. 2001; Finckh et al. 2005; Godbolt et al. 2006

(continued)

Table 1 (continued)

Mutation	Phenotype	Average age of onset	Effects on Aβ production	References
V717F	AD	41	Transfected cells: COS cells: 4-fold reduction in Aβ40 levels, Ab42 levels remain unchanged. HEK-293 cells: ~1.5-fold reduction in Aβ40 and ~2-fold increase in Aβ42 levels.	Murrell et al. 1991; Finckh et al. 2005
V717G	AD		Slight Reduction in Aβ40 levels and 1.5-fold increase in Aβ42 levels.	Chartier-Harlin et al. 1991
L723P Australian mutation	AD	56	Transfected cells: ~2-fold increase in Aβ42/Aβ40 ratio Mouse primary neurons: ~3-fold increase in Aβ42/Aβ40 ratio	Kwok et al. 1998; Kwok et al. 2000
L724N Belgian mutation	AD	55	Transfected cells: ~ 2.5-fold increase in Aβ42/Aβ40 ratio. Mouse primary neuronal cells: ~2.5-fold increase in Aβ42/Aβ40 ratio.	Theuns et al. 2006

date, 26 mutations in 75 families have been identified (Alzheimer Disease and Frontotemporal Dementia Mutation Database) accounting for ~3% of all EOFAD cases. Although APP mutations have only been identified in a small number of EOFAD families, they have profound effects on APP processing (Table 1). All EOFAD-linked APP mutations are strategically located at the N- and C-termini of the Aβ domain (Fig. 2), thus providing insight into why they result in altered Aβ levels as they would alter enzyme cleavage sites.

As shown in Table 1, although the majority of APP mutations result in AD, some mutations, such as the E665D, A673T, H677R, and A713V, do not segregate with AD and are not pathogenic (Peacock et al. 1993, 1994; Janssen et al. 2003); indeed, the A713V mutation was detected in one patient with schizophrenia and in an aged unaffected individual (Jones et al. 1992; Forsell and Lannfelt 1995). Other mutations are associated with cerebral hemorrhage/cerebral amyloid angiopathy in the presence or absence of AD (see Table 1). The Dutch mutation (E693Q) results in hereditary cerebral hemorrhage with amyloidosis–Dutch type phenotype (HCHWA-D) and is characterized by severe cerebral amyloid angiopathy with haemorrhagic strokes and

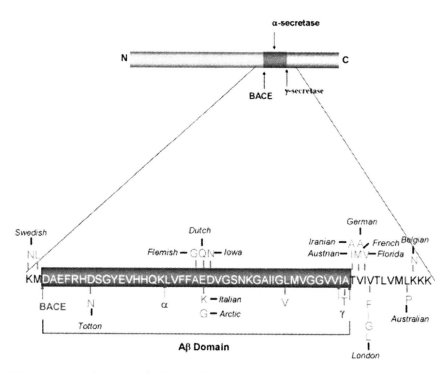

Fig. 2 Location of mutations within the APP molecule. A schematic diagram showing the location of all pathogenic mutations identified to date in the APP molecule. All mutations are located within/surrounding the Aβ domain (*red box*). Some are at or in close proximity to the BACE and γ-secretase enzyme cleavage sites. The majority of mutations are denoted by the ethnic background of the families in which the mutations were identified

dementia (Levy et al. 1990; Van Broeckhoven et al. 1990; Fernandez-Madrid et al. 1991). In contrast to the majority of other APP mutations, cells expressing the Dutch mutation have been shown to have reduced levels of Aβ (Nilsberth et al. 2001). However, this mutation results in enhanced Aβ oligomer and aggregate formation generating protofibrils in vitro (Walsh et al. 1997; Sian et al. 2000; Murakami et al. 2003), which have been shown to induce oxidative stress and death in neuronal cells and electrophysiological changes indicative of neuronal injury (Walsh et al. 1999; Hartley et al. 1999). Furthermore, in vitro evidence suggests that Aβ generated from the Dutch mutation has a greater resistance to degrading enzymes such as insulin degrading enzyme and neprilysin (Morelli et al. 2003; Tsubuki et al. 2003), possibly due to the enhanced ability of this mutation to induce Aβ aggregation.

Similar characteristics are observed for two other mutations (Arctic and Italian mutations). These mutations show reduced levels of Aβ40 and Aβ42 in media from mutation-transfected cells and in plasma from mutation carriers (Nilsberth et al. 2001). Similar to the Dutch mutation, the Italian mutation (E693K) is clinically characterized by accumulation of amyloid within the cerebral blood vessels, typical of congophillic amyloid angiopathy (CAA). However, the Artic mutation (E693G) produces a classical AD phenotype of senile plaques and neurofibrillary tangles, and unlike the other intra-Aβ mutations (Italian, Dutch, and Flemish), CAA is absent (Nilsberth et al. 2001). In addition, the Arctic mutations have been shown to accelerate aggregation of Aβ compared to the wild-type Aβ peptide (Walsh et al. 1997, 2001; Nilsberth et al. 2001). The phenotype seen in subjects with the Arctic, Italian, or Dutch mutations, and its effects on Aβ accumulation and aggregation, provides insight into how Aβ modification may lead to disease and may reflect a similar mechanism for late onset AD, where Aβ is considered an important factor.

Duplication of the APP gene has also been associated with AD. Recent reports have shown duplication of the APP locus on chromosome 21 in 1 out of 10 Dutch families and 5 out of 65 French families with autosomal dominant early onset AD and CAA (Rovelet-Lecrux et al. 2006; Sleegers et al. 2006). While this represents a small number of families with early onset AD, much larger numbers need to be screened for an accurate frequency to be determined. Rovelet-Lecrux et al. (2006) further showed extracellular and vascular deposition of amyloid, with increased CSF levels of Aβ40 and Aβ42 in individuals with a duplication in the APP locus. These families did not have any clinical features suggesting Down syndrome, such as mental retardation prior to dementia, and were clinically diagnosed with progressive dementia of Alzheimer type with a mean age of onset of 52. Conflicting results, however, were reported in the two independent studies regarding whether other gene loci on chromosome 21 were also duplicated. Rovelet-Lecrux et al. (2006) showed in the French families that larger genomic regions (ranging from 0.58 to 6.37 Mb) containing 5–12 genes were duplicated. In contrast, in the Dutch family, Sleegers et al. (2006) found the duplication in the APP locus (extending maximally over 0.7 Mb). However, five other genes are located on this region of the chromosome and the authors could not rule out that these genes are also duplicated and could contribute to Aβ accumulation and aggregation. Nevertheless, both studies detected duplications in the APP locus in the absence of mutations in the known AD genes (APP, PSEN1/PSEN2) suggesting

Molecular Genetics of Alzheimer's Disease 239

that investigating duplication of the APP locus is warranted when mutations are excluded as the cause of AD in early onset familial cases.

Although APP mutations have been identified in familial AD cases, they only account for a small percentage of ~3% (Ercelen and Mercan 2005) of all EOFAD cases. The vast majority of ~50–70% (Ercelen and Mercan 2005) are caused by mutations in the presenilins.

2.2 The Presenilin Genes

Genetic linkage studies of the AD3 locus mapped to chromosome 14q24.3 (Schellenberg et al. 1992; St George-Hyslop et al. 1992). Sherrington et al. (1995) located the presenilin 1 (*PSEN1*) gene in two clusters of the AD3 locus. A second gene was mapped to the AD4 locus on chromosome 1q31.42, termed presenilin 2 (*PSEN2*) in which mutations were first identified in a Volga-German-kindred containing seven related families, clinically diagnosed with autosomal dominant EOAD (Levy-Lahad et al. 1995; Rogaev et al. 1995). Exon trapping experiments showed that the presenilin genes consist of 12 exons, of which 10 comprise the coding sequence (Sherrington et al. 1995; Cruts et al. 1995; Prihar et al. 1996). The presenilin genes can undergo exon splicing and nucleotide insertion and deletion generating alternative transcripts. These transcripts can result in inactive, shortened peptides, proteins containing additional amino acid insertions and proteins lacking certain domains. These transcripts have been associated with familial and sporadic AD, spastic paraparesis, frontal temporal dementia and AD with lewy body type dementia (further discussed below in Sect. 2.2.2).

The presenilin genes encode transmembrane proteins consisting of 6–9 hydrophobic transmembrane domains (Sherrington et al. 1995; Rogaev et al. 1995; Lehmann et al. 1997). The presenilin proteins (PS1 and PS2) are multifunctional proteins involved in a wide range of cellular processes, including signal transduction, cell adhesion, protein trafficking, tau phosphorylation, unfolded protein response, apoptosis, and calcium regulation (recently reviewed in Parks and Curtis 2007). However, it is the presenilins' role in the proteolytic cleavage of APP to generate Aβ that has received the most attention. Presenilin 1 (PS1) is a critical component of the γ-secretase enzyme complex that is responsible for Aβ production. Initial evidence showing that Aβ production was reduced in PS knockout mice (De Strooper et al. 1998; Herreman et al. 1999), mutagenesis of critical residues result in reduced γ-secretase activity (Wolfe et al. 1999) and γ-secretase inhibitors affinity label and bind PS1 (Li et al. 2000; Esler et al. 2000; Seiffert et al. 2000) all suggested that PS1 was the γ-secretase enzyme itself. However, other proteins including nicastrin, anterior pharynx defective 1 (APH-1a/b) and presenilin enhancer (PEN-2) have been identified, which interact with the presenilins to form the multi-subunit γ-secretase enzyme complex. The roles of these proteins within the complex are slowly becoming apparent (recently reviewed in Verdile et al. 2007).

Three-dimensional image reconstructions of the γ-secretase enzyme purified from cells overexpressing the enzyme components have provided insight into the structure of this multi-subunit complex (Ogura et al. 2006; Lazarov et al. 2006). However, there are many facets of γ-secretase activity that are yet to be identified, such as the domain responsible for the actual catalytic activity, the mechanisms of enzymatic activity within the lipid bi-layer, and the existence of single or multiple γ-secretase complexes. Further investigation using appropriate protein expression models and reconstitution studies such as those used to validate the protein components of γ-secretase complex is required to provide some insight into the processes that underlie the catalytic activity of this enzyme. Understanding the exact mechanism(s) of enzyme activity would provide a site that could be a potential target to develop agents that specifically modulate Aβ levels without altering the other known (and unknown) activities of the γ-secretase enzyme.

2.2.1 Distinct Functions for PS1 and PS2

The presenilin genes share 67% sequence homology and have some structural similarities (Levy-Lahad et al. 1995; Rogaev et al. 1995). In particular, high sequence homology is observed within the predicted transmembrane domains within the protein, whereas fewer similarities are apparent in the N-terminus and in the hydrophilic loop (Levy-Lahad et al. 1995; Rogaev et al. 1995). However, several lines of evidence suggest that PS1 and PS2 may have distinct functions. Mice lacking PS1 die before birth and have severe skeletal and brain deformities, whilst those lacking PS2 only develop a mild pulmonary fibrosis and hemorrhage with age (Shen et al. 1997; Wong et al. 1997, Herreman et al. 1999; Rozmahel et al. 2002). Compared to neuronal cultures isolated from PS2 ablated mice, those isolated from PS1 knockout mice exhibit lower Aβ production (Herreman et al. 1999; Lai et al. 2003). In addition, PS2 and PS1 transgenic mice have differential effects on γ-secretase activity (Mastrangelo et al. 2005). Here, the authors reported that, although mutations in PS1 lead to higher brain levels of Aβ40 and Aβ42 than PS2 mutations, compared to PS2 wild-type mice, those mice harboring PS2 mutations had higher brain levels of Aβ42, whilst Aβ40 brain levels remain unchanged (Mastrangelo et al. 2005). These studies suggest that distinct PS1-containing and PS2-containing complexes may exist. If this is the case, evidence to date suggests that PS2-containing complexes have different functions and have less γ-secretase processing power than PS1-containing complexes. This may also hold true for the function of γ-secretase to generate fragments involved in other cellular processes and may account for the phenotypic differences observed in PS2 and PS1 knockout mice. Further in vitro/protein expression studies may provide some insight into the functional differences of PS1 and PS2.

2.2.2 Presenilin Mutations

The vast majority (~50–70%) of early onset familial AD (EOFAD) cases are caused by mutations in the presenilins. Mutations in PS1 account for the majority

Molecular Genetics of Alzheimer's Disease 241

of EFOAD with 162 mutations in 356 families identified to date, whereas 10 PS2 mutations in 18 families have been reported (Alzheimer Disease and Frontotemporal Dementia Mutation Database). In addition, those cases identified with PS2 mutations have later ages of onset of AD with 4 out of 10 mutations (N141I; V148I; G228L; M239V) associated with an average age of onset considered to be late onset AD (70–73 years) (Levy-Lahad et al. 1995; Rogaev et al. 1995; Lao et al. 1998; Zekanowski et al. 2003; Finckh et al. 2005). In contrast, mutations in PS1 are associated with earlier ages of onset ranging from 24 to 59 years old (Fig. 3 a, b). Although the mutations are located throughout the PS1 molecule (Fig. 3b), mutations appear to be concentrated around or within the transmembrane regions of the molecule (Fig. 3a, b), suggesting that these regions may be critical for protein function.

The transmembrane regions are thought to be critical to the activity of the PS1 as mutagenesis of conserved aspartate residues (D257; D385; critical for aspartyl protease activity) are located within transmembrane domains 6 and 7 reduces γ-secretase activity (Wolfe et al. 1999) and bind transition state γ-secretase inhibitors (Li et al. 2000; Esler et al. 2000; Seiffert et al. 2000). Recent in vitro studies have also suggested that the transmembrane domains 6 and 7 form a hydrophilic pore within the membrane to enable catalytic water to enter an otherwise hydrophobic environment and thus allow proteolytic enzyme activity (Tolia et al. 2006; Sato et al. 2006). It would be conceivable that mutations in the transmembrane domains alter the formation of the hydrophilic pore thereby altering γ-secretase activity. This would suggest that the vast number of mutations may lead to a loss of function.

For many years, it was thought that PS1 mutations led to an increase in the more pathogenic Aβ42 or an increase in the Aβ42/40 ratio in vitro and in vivo (Scheuner et al. 1996; Borchelt et al. 1996; Lemere et al. 1996; Duff et al. 1996; Ishii et al. 1997; Citron et al. 1997; Xia et al. 1997; Murayama et al. 1999). However, this notion is changing with a loss of function explaining an increase in Aβ42 production (Wolfe 2007; De Strooper 2007). Some evidence for presenilin mutations causing a loss of function has come from studies that have showed that presenilin mutations result in the loss of Notch signaling (Levitan et al. 1996; Baumeister et al. 1997, which is one of a number of other γ-secretase substrates that include syndecan and N-cadherin (Song et al. 1999; Baki et al. 2001; Schroeter et al. 2003; Bentahir et al. 2006).

Recent experiments re-evaluated the effects of clinical PS1 mutations on APP proteolytic processing expressed in cells deficient of wild-type PS1 or in stably transfected cell lines (Bentahir et al. 2006; Kumar-Singh et al. 2006). These studies measured the absolute levels of Aβ40 and Aβ42 generated from the cells and confirmed that the majority of the PS1/PS2 mutations resulted in an increase in Aβ42/Aβ40 ratio. However, these studies also showed that some mutations led to a reduction in the levels of Aβ40 and two additional fragments generated from γ-secretase enzyme activity [Notch and APP intracellular domains, (NICD and AICD, respectively)] and accumulation of APP-C99 fragments, similar to that observed in cells lacking presenilins (Bentahir et al. 2006; Kumar-Singh et al. 2006). These studies suggest these mutations maybe altering the "normal" function of presenilins to generate Aβ40, NICD, or AICD, favoring the production of Aβ42.

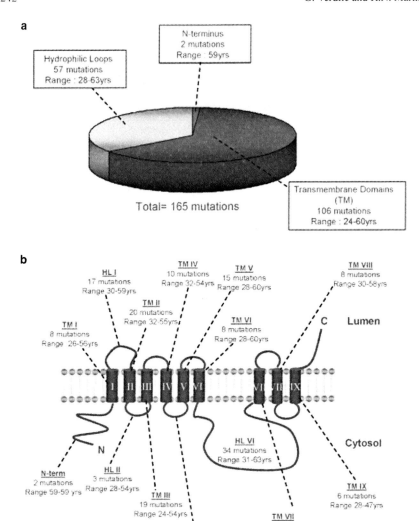

Fig. 3 Number and location of mutations with the PS1 molecule. **a** A pie graph representing the total number of mutations, the number within the N-terminus, hydrophilic loop, and transmembrane domains and the range of ages of onset associated with the mutations. **b** Location of all mutations (and their associated ages of onset) within the PS1 molecule, grouped according to those located within the N-terminus, transmembrane domains and hydrophilic loop. Data obtained from the Alzheimer Disease and Frontotemporal Dementia Mutation Database – http://www.molgen.ua.ac.be/admutations)

The majority of PS1 mutations are missense mutations, resulting in single amino acid substitution. However, mutations resulting in complete exon deletion (e.g., PS1Δ9 mutation) or shifts in the gene open reading frames as a result of single-

Molecular Genetics of Alzheimer's Disease

codon insertions or deletions, have been reported. The PS1Δ9 mutation, first identified in a British FAD family (Perez-Tur et al. 1995), causes a genomic deletion of exon9, resulting in in-frame skipping of exon 9 and an amino acid change at the splice junction of exon 8 and 10. This mutation, results in the removal of the protease cleavage site within PS1. Consequently, the Δ9 mutation cannot be proteolytically cleaved into its constitutive active N-and C-terminal fragments (Thinakaran et al. 1996) and remains as a full-length protein. Similar to other presenilin mutations, the deletion of exon 9 results in increased Aβ42 levels in transfected cell lines (Steiner et al. 1998) and in human brain tissue from FAD patients harboring the exon 9 deletion (Ishii et al. 1997). These results suggest that this deletion mutation is functionally active, although it cannot be processed. Steiner et al. (1999) have shown that indeed the biological activity of the exon 9 deletion is independent of its inability to be processed. Instead, a point mutation (S290C) occurring at the aberrant exon 8/10 splice junction was found to be responsible for the increased Aβ levels.

The PS1Δ9 mutation is also one of few that together with causing an AD phenotype is also associated with familial spastic paraparesis (Kwok et al. 1997; Crook et al. 1998; Prihar et al. 1999; Verkkoniemi et al. 2000; Hiltunen et al. 2000; Houlden et al. 2000; Smith et al. 2001; Halliday et al. 2005), a neurodegenerative disorder of the central motor system. The observation that this PS1 mutation causes another neurodegenerative disorder outlines the phenotypic spectrum of PS1 mutations, which extends beyond EOFAD.

Presenilin mutations can also generate alternative gene transcripts. The PSEN1Intron4; InsTAC mutation is one example that occurs at the splice donor consensus site of intron 4 of the PSEN1 gene and generates three different transcripts (Tysoe et al. 1998; De Jonghe et al. 1999). These transcripts can either result from deletion or insertion of a codon resulting in shifted reading frames, which in turn results in either premature stop codons (in the case for deletions) or longer transcripts (Tysoe et al. 1998; De Jonghe et al. 1999). Although shortened RNA transcripts are generated from deletion mutants, it is predicted to produce a protein truncated at its C-term (~7 kDa). These truncated proteins are undetectable in brain homogenates or lymphoblasts from carriers of the mutation (De Jonghe et al. 1999). It is predicted that this deletion mutation results in inactive proteins since overexpressing the mutation in cells does not result in the increased secretion of Aβ42 (De Jonghe et al. 1999). However, it is noted that the intron 4 mutation generates another transcript resulting from the insertion of a Thr between codons 113 and 114 (PSEN1 T113–114ins) (De Jonghe et al. 1999). Expressing this insertion mutation in cells results in ~a 3-fold increase in Aβ42 levels and is most likely to account for the AD pathology observed in carriers of the intron 4 mutation (De Jonghe et al. 1999).

Alternate presenilin transcripts and truncated proteins have also been observed in other neurodegenerative diseases. A mutation (InsR352) that results in the insertion of an arginine in the coding region of PS1 has been associated with rare forms of familial frontotemporal dementia (FTD) not associated with dominant mutations in the gene encoding the microtubule-associated protein, tau (MAPT) (Rogaeva et al. 2001; Tang-Wai et al. 2002; Amtul et al. 2002; Dermaut et al. 2004). This mutation has been shown to decrease Aβ production and Notch cleavage in vitro (Amtul et al. 2002), indicative of reduced γ-secretase activity. Another PS1 mutation (G183V) was identified in a patient with Picks Disease, a pathological sub-type familial FTD that is, neuropathologically, characterized by neurofibrillary tangles and ovoid "pick" intraneuronal inclusion bodies. This PS1 mutation affects the splice donor site of exon 6 resulting in the loss of this exon from some transcripts (Dermaut et al. 2004). The levels of Aβ42 were shown to only modestly increase in cells expressing the mutation, and no amyloid plaques were observed in the brains from carriers of the mutation (Dermaut et al. 2004). It is conceivable that similar to the deletion mutant located in intron 4 described above, the InsR352 and G183V mutations may represent a loss of PS1 function or a dominant negative effect. In addition, it is not known whether alternative PS1 products are present in the brains of carriers of these mutations. Although alternative PS1 protein products have been observed in brains from individuals with sporadic FTD (Evin et al. 2002), it remains unclear whether these products reflect alternate proteolytic processing of PS1, splicing in/out of codons, or post-translational modifications. Nevertheless, alternative transcripts do exist in neurodegenerative diseases, and their function and role in disease warrants further clarification.

2.2.3 Presenilin Polymorphisms

Although the vast majority of presenilin mutations show 100% penetrance, there are mutations that do not either segregate with disease within family members and/ or appear to be risk factors rather than autosomal dominant. An example of the latter is the PS1 E318G mutation. The pathogenicity of this mutation has been questioned since it is present in nondemented control individuals or early and late onset AD patients without a known familial history (Sandbrink and Beyreuther 1996; Reznik-Wolf et al. 1996; Helisalmi et al. 2000) as well as in FAD patients (Forsell et al. 1997; Cruts et al. 1998; Taddei et al. 2002). Interestingly, the presence of the E318G mutation has been associated with significantly lower cognitive performance, compared to controls, and in nondemented subjects with complaints of memory impairments, suggesting that this mutation is associated with the risk of cognitive decline (Laws et al. 2002a). The diverse effects of this mutation have led to suggestions that the E318G mutation may be either a polymorphism (Mattila et al. 1998; Aldudo et al. 1998; Dermaut et al. 1999), a neutral mutation, a risk factor (Dermaut et al. 1999; Helisalmi et al. 2000; Taddei et al., 2002), or a mutation showing incomplete penetrance (Taddei et al. 2002). Furthermore, it has been suggested that, alternatively, the E318G mutation might be in linkage disequilibrium, with a pathogenic change somewhere else in the PS1 gene or in close proximity to the PS1 gene (Helisalmi et al. 2000).

Polymorphisms have also been identified in the regulatory region of PSEN1. Van Duijin and colleagues were the first to show a significant association between the regulatory region of presenilin and non-familial EOAD in a Dutch population-based

Molecular Genetics of Alzheimer's Disease   245

study (van Duijn et al. 1999). Polymorphisms within the promoter region of PSEN1 have been shown to increase the risk for LOAD (−4,752 C/T; OR= 1.59; Matsubara-Tsutsui et al. 2002) and EOAD (−48C/T; OR= 1.6- ~3; Theuns et al. 2000; Lambert et al. 2001). Furthermore, the −48 CC genotype from the −48 C/T polymorphism is correlated with an increase in the Aβ 40 and Aβ42 load in the AD brain (Lambert et al. 2001). Together, these studies suggest that genetic variations in the regulatory region leading to altered expression levels in the presenilin proteins may influence Aβ production or accumulation. Evidence to date suggests that promoter polymorphisms in PSEN1 result in a reduction in transcription level and promoter activity (Theuns et al. 2003). Whether this translates to reduced protein levels is yet to be determined. However, evidence that ablation and dose dependent knockdown of PS1 (De Strooper et al. 1998; Luo et al. 2004; Xie et al. 2004) results in a reduction in Aβ production, together with evidence showing a reduction in PS1 protein levels is associated with increased Aβ load in EOAD brains (Verdile et al. 2004b), suggest that variations in PS1 levels may contribute to the pathogenesis of AD.

2.2.4 Genetic Mutations and Polymorphisms in Other Components of the γ-Secretase Complex

Considering the biological functions of the other components of the γ-secretase enzyme complex (nicastrin, APH1, and PEN-2) and their role in AD pathogenesis, it would be expected that these components may be candidate genetic risk factors for AD. However, compared to the PSEN or APP genes, there have been very few clinical mutations or genetic polymorphisms in the other γ-secretase complex components. Initial attempts at sequencing and haplotype analysis of the NCSTN gene in AD and control subjects in a Dutch population revealed a set of 4 single nucleotide polymorphisms (SNPs), 10 intronic SNPs, 3 silent mutations, and 1 missense mutation (Dermaut et al. 2002). The frequency of the missense mutation (N417Y) was similar in AD and controls and thus did not appear to segregate with disease. Similar findings were described in an Italian population where the mutation was detected in patients with sporadic AD, EOFAD, and in a control subject (Confaloni et al. 2003). In addition, expressing the mutation in cells did not result in an increase in Aβ42 levels (Dermaut et al. 2002) suggesting that it does not contribute to AD pathogenesis. The frequency of 1 NCSTN SNP haplotype was ~2.5-fold higher in the EOAD group compared to LOAD and control groups and was further increased (up to 4-fold) in individuals with EOAD not carrying the APOE ε4 allele (Dermaut et al. 2002). These initial reports suggest that SNP haplotypes in the NCSTN gene are risk factors for EOAD. However, subsequent studies in Italian (Orlacchio et al. 2004) and French (Cousin et al. 2003) cohorts did not show associations between NCSTN SNP haplotypes suggesting variations amongst different populations.

In a recent study, genetic analysis of all AD related genes was performed in AD ($n = 83$) and control ($n = 113$) samples from a relatively genetically isolated population from one of Italy's major islands, Sardina (Piscopo et al. 2006). The authors found one PS1 mutation (E318G) in 2 subjects with FAD and a NCSTN haplotype (2 silent mutations and 1 intronic polymorphism), the frequency of which was 2-fold higher in the EOAD compared to controls. In addition, an intronic polymorphism was identified in the PEN-2 gene in which the frequency of the homozygous form was ~ 2-fold higher in the EOAD compared to controls. In another Italian population, 3 SNPs were identified of which one had a higher prevalence in AD subjects carrying the ε4 allele (Sala Frigerio et al. 2005). Similar associations were observed in a Chinese population in which SNPs were associated with the risk of LOAD in APOE ε4 carriers (Jia et al. 2007). Mutations in PEN2 are very rare with only one (D90N) identified to date in a 79-year-old Italian man with AD (Sala Frigerio et al. 2005). It is not known whether this mutation is an autosomal dominant mutation as the genetic profile of family members was not available for segregation analysis. In addition, in vitro experiments showed that the mutation did not affect Aβ42 production (Sala Frigerio et al. 2005), suggesting that it is not pathogenic. However, altered Aβ42 production when this mutation is co-expressed with the other γ-−secretase components cannot be excluded. Genetic association studies in the APH-1A/B genes have shown that polymorphisms do not significantly contribute to AD risk in an Italian cohort and the missense mutations identified do not segregate with disease (Poli et al. 2003). Further studies with other ethnic groups or with groups where samples are derived from genetically isolated populations would clarify the risk profile of NCSTN, PEN-2, or APH-1 polymorphisms and mutations.

3 Apolipoprotein E (APOE ε4): Strongest Genetic Risk Factor for Sporadic AD

Genetic mutations or polymorphisms in APP, presenilins, nicastrin, PEN-2, or APH-1 are associated with a relatively small number of AD cases. The vast majority of AD cases (>90%) are sporadic and are associated with a number of risk factors, including age, and changes in reproductive hormones (oestrogen, testosterone, gonadotropins) during menopause/andropause, as well as those associated with cardiovascular disease, including high LDL levels and low HDL levels, obesity, high blood pressure, and type II diabetes. Head injuries, physical inactivity and factors that promote oxidative stress are also considered important risk factors for AD (recently reviewed in Dosunmu et al. 2007). Of these, the ε4 allele of the apolipoprotein E gene (APOE) is by far the strongest genetic risk factor found in most populations studied and accounts for ~50% of AD cases.

3.1 The Apolipoprotein ε4 Allele

The human *APOE* gene is located on chromosome 19 (19q13.2) (Olaisen et al. 1982; Das et al. 1985) with three major alleles existing, denoted as *APOE* ε2, *APOE* ε3, and *APOE* ε4 (Zannis et al. 1982). Other alleles of *APOE* do exist, namely *APOE* ε1,

Molecular Genetics of Alzheimer's Disease

APOE ε5, and *APOE* ε7, but they are extremely rare and usually associated with hyperlipidaemia and atherosclerosis (Weisgraber et al. 1984, Maeda et al. 1989; Yamamura et al. 1984; Matsunaga et al. 1995). The distribution of the three major alleles in almost all populations is 5–10%, 70–80%, and 10–20% for *APOE* ε2, ε3, and ε4 alleles, respectively (reviewed in Huang 2006). Initial linkage studies of families with LOAD resulted in the disease locus identified on chromosome 19 (Pericak-Vance et al. 1991). Subsequent studies led to the association of the *APOE* ε4 allele with both familial and sporadic LOAD cases (Strittmatter et al. 1993; Poirier et al. 1993b). In addition, the *APOE* ε4 allele is associated with more rapid memory decline in nondemented individuals (Dik et al. 2000; Deary et al. 2002) and preclinical memory in asymptomatic middle-aged individuals (Flory et al. 2000; Caselli et al. 2001).

Many studies in various ethnic populations have shown that the ε4 allele is over-represented in the AD population. In the Caucasian population, for example, the prevalence of the ε4 allele has been shown to increase from 14% in the control population up to 50% in the AD population (Poirier et al. 1993b; Saunders et al. 1993; Martins et al. 1995). In contrast, the frequency of ε2 allele is lower in the AD population compared to controls (Poirier et al. 1993b; Saunders et al. 1993; Martins et al. 1995) and is thought to exert a protective effect. The degree of risk of AD associated with the *APOE* ε4 allele varies with different ethnic backgrounds. A meta-analysis of 40 studies (Farrer et al. 1997) that included African-American, Caucasians, Hispanic, and Japanese ethnic backgrounds revealed that the association between the ε4 allele and AD was highest in the Japanese population (ε4/ε3 OR 5.6; ε4/ε4 OR 33.1) followed by the Caucasians (ε3/ε4, OR 3.2; ε4/ε4, OR 14.9). This cumulative data is consistent with the notion that the *APOE* ε4 allele is the major genetic risk factor for AD across most populations.

Interestingly, Africans in Nigeria and East Africa show no association of *APOE* ε4 with AD (Osuntokun et al. 1995; Sayi et al. 1997), whereas African-Americans do show an association. This marked difference between native Africans and the African-Americans would suggest that high-cholesterol intake of the Western diet would contribute to the increased risk in the African-Americans. Of the three apoE isoforms, apoE4 has the strongest affinity for the LDL receptor, which mediates cholesterol uptake. This feature of apoE4 would be beneficial in the hunter-gatherer society (such as the native Africans), where high energy food sources are scarce, but rather detrimental to the western society where the converse is true. A chronically increased level of dietary cholesterol associated with the ε4 allele is a known risk factor for type-2 diabetes, hypertension, hypercholesterolaemia, and cardiovascular disease. Thus, it is conceivable that the increased risk of AD associated with *APOE* ε4 may also be partly attributed to diet (reviewed in Cacabelos et al. 2003; Laws et al. 2003). Evidence for this notion has been provided by a report that a higher intake of calories and fats may be associated with increased risk of AD in individuals carrying this ε4 allele (Luchsinger et al. 2002). Furthermore, individuals with type-2 diabetes possessing the *APOE* ε4 allele are at twice the risk of developing AD when compared to ε4 carriers who do not have diabetes (Peila et al. 2002). These studies provide evidence in support of an interaction between *APOE* ε4 and dietary factors to enhance the risk of AD.

A gene-dosage effect is also present in LOAD families, in which the risk increases from 20% when no ε4 alleles are present to 90% when two copies are present (Corder et al. 1993; Frisoni et al. 1995). The gene-dosage effect also alters the age of onset of clinical symptoms of the disease. For example, homozygous ε4 individuals have an earlier age of onset compared to non-ε4 or ε4 heterozygous individuals (Corder et al. 1993; St George-Hyslop et al. 1994; Sorbi et al. 1995, 1996). The *APOE* ε2 allele, on the other hand, is associated with a delayed age of onset for AD (Chartier-Harlin et al. 1994; Corder et al. 1994; Talbot et al. 1994; West et al. 1994; Roses et al. 1995), even in the presence of the APP – V717I mutation. However, this is not the case for another APP mutation where the phenotype is associated with a more aggressive form of the disease (Flemish mutation; A692G) (Haan et al. 1994). Similar findings were observed in families with EOFAD associated with PS1 mutations. The majority of these studies showed that the presence of the *APOE* ε4 allele did not influence the progression of disease or age of onset (Lendon et al. 1997; Houlden et al. 1998; Heckmann et al. 2004). One exception is a large study of 114 Colombians carrying the PS1 E280A mutation, which showed that compared to the non-ε4 carriers, those possessing the ε4 allele developed AD at an earlier age (35–39 compared to 40–44) (Pastor et al. 2003). Further studies are warranted to determine whether the AD phenotype caused by PS1/APP mutations is influenced by *APOE* ε4. In addition, the effects of homozygosity on *APOE* ε4 interactions with PS1 or APP mutations may provide a more definitive answer to whether *APOE* plays a role in modulating AD onset in these pedigrees.

3.1.1 APOE and Alzheimer's Pathology

Several studies have reported that amyloid plaque load increases in a gene-dosage-dependent manner, with the *APOE* ε4 homozygous subjects demonstrating significantly increased plaque density compared with heterozygotes (Rebeck et al. 1993; Zubenko et al. 1994; Gearing et al. 1995; Hyman et al. 1995; Nagy et al. 1995; Oyama et al. 1995; Polvikoski et al. 1995; Gomez-Isla et al. 1996; Olichney et al. 1996). However, some groups have failed to confirm this association (Landen et al. 1996; Morris et al. 1996). Whether *APOE* ε4 genotype is associated with NFT density remains inconclusive. Evidence has been presented in support of increased NFTs in *APOE* ε4 homozygous patients compared to non-*APOE* ε4 individuals (Schmechel et al. 1993; Nagy et al. 1995; Ohm et al. 1995; Polvikoski et al. 1995; Beffert and Poirier 1996). However, a number of studies have reported no correlation between the amount of NFT and *APOE* genotype (Berr et al. 1994; Oyama et al. 1995; Gomez-Isla et al. 1996; Itoh and Yamada 1996; Landen et al. 1996; Morris et al. 1996; Olichney et al. 1996; Brat et al. 2001).

Hippocampal atrophy, another neuropathological feature of AD, is more prominent in *APOE* ε4 carriers compared to non-*APOE* ε4 carriers. Brain imaging studies (MRI and CT) have demonstrated that individuals homozygous for the *APOE* ε4 allele have smaller hippocampal volumes than non-*APOE* ε4 subjects (Lehtovirta et al. 1995, 1996), and that this ε4 effect is dose dependent (Geroldi et al. 2000). In

Molecular Genetics of Alzheimer's Disease

nondemented *APOE* ε4 carriers, the hippocampal volume is significantly reduced (Soininen et al. 1995; Plassman et al. 1997) and the rate of volume loss is significantly faster than in non-*APOE* ε4 carriers (Moffat et al. 2000). Taken together, these studies suggest that individuals with the *APOE* ε4 allele may be at greater risk of hippocampal atrophy even prior to onset of the clinical signs of AD.

3.1.2 Interaction of APOE and Sex Hormones in AD

The incidence of AD has been reported to be higher in women than men, and is apparent across all *APOE* genotypes (Jorm et al. 1987). However, this gender difference has been reported to be age-dependent, with women over 90 having a higher incidence of AD (Ruitenberg et al. 2001). This is consistent with the neuropathological findings that neurofibrillary tangles and amyloid senile plaques are more substantial in women with AD (particularly *APOE* ε4$^+$ women) compared to men (Corder et al. 2004). These studies not only outline the relative contribution of gender to AD risk, but also provide a role for sex hormones in modulating AD risk. Evidence has also been provided for the role of oestrogen in AD pathogenesis. Serum oestrogen levels are reduced in AD sufferers (Honjo et al. 1989; Yaffe et al. 2000), and biochemical evidence indicates that oestrogen can modulate Aβ deposition in animal models and levels in cell culture (Xu et al. 1998; Petanceska et al. 2000; Zheng et al. 2002). In addition, upregulation of apoE protein levels at nonphysiological concentrations of oestrogen has been reported (Levin-Allerhand et al. 2001).

The benefits of HRT in the prevention and treatment of AD in women has become controversial. Despite early reports of improved cognition and reduced risk of AD in female HRT users (Barrett-Connor and Kritz-Silverstein 1993; Henderson et al. 1995; Tang and Hasselmo 1996; Baldereschi et al. 1998), other studies have found no benefit of HRT (Brenner et al. 1994; Mulnard et al. 2000; Wang et al. 2000; Yaffe 2001), whilst others have even reported negative effects on cognition (Shumaker et al. 2003; Espeland et al. 2004). Many factors may contribute to the discrepancies between the findings from these reports. These include type of oestrogen used (synthetic, equine, oestrogen/progesterone) (Mulnard et al. 2000; Shumaker et al. 2003; Espeland et al. 2004), treatment regime (lower continuous doses) (Luciano et al. 1993), and time of initiation of HRT. More importantly, *APOE* genotype strongly determines the outcome of HRT on cognition, which was not considered in many of the HRT studies. Studies in which the *APOE* genotype was assessed showed a benefit to *APOE* ε4 carriers, though cognition was enhanced in noncarriers (Yaffe et al. 2000; Burkhardt et al. 2004).

Reduced levels of testosterone in men (albeit a slower decline than oestrogen in women) have been associated with cognitive decline and AD. Several studies have reported that, compared to controls, men with AD and other dementias have lower serum (Bowen et al. 2000; Hogervorst et al. 2001; Hogervorst et al. 2003) and brain levels (Rosario et al. 2004) of testosterone. In addition, similar to oestrogen, testosterone modulates Aβ production in neuronal cells (Gouras et al. 2000) and protects against Aβ-mediated neurotoxicity (Goodenough et al. 2000; Pike 2001; Zhang et al.

2004). Furthermore, we have previously shown that a reduction in testosterone levels is associated with increased plasma Aβ levels in dementia cases and also in men who have undergone chemical castration (Almeida et al. 2004, Gandy et al. 2001; Gillett et al. 2003). These studies provide evidence for a role for testosterone in AD pathogenesis.

The *APOE* ε4 allele also appears to modulate the association between testosterone and the risk of developing dementia in men. We have recently reported an association between serum testosterone concentrations and cognitive performance in healthy elderly men (Burkhardt et al. 2006). This study showed that higher levels of free, bio-available testosterone were associated with better cognitive function only in men who did not possess the *APOE* ε4 genotype. However, in animal studies, the therapeutic benefit of testosterone therapy was present even in the presence of the ε4 allele. One study showed that inhibiting the effects of testosterone by blocking the androgen (AR) in male mice expressing *APOE* ε4 results in the development of prominent deficits in spatial learning and memory (Raber et al. 2002). Interestingly, female *APOE* ε4 mice, when compared to female *APOE* ε3 mice, exhibit such deficits in learning and memory without AR blockade, yet treatment with either testosterone or the nonaromatisable androgen, dihydrotestosterone, attenuates the deficits (Raber et al. 2002). These animal studies provide a precedence to assess testosterone as a substitute for oestrogen in assessing the efficacy of these sex steroids by improving cognition in women. The studies described above also outline the importance of stratifying against ε4 genotype in human association studies and clinical trials to assess therapeutic strategies such as HRT.

3.2 Apolipoprotein E (ApoE) and Its Role in AD

The three major alleles of *ApoE* gene encode for the three major isoforms of the apoE protein, denoted as apoE2, apoE3, and apoE4. These isoforms differ by one amino acid substitution at residues 112 or 158. ApoE3, the most common form, has a cysteine residue at 112 and an arginine at 158. The apoE4 isoform is generated when cysteine-112 is substituted by arginine, and the apoE2 occurs when arginine-158 is substituted by cysteine. The apoE protein is expressed in several organs including the brain where it displays the second highest level of expression (Elshourbagy et al. 1985) and is the predominant apolipoprotein in the CSF (Merched et al. 1997). Both apoE mRNA and protein are predominately localized to astrocytes, oligodendrocytes, and activated microglia (Elshourbagy et al. 1985; Boyles et al. 1985; Diedrich et al. 1991; Poirier et al. 1991; Stone et al. 1997). Although initially thought to be localized to astrocytes, albeit at lower levels, both apoE mRNA and protein expression have been found in neurons (Diedrich et al. 1991; Han et al. 1994; Beffert and Poirier 1996; Bao et al. 1996; Metzger et al. 1996; Xu et al. 1999). The relatively low levels detected in neurons may represent uptake from the extracellular surroundings rather than synthesized endogenously as observed with astrocytes.

Molecular Genetics of Alzheimer's Disease 251

The apoE protein aids in the transport of triglyceride-rich lipoproteins (chylomicrons and VLDL), phospholipid, and cholesterol into cells by mediating the binding, internalization and catabolism of lipoprotein particles by interacting with low-density lipoprotein family of receptors found on liver cells (recently reviewed in Martins et al. 2006). In the CNS, apoE is critical for lipid transport and cholesterol homeostasis within the brain (Rebeck et al. 1993; Bu et al. 1994), and it has been suggested that CNS apoE levels, distribution, and apoE allele status can influence AD risk via fluctuations in CNS cholesterol metabolism (Martins et al. 2006).

3.2.1 Apoe Expression in AD

The risk of sporadic AD is associated not only with the *APOE* ε4 genotype but also with increased levels of apoE in plasma and the brain, which is independent of *APOE* genotype (Taddei et al. 1997; Laws et al. 2002b). Moreover, the *APOE* promoter polymorphisms, which most likely have some regulatory influence on *APOE* expression, have been shown to be associated with AD risk. Studies have shown an increased level of apoE mRNA in the brain of AD patients (Yamada et al. 1995; Yamagata et al. 2001). Most data indicate an increase in the level of both the apoE mRNA and protein, most probably owing to an increased apoE expression within reactive astrocytes, a marker of inflammatory processes occurring in AD (Diedrich et al. 1991; Shao et al. 1997; Martins et al. 2001). The mechanisms underlying associations between increased apoE levels and AD risk remain to be determined. Further, it cannot be ruled out that the combination of the ε4 genotype together with apoE levels may increase the risk synergistically. Evidence for this has come from the finding that a significant increase in the relative quantity of *APOE* ε4 mRNA expression in brains of individuals with AD compared to controls (Lambert et al. 1998).

APOE Polymorphisms and AD Risk

Variations in the regulatory regions of *APOE* (Lambert et al. 1997; Artiga et al. 1998; Bullido et al. 1998) have been shown to be risk factors for AD. Three common polymorphisms have been identified in the *APOE* promoter region, at positions −491, −427, and −219 using denaturing gel electrophoresis (Artiga et al. 1998). Functional effects of the polymorphisms, assayed by transient transfection and EMSA (Electromobility shift Assay) in a human hepatoma cell line, showed that sites at −491 and −219 of the *APOE* promoter produce variations in transcriptional activity of the gene, most probably due to differential binding of nuclear proteins (Artiga et al. 1998). Another polymorphism has also been identified in the *APOE* intron 1 enhancer (IE-1) region (+113) (Mui et al. 1996).

Several studies have investigated whether the *APOE* −491, −219, and −427 polymorphisms are associated with altered risk of AD, with some reports supporting an association (Bullido et al. 1998; Lambert et al. 1998; Laws et al. 1999) and others not (Song et al. 1998; Town et al. 1998; Toji et al. 1999). Many studies have found that −491A/T and −219G/T polymorphisms modify risk for AD, and that the −427 C/T polymorphism was not associated with the disease; however, the role for +113G/C has received less attention. A large study showed an increased risk of AD associated with the −219T allele (Lambert et al. 1998). The study confirmed the deleterious effect of the −219T allele and the protective effect of the −491 T allele, which has also been described recently (Bullido et al. 1998; Lambert et al. 1998). These studies indicated that, in addition to the qualitative effect of the *APOE* ε2/ε3/ε4 alleles on AD occurrence, the qualitative variation of expression of these alleles due to functional *APOE* promoter polymoprhisms are a key determinant of AD development. It has been suggested by a number of studies that the association of the −491 A allele with AD is explained by linkage disequilibrium. However, the −491 A allele has been associated with both *APOE ε4* carriers and noncarriers, suggesting that this genotype causes an independent risk for AD (Town et al. 1998).

APOE Promoter Polymorphism and APOE Expression

It has been reported that single nucleotide changes in a promoter region may effect transcriptional activity mediated by transcription factors (Laws et al. 2003), either by directly altering a transcription factor binding site or by changing the structure of DNA, thereby affecting the access of transcription factors. The four *APOE* proximal promoter polymorphisms, described previously, have the potential of affecting the binding of regulatory proteins to *APOE* promoter. If so, they would be functional polymorphisms and can change susceptibility to AD by regulating apoE levels (Bullido and Valdivieso 2000).

In vitro studies in astrocytoma and HepG2 cell lines have supported this theory (Artiga et al. 1998). Studies have investigated the effect of the −491 polymorphisms on apoE levels in plasma (Laws et al. 1999) and in the brain (Laws et al. 2002b). In the former study, individuals with the −491AA genotype had significantly raised plasma apoE levels regardless of *APOE* genotype and AD status. Levels were increased in −491 AA controls, but were higher in individuals with AD having the same genotype. These findings are consistent with the in vitro transfection studies.

The −219 T/G promoter polymorphism has also been associated with altered levels of apoE (Lambert et al. 2000), where it appears to decrease levels of apoE in the plasma. The effects of this promoter polymorphism appear to be dose dependent, as confirmed by in vitro transfection assays. There are no conclusive results as to the effect of −427 T/C polymorphism on apoE levels, with some reporting a decrease in plasma levels (Artiga et al. 1998; Scacchi et al. 2001) and others reporting no change in levels of apoE (Corbo et al. 2001). At present, there are no functional data available for the +113 promoter polymorphism.

3.3 Mechanisms by Which ApoE May Contribute to AD

There have been a number of mechanisms proposed to underlie the associations between apoE and AD pathology. Some proposed mechanisms are outlined below.

ApoE has been shown to have isoform-specific antioxidant activity and to attenuate cytotoxic effects induced by Aβ or other oxidative stress insults, with apoE2 showing greatest benefit and apoE4 the lowest (Lauderback et al. 2002). In addition, reduced metabolic activity in neurons from the nucleus basalis from individuals with AD has been shown to be apoE4 dependent (Salehi et al. 1998), suggesting that one mechanism by which apoE4 may contribute to neurodegeneration is through impairing metabolic activity within neurons. ApoE also has a role in cognitive impairment, synaptic plasticity, and neurite outgrowth. Studies have shown that compared to wild-type mice, ablation of apoE resulted in impaired learning and memory (Gordon et al. 1996), suggesting that apoE has an important role in neuronal plasticity. Subsequent studies have shown that, compared to *APOE* ε4 transgenic mice, those mice expressing *APOE* ε3 showed increased levels of presynaptic and dendritic protein markers following environmental stimulation indicating enhanced synapse formation (Levi et al. 2005). Further, cultured dorsal ganglia cells treated with apoE3 show enhanced neurite extension and branching compared to exposure to apoE4 (Struble et al. 1999). In addition, the apoE protein also plays a key role in the repair process following neuronal injury, possibly by recycling membrane components from damaged cells (Poirier et al. 1991; Poirier et al. 1993a), whether isoform differences in neuronal repair efficiency occurs remain unclear.

Most attention, however, has been focused on the role of apoE in Aβ metabolism and clearance. ApoE has been shown to be involved in APP processing to generate Aβ. ApoE added to cell cultures results in reduced Aβ secretion and accumulation of APP-C-terminal fragment, all features indicative of γ-secretase inhibition (Irizarry et al. 2004). Interestingly, apoE did not alter γ-secretase cleavage of Notch, suggesting that its effect on γ-secretase enzyme activity is specific to APP. Whether apoE interacts with or modulates PS1 or the other enzyme components is yet to be determined. However, apoE has been shown to bind the N-terminus of APP and is thought to influence the maturation and secretion of APP (Hass et al. 2001). Although studies have shown apoE to mediate Aβ production, its interaction with Aβ has been the most widely studied.

ApoE also plays a critical role in Aβ clearance, possibly mediated by ApoE-Aβ interactions. In vivo evidence has been provided to show that apoE is associated with amyloid plaques (Namba et al. 1991; Wisniewski and Frangione 1992; Strittmatter et al. 1993). Indeed, dimeric complexes between Aβ and apoE have been identified in AD brain (Permanne et al. 1997). In addition, evidence has been provided in vitro to show that apoE can accelerate Aβ aggregation and fibril formation (Castano et al. 1995) suggesting that apoE may be involved in plaque formation. Initial interaction studies showed that purified apoE forms SDS and guanidine hydrochloride-stable complexes with Aβ peptides, with apoE4 complexes forming more rapidly and effectively (Sanan et al. 1994; Strittmatter et al. 1993; Cho et al. 2001). However, it was found in subsequent studies that delipidation and denaturation of apoE altered its function with respect to forming complexes with Aβ (LaDu et al. 1995; Yang et al. 1997). These studies showed that although apoE2 and E3 bind Aβ avidly, apoE4 did not bind at all (LaDu et al. 1995; Yang et al. 1997). We subsequently showed that the formation of apoE-Aβ complexes and subsequent uptake is promoted by apoE2 and apoE3 but not apoE4 (Yang et al. 1999). Our in vivo studies on peripheral clearance of Aβ in apoE +/+ or −/− mice have shown that apoE is required for the clearance of Aβ from the kidney and liver (Hone et al. 2003). In cell culture studies, we have also shown that Aβ enhanced the binding of apoE2 lipoprotein-like particles to fibroblasts in culture, whilst markedly reducing the binding of apoE3 and apoE4 (Hone et al. 2005).

Taken together, these findings suggest that apoE4 may contribute to AD pathology by its poor binding to Aβ and thereby not facilitating its uptake and clearance from the cell. We have observed that Aβ is cleared more efficiently from the periphery in E2 and E3 knockin mice but not E4 mice (unpublished results), suggesting that Aβ may be cleared in an isoform-specific manner. However, the mechanism(s) of apoE mediated clearance of Aβ remains unclear, and it still remains to be determined whether apoE4 acts at the level of the periphery or the brain.

4 Other Genetic Risk Factors

AD is a genetically complex, heterogeneous disorder with many cases not accounted for by presenilin/APP mutation or the presence of the *APOE* ε4 allele. Genetic association studies have shown linkage to every chromosome (Fig. 4; Alzgene Database, Bertram et al. 2007). However, the majority of these genetic association studies have shown weak associations that vary across ethnic populations. A recent study conducted meta-analysis for each polymorphism with genotypes from all genetic association studies was published in the AD field (789 studies total, 127 polymorphisms across 69 putative AD risk genes; Bertram et al. 2007). The results of the analysis showed that, in addition to the *APOE* ε4 allele, 20 polymorphisms in 13 genes revealed a significant but modest odds ratio showing average 'risk' effects of 1.25 and average 'protective' effects of 0.82 (Table 2).

Genetic linkage studies continue to identify gene loci on chromosomes that have the strongest associations with AD. Chromosome 10 contains the majority of genes that have been associated with AD (Fig. 4) and, interestingly, genetic linkage analysis

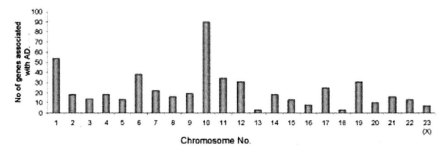

Fig. 4 Number of genes that have been reported to have some associations with AD. *Bar graph* representing the number of genes within all 23 chromosomes that have been reported to have associations with AD. Data obtained from Algene Database (www.alzforum.org/res/com/gen/alzgene/default.asp and Bertram et al. (2007))

Table 2 Genes (other than APOE) and their polymorphisms found to have significant associations with AD following a meta-analysis undertaken by Bertram and colleagues (2007). Modified from Bertram et al. 2007

Gene (Chromosome)	Function	Polymorphism(s)[a]	OR (95% c.i.) (P-value)	AD vs C cases (number of independent samples)
ACE (17) [Angiotensin I converting enzyme (peptidyl-dipeptidase A) 1]	Catalyzes the conversion of angiotensin I into a the active, potent vasopressor angiotensin II.	Intron 16 (ins/del)	1.11 (1.02–1.20) (p=0.01)	AD: 5,778 C: 12,265 (34)
		rs1800764	0.79 (0.68–0.92) (p=0.002)	AD: 818 C: 747 (33)
		rs4291	0.82 (0.70–0.96) (p=0.01)	AD: 812 C: 749 (4)
		rs4343	0.83 (0.72–0.96) (p=0.05)	AD: 824 C: 745 (4)
APOC1 (19) (Apolipoprotein C1)	APOC1 is expressed primarily in the liver, and it is activated when monocytes differentiate into macrophages.	HpaI (ins/del)	2.07 (1.67–2.57) (p=2.4 x10^{11})	AD: 1,668 C: 2,120 (13)
CHRNB2 (1) [cholinergic receptor, nicotinic, beta polypeptide 2 (neuronal)]	CHRNB2 belongs to a superfamily of ligand-gated ion channels which allow the flow of sodium and potassium across the plasma membrane in response to ligands such as acetylcholine and nicotine.	Rs4845378	0.67 (0.50–0.90) (p=0.007)	AD: 576 C: 787 (4)
CST3 (20) (cystatin C)	The cystatin superfamily encompasses proteins that contain multiple cystatin-like sequences having protease inhibitory function.	5′ UTR -157	1.23 (1.03–1.47) (p=0.02)	AD: 983 C: 799 (4)
		5′ UTR -72	1.28 (1.05–1.56) (p=0.01)	AD: 804 C: 571 (3)
		Rs1064039 (A25T)	1.15 (1.02–1.31) (p=0.02)	AD:1,877 C: 1544 (9)
ESR1 (6) (estrogen receptor 1)	A ligand-activated transcription factor composed of several domains important for hormone binding, DNA binding, and activation of transcription.	Pvu11	1.18 (1.00–1.39) (p=0.05)	AD: 1512 C: 7,194 (11)
		XbaI	1.20 (1.02–1.41) (p=0.03)	AD: 1,715 C: 7,377 (12)
GAPDHS (19) (glyceraldehyde-3-phosphate dehydrogenase, spermatogenic)	Encodes a protein belonging to the glyceraldehyde-3-phosphate dehydrogenase family of enzymes that play an important role in carbohydrate metabolism.	rs12984928	0.82 (0.71–0.95) (p=0.007)	AD: 1,014 C: 1,109 (3)

(continued)

Table 2 (continued)

Gene (Chromosome)	Function	Polymorphism(s)[a]	OR (95% c.i.) (P-value)	AD vs C cases (number of independent samples)
		rs4806173	0.82 (0.71–0.94) (p=0.004)	AD:1,015 C: 1,106 (3)
IDE (10) (insulin degrading enzyme)	Belongs to a protease family responsible for intercellular peptide signalling.	rs2251101	0.89 (0.79–1.00) (p=0.05)	AD:2,033 C: 1.902
MTHFR (1) [5,10-methylenetetrahydrofolate reductase (NADPH)]	Catalyzes the conversion of 5,10-methylenetetrahydrofolate to 5-methyltetrahydrofolate, a cosubstrate for homocysteine remethylation to methionine.	A1298C	0.52 (0.73–1.00) (p=0.05)	AD: 735 C:837 (4)
NCSTN (1) (Nicastrin)	Encodes a Type I transmembrane glycoprotein that is an integral component of the multimeric γ-secretase complex.	119 intron 16	1.38 (1.03–1.83) (p=0.03)	AD:952 C:870 (3)
PRNP (20) (prion protein)	Encodes a membrane glycosylphosphatidy linositol-anchored glycoprotein that tends to aggregate into rod-like structures.	Rs1799990 (M129V)	0.89 (0.81–0.98) (p=0.02)	AD:2,270 C: 2,968
PSEN1(14) (presenilin 1)	Encodes a multipass transmembrane protein that is a critical component of the multi-subunit γ-secretase complex	Rs165932 (intron 8)	0.92 (0.86–0.98) (p=0.009)	AD: 5,410 C: 5179 (34)
TF (3) (transferrin)	Involved in transporting iron from the intestine, reticuloendothelial system, and liver parenchymal cells to all proliferating cells in the body.	Rs1049296 (P570S)	1.26 (1.09–1.45) (p=0.002)	AD:1,366 C: 3,934 (11)
TFAM (10) (transcription factor A, mitochondrial)	A mitochondrial transcription factor that is a key activator of mitochondrial transcription as well as a participant in mitochondrial genome replication.	Rs2306604	0.78 (0.62–0.98) (p=0.03)	AD:372 C:294 (3)
TNF (6) (tumor necrosis factor alpha)	A multifunctional proinflammatory cytokine that belongs to the tumor necrosis factor (TNF)	Rs4647198	1.37 (1.05–1.78) (p=0.02)	AD: 389 C: 382 (3)

[a]Polymophisms identified across all ethnicities

has identified a potential AD gene located at or near the insulin degrading enzyme (10q23–25; *IDE*). This gene encodes for the IDE protein, which is involved in the cellular processing of insulin but also has a role in the degradation of Aβ (reviewed in Qiu and Folstein 2006). Further genetic analysis is being undertaken to determine the potential for this locus as another AD related gene (Bertram et al. 2000; Ertekin-Taner et al. 2000). Chromosome 12 also contains a potential AD gene (12p13.3; Alpha 2 macroglobulin; *A2M*) in which gene variants are associated with LOAD (Blacker et al. 1998; Dodel et al. 2000). In addition, recently a gene located on chromosome 11 (11q24.1; *SORL1*) has also been associated with LOAD in Caucasian, Hispanic, and multiethnic study populations (Rogaeva et al. 2007; Lee et al. 2007). The gene encodes for the Sortilin-related receptor 1 (SORL1), which has a role in the intracellular trafficking of APP (Rogaeva et al. 2007). Identification of polymorphisms or mutations within this gene will provided insight into the role of SORL1 in AD pathogenesis.

5   Concluding Remarks

The genetic mutations and polymorphisms identified to date have provided significant insight into the molecular mechanisms that underlie AD pathogenesis. APP and presenilin mutations have provided information on the enzymatic generation of Aβ, which plays a key role in AD pathogenesis. Further, the identification of *APOE* allelic variants and polymorphisms has provided insight into the clearance of accumulated Aβ from the brain. This information provided by AD-related genes identified thus far have led to the development of therapeutic strategies aimed at effectively treating the disease, some of which are currently being assessed in human clinical trials (recently reviewed in Christensen 2007). In addition, identifying genetic determinants and risk factors for the disease and combining them with neuropsychological assessment, bio-markers, and brain imaging, will provide a basis for improved, early diagnosis of the disease, thus allowing intervention prior to onset of clinical features of dementia.

References

Aldudo J, Bullido MJ, Frank A, Valdivieso F (1998) Missense mutation E318G of the presenilin-1 gene appears to be a nonpathogenic polymorphism. Ann Neurol 44:985–986

Almeida OP, Waterreus A, Spry N, Flicker L, Martins RN (2004) One year follow-up study of the association between chemical castration, sex hormones, beta-amyloid, memory and depression in men. Psychoneuroendocrinology 29:1071–1081

Alzheimer A, Stelzmann RA, Schnitzlein HN, Murtagh FR (1995) An english translation of Alzheimer's 1907 paper, "Uber eine eigenartige Erkankung der Hirnrinde". Clin Anat 8:429–431

Amtul Z, Lewis PA, Piper S, Crook R, Baker M, Findlay K, Singleton A, Hogg M, Younkin L, Younkin SG, Hardy J, Hutton M, Boeve BF, Tang-Wai D, Golde TE (2002) A presenilin 1 mutation associated with familial frontotemporal dementia inhibits gamma-secretase cleavage of APP and notch. Neurobiol Dis 9:269–273

Ancolio K, Dumanchin C, Barelli H, Warter JM, Brice A, Campion D, Frebourg T, Checler F (1999) Unusual phenotypic alteration of beta amyloid precursor protein (betaAPP) maturation by a new Val-715 –> Met betaAPP-770 mutation responsible for probable early-onset Alzheimer's disease. Proc Natl Acad Sci U S A 96:4119–4124

Armstrong J, Boada M, Rey MJ, Vidal N, Ferrer I (2004) Familial Alzheimer disease associated with A713T mutation in APP. Neurosci Lett 370:241–243

Artiga MJ, Bullido MJ, Sastre I, Recuero M, Garcia MA, Aldudo J, Vazquez J, Valdivieso F (1998) Allelic polymorphisms in the transcriptional regulatory region of apolipoprotein E gene. FEBS Lett 421:105–108

Baki L, Marambaud P, Efthimiopoulos S, Georgakopoulos A, Wen P, Cui W, Shioi J, Koo E, Ozawa M, Friedrich VL Jr, Robakis NK (2001) Presenilin-1 binds cytoplasmic epithelial cadherin, inhibits cadherin/p120 association, and regulates stability and function of the cadherin/catenin adhesion complex. Proc Natl Acad Sci U S A 98:2381–2386

Baldereschi M, Di Carlo A, Lepore V, Bracco L, Maggi S, Grigoletto F, Scarlato G, Amaducci L (1998) Estrogen-replacement therapy and Alzheimer's disease in the Italian Longitudinal Study on Aging. Neurology 50:996–1002

Bao F, Arai H, Matsushita S, Higuchi S, Sasaki H (1996) Expression of apolipoprotein E in normal and diverse neurodegenerative disease brain. Neuroreport 7:1733–1739

Barrett-Connor E, Kritz-Silverstein D (1993) Estrogen replacement therapy and cognitive function in older women. JAMA 269:2637–2641

Baumeister R, Leimer U, Zweckbronner I, Jakubek C, Grunberg J, Haass C (1997) Human presenilin-1, but not familial Alzheimer's disease (FAD) mutants, facilitate Caenorhabditis elegans Notch signalling independently of proteolytic processing. Genes Funct 1:149–159

Beffert U, Poirier J (1996) Apolipoprotein E, plaques, tangles and cholinergic dysfunction in Alzheimer's disease. Ann N Y Acad Sci 777:166–174

Bentahir M, Nyabi O, Verhamme J, Tolia A, Horre K, Wiltfang J, Esselmann H, De Strooper B (2006) Presenilin clinical mutations can affect gamma-secretase activity by different mechanisms. J Neurochem 96:732–742

Berr C, Hauw JJ, Delaere P, Duyckaerts C, Amouyel P (1994) Apolipoprotein E allele epsilon 4 is linked to increased deposition of the amyloid beta-peptide (A-beta) in cases with or without Alzheimer's disease. Neurosci Lett 178:221–224

Bertram L, Blacker D, Crystal A, Mullin K, Keeney D, Jones J, Basu S, Yhu S, Guenette S, McInnis M, Go R, Tanzi R (2000) Candidate genes showing no evidence for association or linkage with Alzheimer's disease using family-based methodologies. Exp Gerontol 35:1353–1361

Bertram L, McQueen MB, Mullin K, Blacker D, Tanzi RE (2007) Systematic meta-analyses of Alzheimer disease genetic association studies: the AlzGene database. Nat Genet 39:17–23

Blacker D, Wilcox MA, Laird NM, Rodes L, Horvath SM, Go RC, Perry R, Watson B Jr, Bassett SS, McInnis MG, Albert MS, Hyman BT, Tanzi RE (1998) Alpha-2 macroglobulin is genetically associated with Alzheimer disease. Nat Genet 19:357–360

Borchelt DR, Thinakaran G, Eckman CB, Lee MK, Davenport F, Ratovitsky T, Prada CM, Kim G, Seekins S, Yager D, Slunt HH, Wang R, Seeger M, Levey AI, Gandy SE, Copeland NG, Jenkins NA, Price DL, Younkin SG, Sisodia SS (1996) Familial Alzheimer's disease-linked presenilin 1 variants elevate Abeta1–42/1–40 ratio in vitro and in vivo. Neuron 17:1005–1013

Bowen RL, Isley JP, Atkinson RL (2000) An association of elevated serum gonadotropin concentrations and Alzheimer disease? J Neuroendocrinol 12:351–354

Boyles JK, Pitas RE, Wilson E, Mahley RW, Taylor JM (1985) Apolipoprotein E associated with astrocytic glia of the central nervous system and with nonmyelinating glia of the peripheral nervous system. J Clin Invest 76:1501–1513

Brat DJ, Gearing M, Goldthwaite PT, Wainer BH, Burger PC (2001) Tau-associated neuropathology in ganglion cell tumours increases with patient age but appears unrelated to ApoE genotype. Neuropathol Appl Neurobiol 27:197–205

Molecular Genetics of Alzheimer's Disease

Brenner DE, Kukull WA, Stergachis A, van Belle G, Bowen JD, McCormick WC, Teri L, Larson EB (1994) Postmenopausal estrogen replacement therapy and the risk of Alzheimer's disease: a population-based case-control study. Am J Epidemiol 140:262–267

Brooks WS, Martins RN, De Voecht J, Nicholson GA, Schofield PR, Kwok JB, Fisher C, Yeung LU, Van Broeckhoven C (1995) A mutation in codon 717 of the amyloid precursor protein gene in an Australian family with Alzheimer's disease. Neurosci Lett 199:183–186

Bu G, Maksymovitch EA, Nerbonne JM, Schwartz AL (1994) Expression and function of the low density lipoprotein receptor-related protein (LRP) in mammalian central neurons. J Biol Chem 269:18521–18528

Bullido MJ, Artiga MJ, Recuero M, Sastre I, Garcia MA, Aldudo J, Lendon C, Han SW, Morris JC, Frank A, Vazquez J, Goate A, Valdivieso F (1998) A polymorphism in the regulatory region of APOE associated with risk for Alzheimer's dementia. Nat Genet 18:69–71

Bullido MJ, Valdivieso F (2000) Apolipoprotein E gene promoter polymorphisms in Alzheimer's disease. Microsc Res Tech 50:261–267

Burkhardt MS, Foster JK, Clarnette RM, Chubb SA, Bruce DG, Drummond PD, Martins RN, Yeap BB (2006) Interaction between testosterone and apolipoprotein E epsilon4 status on cognition in healthy older men. J Clin Endocrinol Metab 91:1168–1172

Burkhardt MS, Foster JK, Laws SM, Baker LD, Craft S, Gandy SE, Stuckey BG, Clarnette R, Nolan D, Hewson-Bower B, Martins RN (2004) Oestrogen replacement therapy may improve memory functioning in the absence of APOE epsilon4. J Alzheimers Dis 6:221–228

Burns A (2002) Meaningful treatment outcomes in Alzheimer's disease. J Neurol Neurosurg Psychiatry 73:471–472

Cacabelos R, Fernandez-Novoa L, Lombardi V, Corzo L, Pichel V, Kubota Y (2003) Cerebrovascular risk factors in Alzheimer's disease: brain hemodynamics and pharmacogenomic implications. Neurol Res 25:567–580

Cai XD, Golde TE, Younkin SG (1993) Release of excess amyloid beta protein from a mutant amyloid beta protein precursor. Science 259:514–516

Campion D, Brice A, Hannequin D, Charbonnier F, Dubois B, Martin C, Michon A, Penet C, Bellis M, Calenda A, Martinez M, Agid Y, Clerget-Darpoux F, Frebourg T (1996) No founder effect in three novel Alzheimer's disease families with APP 717 Val–>Ile mutation. Clerget-darpoux. French Alzheimer's Disease Study Group. J Med Genet 33:661–664

Campion D, Dumanchin C, Hannequin D, Dubois B, Belliard S, Puel M, Thomas-Anterion C, Michon A, Martin C, Charbonnier F, Raux G, Camuzat A, Penet C, Mesnage V, Martinez M, Clerget-Darpoux F, Brice A, Frebourg T (1999) Early-onset autosomal dominant Alzheimer disease: prevalence, genetic heterogeneity, and mutation spectrum. Am J Hum Genet 65:664–670

Carter DA, Desmarais E, Bellis M, Campion D, Clerget-Darpoux F, Brice A, Agid Y, Jaillard-Serradt A, Mallet J (1992) More missense in amyloid gene. Nat Genet 2:255–256

Caselli RJ, Osborne D, Reiman EM, Hentz JG, Barbieri CJ, Saunders AM, Hardy J, Graff-Radford NR, Hall GR, Alexander GE (2001) Preclinical cognitive decline in late middle-aged asymptomatic apolipoprotein E-e4/4 homozygotes: a replication study. J Neurol Sci 189:93–98

Castano EM, Prelli F, Wisniewski T, Golabek A, Kumar RA, Soto C, Frangione B (1995) Fibrillogenesis in Alzheimer's disease of amyloid beta peptides and apolipoprotein E. Biochem J 306(Pt 2):599–604

Chartier-Harlin MC, Crawford F, Houlden H, Warren A, Hughes D, Fidani L, Goate A, Rossor M, Roques P, Hardy J, et al (1991) Early-onset Alzheimer's disease caused by mutations at codon 717 of the beta-amyloid precursor protein gene. Nature 353:844–846

Chartier-Harlin MC, Parfitt M, Legrain S, Perez-Tur J, Brousseau T, Evans A, Berr C, Vidal O, Roques P, Gourlet V, et al (1994) Apolipoprotein E, epsilon 4 allele as a major risk factor for sporadic early and late-onset forms of Alzheimer's disease: analysis of the 19q13.2 chromosomal region. Hum Mol Genet 3:569–574

Cho HS, Hyman BT, Greenberg SM, Rebeck GW (2001) Quantitation of apoE domains in Alzheimer disease brain suggests a role for apoE in Abeta aggregation. J Neuropathol Exp Neurol 60:342–349

Christensen DD (2007) Alzheimer's disease: progress in the development of anti-amyloid disease-modifying therapies. CNS Spectr 12:113–116, 119–123

Citron M, Oltersdorf T, Haass C, McConlogue L, Hung AY, Seubert P, Vigo-Pelfrey C, Lieberburg I, Selkoe DJ (1992) Mutation of the beta-amyloid precursor protein in familial Alzheimer's disease increases beta-protein production. Nature 360:672–674

Citron M, Westaway D, Xia W, Carlson G, Diehl T, Levesque G, Johnson-Wood K, Lee M, Seubert P, Davis A, Kholodenko D, Motter R, Sherrington R, Perry B, Yao H, Strome R, Lieberburg I, Rommens J, Kim S, Schenk D, Fraser P, St George-Hyslop P, Selkoe DJ (1997) Mutant presenilins of Alzheimer's disease increase production of 42-residue amyloid beta-protein in both transfected cells and transgenic mice. Nat Med 3:67–72

Confaloni A, Terreni L, Piscopo P, Crestini A, Campeggi LM, Frigerio CS, Blotta I, Perri M, Di Natale M, Maletta R, Marcon G, Franceschi M, Bruni AC, Forloni G, Cantafora A (2003) Nicastrin gene in familial and sporadic Alzheimer's disease. Neurosci Lett 353:61–65

Corbo RM, Scacchi R, Vilardo T, Ruggeri M (2001) Polymorphisms in the apolipoprotein E gene regulatory region in relation to coronary heart disease and their effect on plasma apolipoprotein E. Clin Chem Lab Med 39:2–6

Corder EH, Ghebremedhin E, Taylor MG, Thal DR, Ohm TG, Braak H (2004) The biphasic relationship between regional brain senile plaque and neurofibrillary tangle distributions: modification by age, sex, and APOE polymorphism. Ann N Y Acad Sci 1019:24–28

Corder EH, Saunders AM, Risch NJ, Strittmatter WJ, Schmechel DE, Gaskell PC Jr, Rimmler JB, Locke PA, Conneally PM, Schmader KE, et al (1994) Protective effect of apolipoprotein E type 2 allele for late onset Alzheimer disease. Nat Genet 7:180–184

Corder EH, Saunders AM, Strittmatter WJ, Schmechel DE, Gaskell PC, Small GW, Roses AD, Haines JL, Pericak-Vance MA (1993) Gene dose of apolipoprotein E type 4 allele and the risk of Alzheimer's disease in late onset families. Science 261:921–923

Cork LC (1990) Neuropathology of Down syndrome and Alzheimer disease. Am J Med Genet Suppl 7:282–286

Cousin E, Hannequin D, Mace S, Dubois B, Ricard S, Genin E, Brun C, Chansac C, Pradier L, Frebourg T, Brice A, Campion D, Deleuze JF (2003) No replication of the association between the Nicastrin gene and familial early-onset Alzheimer's disease. Neurosci Lett 353:153–155

Crook R, Verkkoniemi A, Perez-Tur J, Mehta N, Baker M, Houlden H, Farrer M, Hutton M, Lincoln S, Hardy J, Gwinn K, Somer M, Paetau A, Kalimo H, Ylikoski R, Poyhonen M, Kucera S, Haltia M (1998) A variant of Alzheimer's disease with spastic paraparesis and unusual plaques due to deletion of exon 9 of presenilin 1. Nat Med 4:452–455

Cruts M, Backhovens H, Theuns J, Clark RF, Le Paslier D, Weissenbach J, Goate AM, Martin JJ, Van Broeckhoven C (1995) Genetic and physical characterization of the early-onset Alzheimer's disease AD3 locus on chromosome 14q24.3. Hum Mol Genet 4:1355–1364

Cruts M, Dermaut B, Rademakers R, Van den Broeck M, Stogbauer F, Van Broeckhoven C (2003) Novel APP mutation V715A associated with presenile Alzheimer's disease in a German family. J Neurol 250:1374–1375

Cruts M, van Duijn CM, Backhovens H, Van den Broeck M, Wehnert A, Serneels S, Sherrington R, Hutton M, Hardy J, St George-Hyslop PH, Hofman A, Van Broeckhoven C (1998) Estimation of the genetic contribution of presenilin-1 and -2 mutations in a population-based study of presenile Alzheimer disease. Hum Mol Genet 7:43–51

Das HK, McPherson J, Bruns GA, Karathanasis SK, Breslow JL (1985) Isolation, characterization, and mapping to chromosome 19 of the human apolipoprotein E gene. J Biol Chem 260:6240–6247

De Jonghe C, Cruts M, Rogaeva EA, Tysoe C, Singleton A, Vanderstichele H, Meschino W, Dermaut B, Vanderhoeven I, Backhovens H, Vanmechelen E, Morris CM, Hardy J, Rubinsztein DC, St George-Hyslop PH, Van Broeckhoven C (1999) Aberrant splicing in the presenilin-1 intron 4 mutation causes presenile Alzheimer's disease by increased Abeta42 secretion. Hum Mol Genet 8:1529–1540

De Jonghe C, Esselens C, Kumar-Singh S, Craessaerts K, Serneels S, Checler F, Annaert W, Van Broeckhoven C, De Strooper B (2001) Pathogenic APP mutations near the gamma-secretase

Molecular Genetics of Alzheimer's Disease

cleavage site differentially affect Abeta secretion and APP C-terminal fragment stability. Hum Mol Genet 10:1665–1671

De Jonghe C, Zehr C, Yager D, Prada CM, Younkin S, Hendriks L, Van Broeckhoven C, Eckman CB (1998) Flemish and Dutch mutations in amyloid beta precursor protein have different effects on amyloid beta secretion. Neurobiol Dis 5:281–286

De Strooper B (2007) Loss-of-function presenilin mutations in Alzheimer disease. Talking Point on the role of presenilin mutations in Alzheimer disease. EMBO Rep 8:141–146

De Strooper B, Saftig P, Craessaerts K, Vanderstichele H, Guhde G, Annaert W, Von ura K, Van Leuven F (1998) Deficiency of presenilin-1 inhibits the normal cleavage of amyloid precursor protein. Nature 391:387–390

Deary IJ, Whiteman MC, Pattie A, Starr JM, Hayward C, Wright AF, Carothers A, Whalley LJ (2002) Cognitive change and the APOE epsilon 4 allele. Nature 418:932

Dermaut B, Cruts M, Slooter AJ, Van Gestel S, De Jonghe C, Vanderstichele H, Vanmechelen E, Breteler MM, Hofman A, van Duijn CM, Van Broeckhoven C (1999) The Glu318Gly substitution in presenilin 1 is not causally related to Alzheimer disease. Am J Hum Genet 64:290–292

Dermaut B, Kumar-Singh S, Engelborghs S, Theuns J, Rademakers R, Saerens J, Pickut BA, Peeters K, van den Broeck M, Vennekens K, Claes S, Cruts M, Cras P, Martin JJ, Van Broeckhoven C, De Deyn PP (2004) A novel presenilin 1 mutation associated with Pick's disease but not beta-amyloid plaques. Ann Neurol 55:617–626

Dermaut B, Theuns J, Sleegers K, Hasegawa H, Van den Broeck M, Vennekens K, Corsmit E, St George-Hyslop P, Cruts M, van Duijn CM, Van Broeckhoven C (2002) The gene encoding nicastrin, a major gamma-secretase component, modifies risk for familial early-onset Alzheimer disease in a Dutch population-based sample. Am J Hum Genet 70:1568–1574

Diedrich JF, Minnigan H, Carp RI, Whitaker JN, Race R, Frey W II, Haase AT (1991) Neuropathological changes in scrapie and Alzheimer's disease are associated with increased expression of apolipoprotein E and cathepsin D in astrocytes. J Virol 65:4759–4768

Dik MG, Jonker C, Bouter LM, Geerlings MI, van Kamp GJ, Deeg DJ (2000) APOE-epsilon4 is associated with memory decline in cognitively impaired elderly. Neurology 54:1492–1497

Dodel RC, Du Y, Bales KR, Gao F, Eastwood B, Glazier B, Zimmer R, Cordell B, Hake A, Evans R, Gallagher-Thompson D, Thompson LW, Tinklenberg JR, Pfefferbaum A, Sullivan EV, Yesavage J, Alstiel L, Gasser T, Farlow MR, Murphy GM Jr, Paul SM (2000) Alpha2 macroglobulin and the risk of Alzheimer's disease. Neurology 54:438–442

Dosunmu R, Wu J, Basha MR, Zawia NH (2007) Environmental and dietary risk factors in Alzheimer's disease. Expert Rev Neurother 7:887–900

Duff K, Eckman C, Zehr C, Yu X, Prada CM, Perez-tur J, Hutton M, Buee L, Harigaya Y, Yager D, Morgan D, Gordon MN, Holcomb L, Refolo L, Zenk B, Hardy J, Younkin S (1996) Increased amyloid-beta42(43) in brains of mice expressing mutant presenilin 1. Nature 383:710–713

Eckman CB, Mehta ND, Crook R, Perez-tur J, Prihar G, Pfeiffer E, Graff-Radford N, Hinder P, Yager D, Zenk B, Refolo LM, Prada CM, Younkin SG, Hutton M, Hardy J (1997) A new pathogenic mutation in the APP gene (I716V) increases the relative proportion of A beta 42(43). Hum Mol Genet 6:2087–2089

Edwards-Lee T, Ringman JM, Chung J, Werner J, Morgan A, St George-Hyslop P, Thompson P, Dutton R, Mlikotic A, Rogaeva E, Hardy J (2005) An African American family with early-onset Alzheimer disease and an APP (T714I) mutation. Neurology 64:377–379

Elshourbagy NA, Liao WS, Mahley RW, Taylor JM (1985) Apolipoprotein E mRNA is abundant in the brain and adrenals, as well as in the liver, and is present in other peripheral tissues of rats and marmosets. Proc Natl Acad Sci U S A 82:203–207

Ercelen N, Mercan S (2005) Alzheimer's disease and genes Advances in Molecular. Medicine 1:155–164

Ertekin-Taner N, Graff-Radford N, Younkin LH, Eckman C, Baker M, Adamson J, Ronald J, Blangero J, Hutton M, Younkin SG (2000) Linkage of plasma Abeta42 to a quantitative locus on chromosome 10 in late-onset Alzheimer's disease pedigrees. Science 290:2303–2304

Esler WP, Kimberly WT, Ostaszewski BL, Diehl TS, Moore CL, Tsai JY, Rahmati T, Xia W, Selkoe DJ, Wolfe MS (2000) Transition-state analogue inhibitors of gamma-secretase bind directly to presenilin-1. Nat Cell Biol 2:428–434

Espeland MA, Rapp SR, Shumaker SA, Brunner R, Manson JE, Sherwin BB, Hsia J, Margolis KL, Hogan PE, Wallace R, Dailey M, Freeman R, Hays J (2004) Conjugated equine estrogens and global cognitive function in postmenopausal women: Women's Health Initiative Memory Study. Jama 291:2959–2968

Evin G, Smith MJ, Tziotis A, McLean C, Canterford L, Sharples RA, Cappai R, Weidemann A, Beyreuther K, Cotton RG, Masters CL, Culvenor JG (2002) Alternative transcripts of presenilin-1 associated with frontotemporal dementia. Neuroreport 13:917–921

Farrer LA, Cupples LA, Haines JL, Hyman B, Kukull WA, Mayeux R, Myers RH, Pericak-Vance MA, Risch N, van Duijn CM (1997) Effects of age, sex, and ethnicity on the association between apolipoprotein E genotype and Alzheimer disease. A meta-analysis. APOE and Alzheimer Disease Meta Analysis Consortium. JAMA 278:1349–1356

Felsenstein KM, Ingalls KM, Hunihan LW, Roberts SB (1994) Reversal of the Swedish familial Alzheimer's disease mutant phenotype in cultured cells treated with phorbol 12,13-dibutyrate. Neurosci Lett 174:173–176

Fernandez-Madrid I, Levy E, Marder K, Frangione B (1991) Codon 618 variant of Alzheimer amyloid gene associated with inherited cerebral hemorrhage. Ann Neurol 30:730–733

Ferri CP, Prince M, Brayne C, Brodaty H, Fratiglioni L, Ganguli M, Hall K, Hasegawa K, Hendrie H, Huang Y, Jorm A, Mathers C, Menezes PR, Rimmer E, Scazufca M (2005) Global prevalence of dementia: a Delphi consensus study. Lancet 366:2112–2117

Finckh U, Kuschel C, Anagnostouli M, Patsouris E, Pantes GV, Gatzonis S, Kapaki E, Davaki P, Lamszus K, Stavrou D, Gal A (2005) Novel mutations and repeated findings of mutations in familial Alzheimer disease. Neurogenetics 6:85–89

Finckh U, Muller-Thomsen T, Mann U, Eggers C, Marksteiner J, Meins W, Binetti G, Alberici A, Hock C, Nitsch RM, Gal A (2000) High prevalence of pathogenic mutations in patients with early-onset dementia detected by sequence analyses of four different genes. Am J Hum Genet 66:110–117

Flory JD, Manuck SB, Ferrell RE, Ryan CM, Muldoon MF (2000) Memory performance and the apolipoprotein E polymorphism in a community sample of middle-aged adults. Am J Med Genet 96:707–711

Forsell C, Froelich S, Axelman K, Vestling M, Cowburn RF, Lilius L, Johnston JA, Engvall B, Johansson K, Dahlkild A, Ingelson M, St George-Hyslop PH, Lannfelt L (1997) A novel pathogenic mutation (Leu262Phe) found in the presenilin 1 gene in early-onset Alzheimer's disease. Neurosci Lett 234:3–6

Forsell C, Lannfelt L (1995) Amyloid precursor protein mutation at codon 713 (Ala–>Val) does not cause schizophrenia: non-pathogenic variant found at codon 705 (silent). Neurosci Lett 184:90–93

Frisoni GB, Govoni S, Geroldi C, Bianchetti A, Calabresi L, Franceschini G, Trabucchi M (1995) Gene dose of the epsilon 4 allele of apolipoprotein E and disease progression in sporadic late-onset Alzheimer's disease. Ann Neurol 37:596–604

Gandy S, Almeida OP, Fonte J, Lim D, Waterrus A, Spry N, Flicker L, Martins RN (2001) Chemical andropause and amyloid-beta peptide. JAMA 285:2195–2196

Gandy S, Petanceska S (2001) Regulation of alzheimer beta-amyloid precursor trafficking and metabolism. Adv Exp Med Biol 487:85–100

Gearing M, Schneider JA, Rebeck GW, Hyman BT, Mirra SS (1995) Alzheimer's disease with and without coexisting Parkinson's disease changes: apolipoprotein E genotype and neuropathologic correlates. Neurology 45:1985–1990

Geroldi C, Laakso MP, DeCarli C, Beltramello A, Bianchetti A, Soininen H, Trabucchi M, Frisoni GB (2000) Apolipoprotein E genotype and hippocampal asymmetry in Alzheimer's disease: a volumetric MRI study. J Neurol Neurosurg Psychiatry 68:93–96

Gillett MJ, Martins RN, Clarnette RM, Chubb SA, Bruce DG, Yeap BB (2003) Relationship between testosterone, sex hormone binding globulin and plasma amyloid beta peptide 40 in older men with subjective memory loss or dementia. J Alzheimers Dis 5:267–269

Molecular Genetics of Alzheimer's Disease   263

Goate A, Chartier-Harlin MC, Mullan M, Brown J, Crawford F, Fidani L, Giuffra L, Haynes A, Irving N, James L, et al (1991) Segregation of a missense mutation in the amyloid precursor protein gene with familial Alzheimer's disease. Nature 349:704–706

Godbolt AK, Beck JA, Collinge JC, Cipolotti L, Fox NC, Rossor MN (2006) A second family with familial AD and the V717L APP mutation has a later age at onset. Neurology 66:611–612

Gomez-Isla T, West HL, Rebeck GW, Harr SD, Growdon JH, Locascio JJ, Perls TT, Lipsitz LA, Hyman BT (1996) Clinical and pathological correlates of apolipoprotein E epsilon 4 in Alzheimer's disease. Ann Neurol 39:62–70

Goodenough S, Engert S, Behl C (2000) Testosterone stimulates rapid secretory amyloid precursor protein release from rat hypothalamic cells via the activation of the mitogen-activated protein kinase pathway. Neurosci Lett 296:49–52

Gordon I, Genis I, Grauer E, Sehayek E, Michaelson DM (1996) Biochemical and cognitive studies of apolipoprotein-E-deficient mice. Mol Chem Neuropathol 28:97–103

Gouras GK, Xu H, Gross RS, Greenfield JP, Hai B, Wang R, Greengard P (2000) Testosterone reduces neuronal secretion of Alzheimer's beta-amyloid peptides. Proc Natl Acad Sci U S A 97:1202–1205

Grabowski TJ, Cho HS, Vonsattel JP, Rebeck GW, Greenberg SM (2001) Novel amyloid precursor protein mutation in an Iowa family with dementia and severe cerebral amyloid angiopathy. Ann Neurol 49:697–705

Haan J, Van Broeckhoven C, van Duijn CM, Voorhoeve E, van Harskamp F, van Swieten JC, Maat-Schieman ML, Roos RA, Bakker E (1994) The apolipoprotein E epsilon 4 allele does not influence the clinical expression of the amyloid precursor protein gene codon 693 or 692 mutations. Ann Neurol 36:434–437

Haass C, Hung AY, Selkoe DJ, Teplow DB (1994) Mutations associated with a locus for familial Alzheimer's disease result in alternative processing of amyloid beta-protein precursor. J Biol Chem 269:17741–17748

Halliday GM, Song YJ, Lepar G, Brooks WS, Kwok JB, Kersaitis C, Gregory G, Shepherd CE, Rahimi F, Schofield PR, Kril JJ (2005) Pick bodies in a family with presenilin-1 Alzheimer's disease. Ann Neurol 57:139–143

Han SH, Einstein G, Weisgraber KH, Strittmatter WJ, Saunders AM, Pericak-Vance M, Roses AD, Schmechel DE (1994) Apolipoprotein E is localized to the cytoplasm of human cortical neurons: a light and electron microscopic study. J Neuropathol Exp Neurol 53:535–544

Hartley DM, Walsh DM, Ye CP, Diehl T, Vasquez S, Vassilev PM, Teplow DB, Selkoe DJ (1999) Protofibrillar intermediates of amyloid beta-protein induce acute electrophysiological changes and progressive neurotoxicity in cortical neurons. J Neurosci 19:8876–8884

Hass S, Weidemann A, Utermann G, Baier G (2001) Intracellular apolipoprotein E affects Amyloid Precursor Protein processing and amyloid Abeta production in COS-1 cells. Mol Genet Genomics 265:791–800

Heckmann JM, Low WC, de Villiers C, Rutherfoord S, Vorster A, Rao H, Morris CM, Ramesar RS, Kalaria RN (2004) Novel presenilin 1 mutation with profound neurofibrillary pathology in an indigenous Southern African family with early-onset Alzheimer's disease. Brain 127:133–142

Helisalmi S, Hiltunen M, Mannermaa A, Koivisto AM, Lehtovirta M, Alafuzoff I, Ryynanen M, Soininen H (2000) Is the presenilin-1 E318G missense mutation a risk factor for Alzheimer's disease? Neurosci Lett 278:65–68

Henderson AS, Easteal S, Jorm AF, Mackinnon AJ, Korten AE, Christensen H, Croft L, Jacomb PA (1995) Apolipoprotein E allele epsilon 4, dementia, and cognitive decline in a population sample. Lancet 346:1387–1390

Hendriks L, van Duijn CM, Cras P, Cruts M, Van Hul W, van Harskamp F, Warren A, McInnis MG, Antonarakis SE, Martin JJ, et al (1992) Presenile dementia and cerebral haemorrhage linked to a mutation at codon 692 of the beta-amyloid precursor protein gene. Nat Genet 1:218–221

Herreman A, Hartmann D, Annaert W, Saftig P, Craessaerts K, Serneels L, Umans L, Schrijvers V, Checler F, Vanderstichele H, Baekelandt V, Dressel R, Cupers P, Huylebroeck D, Zwijsen

A, Van Leuven F, De Strooper B (1999) Presenilin 2 deficiency causes a mild pulmonary phenotype and no changes in amyloid precursor protein processing but enhances the embryonic lethal phenotype of presenilin 1 deficiency. Proc Natl Acad Sci U S A 96:11872–11877

Hiltunen M, Helisalmi S, Mannermaa A, Alafuzoff I, Koivisto AM, Lehtovirta M, Pirskanen M, Sulkava R, Verkkoniemi A, Soininen H (2000) Identification of a novel 4.6-kb genomic deletion in presenilin-1 gene which results in exclusion of exon 9 in a Finnish early onset Alzheimer's disease family: an Alu core sequence-stimulated recombination? Eur J Hum Genet 8:259–266

Hogervorst E, Combrinck M, Smith AD (2003) Testosterone and gonadotropin levels in men with dementia. Neuro Endocrinol Lett 24:203–208

Hogervorst E, Williams J, Budge M, Barnetson L, Combrinck M, Smith AD (2001) Serum total testosterone is lower in men with Alzheimer's disease. Neuro Endocrinol Lett 22:163–168

Hone E, Martins IJ, Fonte J, Martins RN (2003) Apolipoprotein E influences amyloid-beta clearance from the murine periphery. J Alzheimers Dis 5:1–8

Hone E, Martins IJ, Jeoung M, Ji TH, Gandy SE, Martins RN (2005) Alzheimer's disease amyloid-beta peptide modulates apolipoprotein E isoform specific receptor binding. J Alzheimers Dis 7:303–314

Honjo H, Ogino Y, Naitoh K, Urabe M, Kitawaki J, Yasuda J, Yamamoto T, Ishihara S, Okada H, Yonezawa T, et al (1989) In vivo effects by estrone sulfate on the central nervous system-senile dementia (Alzheimer's type). J Steroid Biochem 34:521–525

Houlden H, Baker M, McGowan E, Lewis P, Hutton M, Crook R, Wood NW, Kumar-Singh S, Geddes J, Swash M, Scaravilli F, Holton JL, Lashley T, Tomita T, Hashimoto T, Verkkoniemi A, Kalimo H, Somer M, Paetau A, Martin JJ, Van Broeckhoven C, Golde T, Hardy J, Haltia M, Revesz T (2000) Variant Alzheimer's disease with spastic paraparesis and cotton wool plaques is caused by PS-1 mutations that lead to exceptionally high amyloid-beta concentrations. Ann Neurol 48:806–808

Houlden H, Crook R, Backhovens H, Prihar G, Baker M, Hutton M, Rossor M, Martin JJ, Van Broeckhoven C, Hardy J (1998) ApoE genotype is a risk factor in nonpresenilin early-onset Alzheimer's disease families. Am J Med Genet 81:117–121

Huang Y (2006) Apolipoprotein E and Alzheimer disease. Neurology 66:S79–85

Hyman BT, West HL, Rebeck GW, Lai F, Mann DM (1995) Neuropathological changes in Down's syndrome hippocampal formation. Effect of age and apolipoprotein E genotype. Arch Neurol 52:373–378

Irizarry MC, Deng A, Lleo A, Berezovska O, Von Arnim CA, Martin-Rehrmann M, Manelli A, LaDu MJ, Hyman BT, Rebeck GW (2004) Apolipoprotein E modulates gamma-secretase cleavage of the amyloid precursor protein. J Neurochem 90:1132–1143

Ishii K, Ii K, Hasegawa T, Shoji S, Doi A, Mori H (1997) Increased A beta 42(43)-plaque deposition in early-onset familial Alzheimer's disease brains with the deletion of exon 9 and the missense point mutation (H163R) in the PS-1 gene. Neurosci Lett 228:17–20.

Itoh Y, Yamada M (1996) Apolipoprotein E and the neuropathology of dementia. N Engl J Med 334:599–600

Iwatsubo T, Odaka A, Suzuki N, Mizusawa H, Nukina N, Ihara Y (1994) Visualization of A beta 42(43) and A beta 40 in senile plaques with end-specific A beta monoclonals: evidence that an initially deposited species is A beta 42(43). Neuron 13:45–53

Janssen JC, Beck JA, Campbell TA, Dickinson A, Fox NC, Harvey RJ, Houlden H, Rossor MN, Collinge J (2003) Early onset familial Alzheimer's disease: mutation frequency in 31 families. Neurology 60:235–239

Jia L, Ye J, Haiyan LV, Wang W, Zhou C, Zhang X, Xu J, Wang L, Jia J (2007) Genetic association between polymorphisms of Pen2 gene and late onset Alzheimer's disease in the North Chinese population. Brain Res 1141:10–14

Jones CT, Morris S, Yates CM, Moffoot A, Sharpe C, Brock DJ, St Clair D (1992) Mutation in codon 713 of the beta amyloid precursor protein gene presenting with schizophrenia. Nat Genet 1:306–309

Molecular Genetics of Alzheimer's Disease

Jorm AF, Korten AE, Henderson AS (1987) The prevalence of dementia: a quantitative integration of the literature. Acta Psychiatr Scand 76:465–479

Kamino K, Orr HT, Payami H, Wijsman EM, Alonso ME, Pulst SM, Anderson L, O'Dahl S, Nemens E, White JA, et al (1992) Linkage and mutational analysis of familial Alzheimer disease kindreds for the APP gene region. Am J Hum Genet 51:998–1014

Kang J, Lemaire HG, Unterbeck A, Salbaum JM, Masters CL, Grzeschik KH, Multhaup G, Beyreuther K, Muller-Hill B (1987) The precursor of Alzheimer's disease amyloid A4 protein resembles a cell-surface receptor. Nature 325:733–736

Kitaguchi N, Takahashi Y, Tokushima Y, Shiojiri S, Ito H (1988) Novel precursor of Alzheimer's disease amyloid protein shows protease inhibitory activity. Nature 331:530–532

Kitaguchi N, Tokushima Y, Oishi K, Takahashi Y, Shiojiri S, Nakamura S, Tanaka S, Kodaira R, Ito H (1990) Determination of amyloid beta protein precursors harboring active form of proteinase inhibitor domains in cerebrospinal fluid of Alzheimer's disease patients by trypsin-antibody sandwich ELISA. Biochem Biophys Res Commun 166:1453–1459

Kumar-Singh S, Theuns J, Van Broeck B, Pirici D, Vennekens K, Corsmit E, Cruts M, Dermaut B, Wang R, Van Broeckhoven C (2006) Mean age-of-onset of familial alzheimer disease caused by presenilin mutations correlates with both increased Abeta42 and decreased Abeta40. Hum Mutat 27:686–695

Kwok, J. Q. X., Li, M. Hallupp, L. Milward, S. Whyte, P. Schofield (1998) Novel familial early-onset Alzheimer's disease mutation (Leu723Pro) in amyloid precursor protein (APP) gene increases production of 42(43) amino acid isoform of amyloid beta peptide. Neurobiol Aging 19(Suppl 4):S91

Kwok J, Li Q, Hallupp M, Whyte S, Ames D, Beyreuther K, Masters C, Schofield P (2000) Novel Leu723Pro amyloid precursor protein mutation increases amyloid beta42(43) peptide levels and induces apoptosis. Ann Neurol 47:249–253

Kwok JB, Taddei K, Hallupp M, Fisher C, Brooks WS, Broe GA, Hardy J, Fulham MJ, Nicholson GA, Stell R, St George Hyslop PH, Fraser PE, Kakulas B, Clarnette R, Relkin N, Gandy SE, Schofield PR, Martins RN (1997) Two novel (M233T and R278T) presenilin-1 mutations in early-onset Alzheimer's disease pedigrees and preliminary evidence for association of presenilin-1 mutations with a novel phenotype. Neuroreport 8:1537–1542

LaDu MJ, Pederson TM, Frail DE, Reardon CA, Getz GS, Falduto MT (1995) Purification of apolipoprotein E attenuates isoform-specific binding to beta-amyloid. J Biol Chem 270:9039–9042

Lai MT, Chen E, Crouthamel MC, DiMuzio-Mower J, Xu M, Huang Q, Price E, Register RB, Shi XP, Donoviel DB, Bernstein A, Hazuda D, Gardell SJ, Li YM (2003) Presenilin-1 and presenilin-2 exhibit distinct yet overlapping gamma-secretase activities. J Biol Chem 278:22475–22481

Lambert JC, Berr C, Pasquier F, Delacourte A, Frigard B, Cottel D, Perez-Tur J, Mouroux V, Mohr M, Cecyre D, Galasko D, Lendon C, Poirier J, Hardy J, Mann D, Amouyel P, Chartier-Harlin MC (1998) Pronounced impact of Th1/E47cs mutation compared with -491 AT mutation on neural APOE gene expression and risk of developing Alzheimer's disease. Hum Mol Genet 7:1511–1516

Lambert JC, Brousseau T, Defosse V, Evans A, Arveiler D, Ruidavets JB, Haas B, Cambou JP, Luc G, Ducimetiere P, Cambien F, Chartier-Harlin MC, Amouyel P (2000) Independent association of an APOE gene promoter polymorphism with increased risk of myocardial infarction and decreased APOE plasma concentrations-the ECTIM study. Hum Mol Genet 9:57–61

Lambert JC, Mann DM, Harris JM, Chartier-Harlin MC, Cumming A, Coates J, Lemmon H, StClair D, Iwatsubo T, Lendon C (2001) The -48 C/T polymorphism in the presenilin 1 promoter is associated with an increased risk of developing Alzheimer's disease and an increased Abeta load in brain. J Med Genet 38:353–355

Lambert JC, Perez-Tur J, Dupire MJ, Galasko D, Mann D, Amouyel P, Hardy J, Delacourte A, Chartier-Harlin MC (1997) Distortion of allelic expression of apolipoprotein E in Alzheimer's disease. Hum Mol Genet 6:2151–2154

Landen M, Thorsell A, Wallin A, Blennow K (1996) The apolipoprotein E allele epsilon 4 does not correlate with the number of senile plaques or neurofibrillary tangles in patients with Alzheimer's disease. J Neurol Neurosurg Psychiatr 61:352–356

Lao JI, Beyer K, Fernandez-Novoa L, Cacabelos R (1998) A novel mutation in the predicted TM2 domain of the presenilin 2 gene in a Spanish patient with late-onset Alzheimer's disease. Neurogenetics 1:293–296

Lauderback CM, Kanski J, Hackett JM, Maeda N, Kindy MS, Butterfield DA (2002) Apolipoprotein E modulates Alzheimer's Abeta(1–42)-induced oxidative damage to synaptosomes in an allele-specific manner. Brain Res 924:90–97

Laws SM, Clarnette RM, Taddei K, Martins G, Paton A, Almeida OP, Forstl H, Martins RN (2002a) Association between the presenilin-1 mutation Glu318Gly and complaints of memory impairment. Neurobiol Aging 23:55–58

Laws SM, Hone E, Gandy S, Martins RN (2003) Expanding the association between the APOE gene and the risk of Alzheimer's disease: possible roles for APOE promoter polymorphisms and alterations in APOE transcription. J Neurochem 84:1215–123

Laws SM, Hone E, Taddei K, Harper C, Dean B, McClean C, Masters C, Lautenschlager N, Gandy SE, Martins RN (2002b) Variation at the APOE -491 promoter locus is associated with altered brain levels of apolipoprotein E. Mol Psychiatry 7:886–890

Laws SM, Taddei K, Martins G, Paton A, Fisher C, Clarnette R, Hallmayer J, Brooks WS, Gandy SE, Martins RN (1999) The -491AA polymorphism in the APOE gene is associated with increased plasma apoE levels in Alzheimer's disease. Neuroreport 10:879–882

Lazarov VK, Fraering PC, Ye W, Wolfe MS, Selkoe DJ, Li H (2006) Electron microscopic structure of purified, active gamma-secretase reveals an aqueous intramembrane chamber and two pores. Proc Natl Acad Sci U S A 103:6889–6894

Lee JH, Cheng R, Schupf N, Manly J, Lantigua R, Stern Y, Rogaeva E, Wakutani Y, Farrer L, St George-Hyslop P, Mayeux R (2007) The association between genetic variants in SORL1 and Alzheimer disease in an urban, multiethnic, community-based cohort. Arch Neurol 64:501–506

Lehmann S, Chiesa R, Harris DA (1997) Evidence for a six-transmembrane domain structure of presenilin 1. J Biol Chem 272:12047–12051

Lehtovirta M, Laakso MP, Soininen H, Helisalmi S, Mannermaa A, Helkala EL, Partanen K, Ryynanen M, Vainio P, Hartikainen P, et al (1995) Volumes of hippocampus, amygdala and frontal lobe in Alzheimer patients with different apolipoprotein E genotypes. Neuroscience 67:65–72

Lehtovirta M, Soininen H, Laakso MP, Partanen K, Helisalmi S, Mannermaa A, Ryynanen M, Kuikka J, Hartikainen P, Riekkinen PJ, Sr. (1996) SPECT and MRI analysis in Alzheimer's disease: relation to apolipoprotein E epsilon 4 allele. J Neurol Neurosurg Psychiatr 60:644–649

Lemere CA, Lopera F, Kosik KS, Lendon CL, Ossa J, Saido TC, Yamaguchi H, Ruiz A, Martinez A, Madrigal L, Hincapie L, Arango JC, Anthony DC, Koo EH, Goate AM, Selkoe DJ, Arango JC (1996) The E280A presenilin 1 Alzheimer mutation produces increased A beta 42 deposition and severe cerebellar pathology. Nat Med 2:1146–1150

Lendon CL, Myers A, Cumming A, Goate AM, St Clair D (1997) A polymorphism in the presenilin 1 gene does not modify risk for Alzheimer's disease in a cohort with sporadic early onset. Neurosci Lett 228:212–214

Levi O, Jongen-Relo AL, Feldon J, Michaelson DM (2005) Brain area- and isoform-specific inhibition of synaptic plasticity by apoE4. J Neurol Sci 229–230:241–248

Levin-Allerhand J, McEwen BS, Lominska CE, Lubahn DB, Korach KS, Smith JD (2001) Brain region-specific up-regulation of mouse apolipoprotein E by pharmacological estrogen treatments. J Neurochem 79:796–803

Levitan D, Doyle TG, Brousseau D, Lee MK, Thinakaran G, Slunt HH, Sisodia SS, Greenwald I (1996) Assessment of normal and mutant human presenilin function in Caenorhabditis elegans. Proc Natl Acad Sci U S A 93:14940–14944

Levy-Lahad E, Wasco W, Poorkaj P, Romano DM, Oshima J, Pettingell WH, Yu CE, Jondro PD, Schmidt SD, Wang K, et al (1995) Candidate gene for the chromosome 1 familial Alzheimer's disease locus. Science 269:973–977

Levy E, Carman MD, Fernandez-Madrid IJ, Power MD, Lieberburg I, van Duinen SG, Bots GT, Luyendijk W, Frangione B (1990) Mutation of the Alzheimer's disease amyloid gene in hereditary cerebral hemorrhage, Dutch type. Science 248:1124–1126

Molecular Genetics of Alzheimer's Disease

Li QX, Evin G, Small DH, Multhaup G, Beyreuther K, Masters CL (1995) Proteolytic processing of Alzheimer's disease beta A4 amyloid precursor protein in human platelets. J Biol Chem 270:14140–14147

Li YM, Xu M, Lai MT, Huang Q, Castro JL, DiMuzio-Mower J, Harrison T, Lellis C, Nadin A, Neduvelil JG, Register RB, Sardana MK, Shearman MS, Smith AL, Shi XP, Yin KC, Shafer JA, Gardell SJ (2000) Photoactivated gamma-secretase inhibitors directed to the active site covalently label presenilin 1. Nature 405:689–694

Luchsinger JA, Tang MX, Shea S, Mayeux R (2002) Caloric intake and the risk of Alzheimer disease. Arch Neurol 59:1258–1263

Luciano AA, De Souza MJ, Roy MP, Schoenfeld MJ, Nulsen JC, Halvorson CV (1993) Evaluation of low-dose estrogen and progestin therapy in postmenopausal women. A double-blind, prospective study of sequential versus continuous therapy. J Reprod Med 38:207–214

Luo HM, Deng H, Xiao F, Gao Q, Weng W, Zhang PF, Li XG (2004) Down-regulation amyloid beta-protein 42 production by interfering with transcript of presenilin 1 gene with siRNA. Acta Pharmacol Sin 25:1613–1618

Maeda H, Nakamura H, Kobori S, Okada M, Mori H, Niki H, Ogura T, Hiraga S (1989) Identification of human apolipoprotein E variant gene: apolipoprotein E7 (Glu244,245–Lys244,245). J Biochem (Tokyo) 105:51–54

Mann DM (1989) Cerebral amyloidosis, ageing and Alzheimer's disease: a contribution from studies on Down's syndrome. Neurobiol Aging 10:397–399; discussion 412–414

Martins IJ, Hone E, Foster JK, Sunram-Lea SI, Gnjec A, Fuller SJ, Nolan D, Gandy SE, Martins RN (2006) Apolipoprotein E, cholesterol metabolism, diabetes, and the convergence of risk factors for Alzheimer's disease and cardiovascular disease. Mol Psychiatr 11:721–736

Martins RN, Clarnette R, Fisher C, Broe GA, Brooks WS, Montgomery P, Gandy SE (1995) ApoE genotypes in Australia: roles in early and late onset Alzheimer's disease and Down's syndrome. Neuroreport 6:1513–1516

Martins RN, Taddei K, Kendall C, Evin G, Bates KA, Harvey AR (2001) Altered expression of apolipoprotein E, amyloid precursor protein and presenilin-1 is associated with chronic reactive gliosis in rat cortical tissue. Neuroscience 106:557–569

Maruyama K, Usami M, Yamao-Harigaya W, Tagawa K, Ishiura S (1991) Mutation of Glu693 to Gln or Val717 to Ile has no effect on the processing of Alzheimer amyloid precursor protein expressed in COS-1 cells by cDNA transfection. Neurosci Lett 132:97–100

Mastrangelo P, Mathews PM, Chishti MA, Schmidt SD, Gu Y, Yang J, Mazzella MJ, Coomaraswamy J, Horne P, Strome B, Pelly H, Levesque G, Ebeling C, Jiang Y, Nixon RA, Rozmahel R, Fraser PE, St George-Hyslop P, Carlson GA, Westaway D (2005) Dissociated phenotypes in presenilin transgenic mice define functionally distinct gamma-secretases. Proc Natl Acad Sci U S A 102:8972–8977

Matsubara-Tsutsui M, Yamagata H, Morishima A, Nakura J, Mitsuda N, Kamino K, Kondo I, Miki T (2002) The 4,752 C/T polymorphism in the presenilin 1 gene increases the risk of Alzheimer's disease in apolipoprotein E4 carriers. Intern Med 41:823–828

Matsumura Y, Kitamura E, Miyoshi K, Yamamoto Y, Furuyama J, Sugihara T (1996) Japanese siblings with missense mutation (717Val –> Ile) in amyloid precursor protein of early-onset Alzheimer's disease. Neurology 46:1721–1723

Matsunaga A, Sasaki J, Moriyama K, Arakawa F, Takada Y, Nishi K, Hidaka K, Arakawa K (1995) Population frequency of apolipoprotein E5 (Glu3–>Lys) and E7 (Glu244–>Lys, Glu245–>Lys) variants in western Japan. Clin Genet 48:93–99

Mattila KM, Forsell C, Pirttila T, Rinne JO, Lehtimaki T, Roytta M, Lilius L, Eerola A, St George-Hyslop PH, Frey H, Lannfelt L (1998) The Glu318Gly mutation of the presenilin-1 gene does not necessarily cause Alzheimer's disease. Ann Neurol 44:965–967

Merched A, Blain H, Visvikis S, Herbeth B, Jeandel C, Siest G (1997) Cerebrospinal fluid apolipoprotein E level is increased in late-onset Alzheimer's disease. J Neurol Sci 145:33–39

Metzger RE, LaDu MJ, Pan JB, Getz GS, Frail DE, Falduto MT (1996) Neurons of the human frontal cortex display apolipoprotein E immunoreactivity: implications for Alzheimer's disease. J Neuropathol Exp Neurol 55:372–380

Moechars D, Lorent K ,Van Leuven F (1999) Premature death in transgenic mice that overexpress a mutant amyloid precursor protein is preceded by severe neurodegeneration and apoptosis. Neuroscience 91:819–830

Moffat SD, Szekely CA, Zonderman AB, Kabani NJ, Resnick SM (2000) Longitudinal change in hippocampal volume as a function of apolipoprotein. E genotype. Neurology 55:134–136

Morelli L, Llovera R, Gonzalez SA, Affranchino JL, Prelli F, Frangione B, Ghiso J, Castano EM (2003) Differential degradation of amyloid beta genetic variants associated with hereditary dementia or stroke by insulin-degrading enzyme. J Biol Chem 278:23221–23226

Morris CM, Massey HM, Benjamin R, Leake A, Broadbent C, Griffiths M, Lamb H, Brown A, Ince PG, Tyrer S, Thompson P, McKeith IG, Edwardson JA, Perry RH, Perry EK (1996) Molecular biology of APO E alleles in Alzheimer's and non-Alzheimer's dementias. J Neural Transm Suppl 47:205–218

Mui S, Briggs M, Chung H, Wallace RB, Gomez-Isla T, Rebeck GW, Hyman BT (1996) A newly identified polymorphism in the apolipoprotein E enhancer gene region is associated with Alzheimer's disease and strongly with the epsilon 4 allele. Neurology 47:196–201

Mullan M, Houlden H, Windelspecht M, Fidani L, Lombardi C, Diaz P, Rossor M, Crook R, Hardy J, Duff K, et al (1992) A locus for familial early-onset Alzheimer's disease on the long arm of chromosome 14, proximal to the alpha 1-antichymotrypsin gene. Nat Genet 2:340–342

Mulnard RA, Cotman CW, Kawas C, van Dyck CH, Sano M, Doody R, Koss E, Pfeiffer E, Jin S, Gamst A, Grundman M, Thomas R, Thal LJ (2000) Estrogen replacement therapy for treatment of mild to moderate Alzheimer disease: a randomized controlled trial. Alzheimer's Disease Cooperative Study. JAMA 283:1007–1015

Murakami K, Irie K, Morimoto A, Ohigashi H, Shindo M, Nagao M, Shimizu T, Shirasawa T (2003) Neurotoxicity and physicochemical properties of Abeta mutant peptides from cerebral amyloid angiopathy: implication for the pathogenesis of cerebral amyloid angiopathy and Alzheimer's disease. J Biol Chem 278:46179–46187

Murayama O, Tomita T, Nihonmatsu N, Murayama M, Sun X, Honda T, Iwatsubo T, Takashima A (1999) Enhancement of amyloid beta 42 secretion by 28 different presenilin 1 mutations of familial Alzheimer's disease. Neurosci Lett 265:61–63

Murrell J, Farlow M, Ghetti B, Benson MD (1991) A mutation in the amyloid precursor protein associated with hereditary Alzheimer's disease. Science 254:97–99

Murrell JR, Hake AM, Quaid KA, Farlow MR, Ghetti B (2000) Early-onset Alzheimer disease caused by a new mutation (V717L) in the amyloid precursor protein gene. Arch Neurol 57:885–887

Nagy Z, Esiri MM, Jobst KA, Johnston C, Litchfield S, Sim E, Smith AD (1995) Influence of the apolipoprotein E genotype on amyloid deposition and neurofibrillary tangle formation in Alzheimer's disease. Neuroscience 69:757–761

Namba Y, Tomonaga M, Kawasaki H, Otomo E, Ikeda K (1991) Apolipoprotein E immunoreactivity in cerebral amyloid deposits and neurofibrillary tangles in Alzheimer's disease and kuru plaque amyloid in Creutzfeldt-Jakob disease. Brain Res 541:163–166

Naruse S, Igarashi S, Kobayashi H, Aoki K, Inuzuka T, Kaneko K, Shimizu T, Iihara K, Kojima T, Miyatake T, et al (1991) Mis-sense mutation Val—Ile in exon 17 of amyloid precursor protein gene in Japanese familial Alzheimer's disease. Lancet 337:978–979

Nilsberth C, Westlind-Danielsson A, Eckman CB, Condron MM, Axelman K, Forsell C, Stenh C, Luthman J, Teplow DB, Younkin SG, Naslund J, Lannfelt L (2001) The 'Arctic' APP mutation (E693G) causes Alzheimer's disease by enhanced Abeta protofibril formation. Nat Neurosci 4:887–893

Obici L, Demarchi A, de Rosa G, Bellotti V, Marciano S, Donadei S, Arbustini E, Palladini G, Diegoli M, Genovese E, Ferrari G, Coverlizza S, Merlini G (2005) A novel AbetaPP mutation exclusively associated with cerebral amyloid angiopathy. Ann Neurol 58:639–644

Ogura T, Mio K, Hayashi I, Miyashita H, Fukuda R, Kopan R, Kodama T, Hamakubo T, Iwatsubo T, Tomita T, Sato C (2006) Three-dimensional structure of the gamma-secretase complex. Biochem Biophys Res Commun 343:525–534

Molecular Genetics of Alzheimer's Disease 269

Ohm TG, Kirca M, Bohl J, Scharnagl H, Gross W, Marz W (1995) Apolipoprotein E polymorphism influences not only cerebral senile plaque load but also Alzheimer-type neurofibrillary tangle formation. Neuroscience 66:583–587

Olaisen B, Teisberg P, Gedde-Dahl T Jr (1982) The locus for apolipoprotein E (apoE) is linked to the complement component C3 (C3) locus on chromosome 19 in man. Hum Genet 62:233–236

Olichney JM, Hansen LA, Galasko D, Saitoh T, Hofstetter CR, Katzman R, Thal LJ (1996) The apolipoprotein E epsilon 4 allele is associated with increased neuritic plaques and cerebral amyloid angiopathy in Alzheimer's disease and Lewy body variant. Neurology 47:190–196

Orlacchio A, Kawarai T, Polidoro M, Paterson AD, Rogaeva E, Orlacchio A, St George-Hyslop PH, Bernardi G (2004) Lack of association between Alzheimer's disease and the promoter region polymorphisms of the nicastrin gene. Neurosci Lett 363:49–53

Osuntokun BO, Sahota A, Ogunniyi AO, Gureje O, Baiyewu O, Adeyinka A, Oluwole SO, Komolafe O, Hall KS, Unverzagt FW, et al (1995) Lack of an association between apolipoprotein E epsilon 4 and Alzheimer's disease in elderly Nigerians. Ann Neurol 38:463–465

Oyama F, Shimada H, Oyama R, Ihara Y (1995) Apolipoprotein E genotype, Alzheimer's pathologies and related gene expression in the aged population. Brain Res Mol Brain Res 29:92–98

Parks AL, Curtis D (2007) Presenilin diversifies its portfolio. Trends Genet 23:140–150

Pasalar P, Najmabadi H, Noorian AR, Moghimi B, Jannati A, Soltanzadeh A, Krefft T, Crook R, Hardy J (2002) An Iranian family with Alzheimer's disease caused by a novel APP mutation (Thr714Ala). Neurology 58:1574–1575

Pastor P, Roe CM, Villegas A, Bedoya G, Chakraverty S, Garcia G, Tirado V, Norton J, Rios S, Martinez M, Kosik KS, Lopera F, Goate AM (2003) Apolipoprotein Eepsilon4 modifies Alzheimer's disease onset in an E280A PS1 kindred. Ann Neurol 54:163–169

Peacock ML, Murman DL, Sima AA, Warren JT Jr, Roses AD, Fink JK (1994) Novel amyloid precursor protein gene mutation (codon 665Asp) in a patient with late-onset Alzheimer's disease. Ann Neurol 35:432–438

Peacock ML, Warren JT Jr, Roses AD, Fink JK (1993) Novel polymorphism in the A4 region of the amyloid precursor protein gene in a patient without Alzheimer's disease. Neurology 43:1254–1256

Peila R, Rodriguez BL, Launer LJ (2002) Type 2 diabetes, APOE gene, and the risk for dementia and related pathologies: the Honolulu-Asia. Aging Study Diabetes 51:1256–1262

Perez-Tur J, Froelich S, Prihar G, Crook R, Baker M, Duff K, Wragg M, Busfield F, Lendon C, Clark RF, et al (1995) A mutation in Alzheimer's disease destroying a splice acceptor site in the presenilin-1 gene. Neuroreport 7:297–301

Pericak-Vance MA, Bebout JL, Gaskell PC Jr, Yamaoka LH, Hung WY, Alberts MJ, Walker AP, Bartlett RJ, Haynes CA, Welsh KA, et al (1991) Linkage studies in familial Alzheimer disease: evidence for chromosome 19 linkage. Am J Hum Genet 48:1034–1050

Permanne B, Perez C, Soto C, Frangione B, Wisniewski T (1997) Detection of apolipoprotein E/dimeric soluble amyloid beta complexes in Alzheimer's disease brain supernatants. Biochem Biophys Res Commun 240:715–720

Petanceska SS, Nagy V, Frail D, Gandy S (2000) Ovariectomy and 17beta-estradiol modulate the levels of Alzheimer's amyloid beta peptides in brain. Neurology. 54:2212–2217.

Pike CJ (2001) Testosterone attenuates beta-amyloid toxicity in cultured hippocampal neurons. Brain Res 919:160–165

Piscopo P, Manfredi A, Malvezzi-Campeggi L, Crestini A, Spadoni O, Cherchi R, Deiana E, Piras MR, Confaloni A (2006) Genetic study of Sardinian patients with Alzheimer's disease. Neurosci Lett 398:124–128

Plassman BL, Welsh-Bohmer KA, Bigler ED, Johnson SC, Anderson CV, Helms MJ, Saunders AM, Breitner JC (1997) Apolipoprotein E epsilon 4 allele and hippocampal volume in twins with normal cognition. Neurology 48:985–989

Poirier J, Baccichet A, Dea D, Gauthier S (1993a) Cholesterol synthesis and lipoprotein reuptake during synaptic remodelling in hippocampus in adult rats. Neuroscience 55:81–90

Poirier J, Davignon J, Bouthillier D, Kogan S, Bertrand P, Gauthier S (1993b) Apolipoprotein E polymorphism and Alzheimer's disease. Lancet 342:697–699

Poirier J, Hess M, May PC, Finch CE (1991) Astrocytic apolipoprotein E mRNA and GFAP mRNA in hippocampus after entorhinal cortex lesioning. Brain Res Mol Brain Res 11:97–106

Poli M, Gatta LB, Archetti S, Padovani A, Albertini A, Finazzi D (2003) Association analysis between anterior-pharynx defective-1 genes polymorphisms and Alzheimer's disease. Neurosci Lett 350:77–80

Polvikoski T, Sulkava R, Haltia M, Kainulainen K, Vuorio A, Verkkoniemi A, Niinisto L, Halonen P, Kontula K (1995) Apolipoprotein E, dementia, and cortical deposition of beta-amyloid protein. N Engl J Med 333:1242–1247

Ponte P, Gonzalez-DeWhitt P, Schilling J, Miller J, Hsu D, Greenberg B, Davis K, Wallace W, Lieberburg I, Fuller F (1988) A new A4 amyloid mRNA contains a domain homologous to serine proteinase inhibitors. Nature 331:525–527

Prihar G, Fuldner RA, Perez-Tur J, Lincoln S, Duff K, Crook R, Hardy J, Philips CA, Venter C, Talbot C, Clark RF, Goate A, Li J, Potter H, Karran E, Roberts GW, Hutton M, Adams MD (1996) Structure and alternative splicing of the presenilin-2 gene. Neuroreport 7:1680–1684

Prihar G, Verkkoniem A, Perez-Tur J, Crook R, Lincoln S, Houlden H, Somer M, Paetau A, Kalimo H, Grover A, Myllykangas L, Hutton M, Hardy J, Haltia M (1999) Alzheimer disease PS-1 exon 9 deletion defined. Nat Med 5:1090

Qiu WQ, Folstein MF (2006) Insulin, insulin-degrading enzyme and amyloid-beta peptide in Alzheimer's disease: review and hypothesis. Neurobiol Aging 27:190–198

Raber J, Bongers G, LeFevour A, Buttini M, Mucke L (2002) Androgens protect against apolipoprotein E4-induced cognitive deficits. J Neurosci 22:5204–5209

Raux G, Guyant-Marechal L, Martin C, Bou J, Penet C, Brice A, Hannequin D, Frebourg T, Campion D (2005) Molecular diagnosis of autosomal dominant early onset Alzheimer's disease: an update. J Med Genet 42:793–795

Rebeck GW, Reiter JS, Strickland DK, Hyman BT (1993) Apolipoprotein E in sporadic Alzheimer's disease: allelic variation and receptor interactions. Neuron 11:575–580.

Reddy PH (2006) Mitochondrial oxidative damage in aging and Alzheimer's disease: implications for mitochondrially targeted antioxidant therapeutics. J Biomed Biotechnol 2006:31372

Reznik-Wolf H, Treves TA, Davidson M, Aharon-Peretz J, St George Hyslop PH, Chapman J, Korczyn AD, Goldman B, Friedman E (1996) A novel mutation of presenilin 1 in familial Alzheimer's disease in Israel detected by denaturing gradient gel electrophoresis. Hum Genet 98:700–702

Rogaev EI, Sherrington R, Rogaeva EA, Levesque G, Ikeda M, Liang Y, Chi H, Lin C, Holman K, Tsuda T, et al (1995) Familial Alzheimer's disease in kindreds with missense mutations in a gene on chromosome 1 related to the Alzheimer's disease type 3 gene. Nature 376:775–778

Rogaeva E, Meng Y, Lee JH, Gu Y, Kawarai T, Zou F, Katayama T, Baldwin CT, Cheng R, Hasegawa H, Chen F, Shibata N, Lunetta KL, Pardossi-Piquard R, Bohm C, Wakutani Y, Cupples LA, Cuenco KT, Green RC, Pinessi L, Rainero I, Sorbi S, Bruni A, Duara R, Friedland RP, Inzelberg R, Hampe W, Bujo H, Song YQ, Andersen OM, Willnow TE, Graff-Radford N, Petersen RC, Dickson D, Der SD, Fraser PE, Schmitt-Ulms G, Younkin S, Mayeux R, Farrer LA, St George-Hyslop P (2007) The neuronal sortilin-related receptor SORL1 is genetically associated with Alzheimer disease. Nat Genet 39:168–177

Rogaeva EA, Fafel KC, Song YQ, Medeiros H, Sato C, Liang Y, Richard E, Rogaev EI, Frommelt P, Sadovnick AD, Meschino W, Rockwood K, Boss MA, Mayeux R, St George-Hyslop P (2001) Screening for PS1 mutations in a referral-based series of AD cases: 21 novel mutations. Neurology 57:621–625

Roks G, Van Harskamp F, De Koning I, Cruts M, De Jonghe C, Kumar-Singh S, Tibben A, Tanghe H, Niermeijer MF, Hofman A, Van Swieten JC, Van Broeckhoven C, Van Duijn CM (2000) Presentation of amyloidosis in carriers of the codon 692 mutation in the amyloid precursor protein gene (APP692). Brain 123(Pt 10):2130–2140

Molecular Genetics of Alzheimer's Disease

Rosario ER, Chang L, Stanczyk FZ, Pike CJ (2004) Age-related testosterone depletion and the development of Alzheimer disease. JAMA 292:1431–1432

Roses AD, Saunders AM, Alberts MA, Strittmatter WJ, Schmechel D, Gorder E, Pericak-Vance MA (1995) Apolipoprotein E E4 allele and risk of dementia. JAMA 273:374–375; author reply 375–376

Rossi G, Giaccone G, Maletta R, Morbin M, Capobianco R, Mangieri M, Giovagnoli AR, Bizzi A, Tomaino C, Perri M, Di Natale M, Tagliavini F, Bugiani O, Bruni AC (2004) A family with Alzheimer disease and strokes associated with A713T mutation of the APP gene. Neurology 63:910–912

Rovelet-Lecrux A, Hannequin D, Raux G, Le Meur N, Laquerriere A, Vital A, Dumanchin C, Feuillette S, Brice A, Vercelletto M, Dubas F, Frebourg T, Campion D (2006) APP locus duplication causes autosomal dominant early-onset Alzheimer disease with cerebral amyloid angiopathy. Nat Genet 38:24–26

Rozmahel R, Huang J, Chen F, Liang Y, Nguyen V, Ikeda M, Levesque G, Yu G, Nishimura M, Mathews P, Schmidt SD, Mercken M, Bergeron C, Westaway D, St George-Hyslop P (2002) Normal brain development in PS1 hypomorphic mice with markedly reduced gamma-secretase cleavage of betaAPP. Neurobiol Aging 23:187–194

Ruitenberg A, Ott A, van Swieten JC, Hofman A, Breteler MM (2001) Incidence of dementia: does gender make a difference? Neurobiol Aging 22:575–580

Sahasrabudhe SR, Spruyt MA, Muenkel HA, Blume AJ, Vitek MP, Jacobsen JS (1992) Release of amino-terminal fragments from amyloid precursor protein reporter and mutated derivatives in cultured cells. J Biol Chem 267:25602–25608

Sala Frigerio C, Piscopo P, Calabrese E, Crestini A, Malvezzi Campeggi L, Civita di Fava R, Fogliarino S, Albani D, Marcon G, Cherchi R, Piras R, Forloni G, Confaloni A (2005) PEN-2 gene mutation in a familial Alzheimer's disease case. J Neurol 252:1033–1036

Salehi A, Dubelaar EJ, Mulder M, Swaab DF (1998) Aggravated decrease in the activity of nucleus basalis neurons in Alzheimer's disease is apolipoprotein E-type dependent. Proc Natl Acad Sci U S A 95:11445–11449

Sanan DA, Weisgraber KH, Russell SJ, Mahley RW, Huang D, Saunders A, Schmechel D, Wisniewski T, Frangione B, Roses AD, et al (1994) Apolipoprotein E associates with beta amyloid peptide of Alzheimer's disease to form novel monofibrils. Isoform apoE4 associates more efficiently than apoE3. J Clin Invest 94:860–869

Sandbrink R, Beyreuther K (1996) Unraveling the molecular pathway of Alzheimer's disease: research about presenilins gathers momentum. Mol Psychiatr 1:438–444

Sato C, Morohashi Y, Tomita T, Iwatsubo T (2006) Structure of the catalytic pore of gamma-secretase probed by the accessibility of substituted cysteines. J Neurosci 26:12081–12088

Saunders AM, Strittmatter WJ, Schmechel D, George-Hyslop PH, Pericak-Vance MA, Joo SH, Rosi BL, Gusella JF, Crapper-MacLachlan DR, Alberts MJ, et al (1993) Association of apolipoprotein E allele epsilon 4 with late-onset familial and sporadic Alzheimer's disease. Neurology 43:1467–1472

Sayi JG, Patel NB, Premkumar DR, Adem A, Winblad B, Matuja WB, Mtui EP, Gatere S, Friedland RP, Koss E, Kalaria RN (1997) Apolipoprotein E polymorphism in elderly east. Africans East Afr Med J 74:668–670

Scacchi R, Gambina G, Martini MC, Ruggeri M, Ferrari G, Silvestri M, Schiavon R, Corbo RM (2001) Polymorphisms of the apolipoprotein E gene regulatory region and of the LDL receptor gene in late-onset Alzheimer's disease in relation to the plasma lipidic pattern. Dement Geriatr Cogn Disord 12:63–68

Schellenberg GD, Bird TD, Wijsman EM, Orr HT, Anderson L, Nemens E, White JA, Bonnycastle L, Weber JL, Alonso ME, et al (1992) Genetic linkage evidence for a familial Alzheimer's disease locus on chromosome 14. Science 258:668–671

Scheuner D, Eckman C, Jensen M, Song X, Citron M, Suzuki N, Bird TD, Hardy J, Hutton M, Kukull W, Larson E, Levy-Lahad E, Viitanen M, Peskind E, Poorkaj P, Schellenberg G, Tanzi R, Wasco W, Lannfelt L, Selkoe D, Younkin S (1996) Secreted amyloid beta-protein similar to that in the senile plaques of Alzheimer's disease is increased in vivo by the presenilin 1 and 2 and APP mutations linked to familial Alzheimer's disease. Nat Med 2:864–870

Schmechel DE, Saunders AM, Strittmatter WJ, Crain BJ, Hulette CM, Joo SH, Pericak-Vance MA, Goldgaber D, Roses AD (1993) Increased amyloid beta-peptide deposition in cerebral cortex as a consequence of apolipoprotein E genotype in late-onset Alzheimer disease. Proc Natl Acad Sci U S A 90:9649–9653

Schroeter EH, Ilagan MX, Brunkan AL, Hecimovic S, Li YM, Xu M, Lewis HD, Saxena MT, De Strooper B, Coonrod A, Tomita T, Iwatsubo T, Moore CL, Goate A, Wolfe MS, Shearman M, Kopan R (2003) A presenilin dimer at the core of the gamma-secretase enzyme: insights from parallel analysis of Notch 1 and APP proteolysis. Proc Natl Acad Sci U S A 100:13075–13080

Seiffert D, Bradley JD, Rominger CM, Rominger DH, Yang F, Meredith JE Jr, Wang Q, Roach AH, Thompson LA, Spitz SM, Higaki JN, Prakash SR, Combs AP, Copeland RA, Arneric SP, Hartig PR, Robertson DW, Cordell B, Stern AM, Olson RE, Zaczek R (2000) Presenilin-1 and -2 are molecular targets for gamma-secretase inhibitors. J Biol Chem 275:34086–34091

Shao Y, Gearing M, Mirra SS (1997) Astrocyte-apolipoprotein E associations in senile plaques in Alzheimer disease and vascular lesions: a regional immunohistochemical study J Neuropathol. Exp Neurol 56:376–381

Shen J, Bronson RT, Chen DF, Xia W, Selkoe DJ, Tonegawa S (1997) Skeletal and CNS defects in Presenilin-1-deficient mice. Cell 89:629–639.

Sherrington R, Rogaev EI, Liang Y, Rogaeva EA, Levesque G, Ikeda M, Chi H, Lin C, Li G, Holman K, et al (1995) Cloning of a gene bearing missense mutations in early-onset familial Alzheimer's disease. Nature 375:754–760

Shumaker SA, Legault C, Rapp SR, Thal L, Wallace RB, Ockene JK, Hendrix SL, Jones BN III, Assaf AR, Jackson RD, Kotchen JM, Wassertheil-Smoller S, Wactawski-Wende J (2003) Estrogen plus progestin and the incidence of dementia and mild cognitive impairment in post-menopausal women: the Women's Health Initiative Memory Study: a randomized controlled trial. JAMA 289:2651–2662

Sian AK, Frears ER, El-Agnaf OM, Patel BP, Manca MF, Siligardi G, Hussain R, Austen BM (2000) Oligomerization of beta-amyloid of the Alzheimer's and the Dutch-cerebral-haemorrhage types. Biochem J 349:299–308

Sleegers K, Brouwers N, Gijselinck I, Theuns J, Goossens D, Wauters J, Del-Favero J, Cruts M, van Duijn CM, Van Broeckhoven C (2006) APP duplication is sufficient to cause early onset Alzheimer's dementia with cerebral amyloid angiopathy. Brain 129:2977–2983

Smith MJ, Kwok JB, McLean CA, Kril JJ, Broe GA, Nicholson GA, Cappai R, Hallupp M, Cotton RG, Masters CL, Schofield PR, Brooks WS (2001) Variable phenotype of Alzheimer's disease with spastic paraparesis. Ann Neurol 49:125–129

Soininen H, Partanen K, Pitkanen A, Hallikainen M, Hanninen T, Helisalmi S, Mannermaa A, Ryynanen M, Koivisto K, Riekkinen P, Sr. (1995) Decreased hippocampal volume asymmetry on MRIs in nondemented elderly subjects carrying the apolipoprotein E epsilon 4 allele. Neurology 45:391–392

Song W, Nadeau P, Yuan M, Yang X, Shen J, Yankner BA (1999) Proteolytic release and nuclear translocation of Notch-1 are induced by presenilin-1 and impaired by pathogenic presenilin-1 mutations. Proc Natl Acad Sci U S A 96:6959–6963

Song YQ, Rogaeva E, Premkumar S, Brindle N, Kawarai T, Orlacchio A, Yu G, Levesque G, Nishimura M, Ikeda M, Pei Y, O'Toole C, Duara R, Barker W, Sorbi S, Freedman M, Farrer L, St George-Hyslop P (1998) Absence of association between Alzheimer disease and the -491 regulatory region polymorphism of APOE. Neurosci Lett 250:189–192

Sorbi S, Nacmias B, Forleo P, Piacentini S, Amaducci L (1996) Alzheimer's disease and apolipoprotein E in Italy Ann N Y. Acad Sci 777:260–265

Sorbi S, Nacmias B, Forleo P, Piacentini S, Latorraca S, Amaducci L (1995) Epistatic effect of APP717 mutation and apolipoprotein E genotype in familial Alzheimer's disease. Ann Neurol 38:124–127

St George-Hyslop P, Haines J, Rogaev E, Mortilla M, Vaula G, Pericak-Vance M, Foncin JF, Montesi M, Bruni A, Sorbi S, et al (1992) Genetic evidence for a novel familial Alzheimer's disease locus on chromosome 14. Nat Genet 2:330–334

Molecular Genetics of Alzheimer's Disease

St George-Hyslop P, McLachlan DC, Tsuda T, Rogaev E, Karlinsky H, Lippa CF, Pollen D (1994) Alzheimer's disease and possible gene interaction. Science 263:537

St George-Hyslop PH, Tanzi RE, Polinsky RJ, Haines JL, Nee L, Watkins PC, Myers RH, Feldman RG, Pollen D, Drachman D, et al (1987) The genetic defect causing familial Alzheimer's disease maps on chromosome 21. Science 235:885–890

Steiner H, Capell A, Pesold B, Citron M, Kloetzel PM, Selkoe DJ, Romig H, Mendla K, Haass C (1998) Expression of Alzheimer's disease-associated presenilin-1 is controlled by proteolytic degradation and complex formation. J Biol Chem 273:32322–32331

Steiner H, Romig H, Grim MG, Philipp U, Pesold B, Citron M, Baumeister R, Haass C (1999) The biological and pathological function of the presenilin-1 Deltaexon 9 mutation is independent of its defect to undergo proteolytic processing. J Biol Chem 274:7615–7618

Stone DJ, Rozovsky I, Morgan TE, Anderson CP, Hajian H, Finch CE (1997) Astrocytes and microglia respond to estrogen with increased apoE mRNA in vivo and in vitro. Exp Neurol 143:313–318

Strittmatter WJ, Saunders AM, Schmechel D, Pericak-Vance M, Enghild J, Salvesen GS, Roses AD (1993) Apolipoprotein E: high-avidity binding to beta-amyloid and increased frequency of type 4 allele in late-onset familial Alzheimer disease. Proc Natl Acad Sci U S A 90:1977–1981

Struble RG, Short J, Ghobrial M, Nathan BP (1999) Apolipoprotein E immunoreactivity in human and mouse olfactory bulb. Neurosci Lett 267:137–140

Suzuki N, Cheung TT, Cai XD, Odaka A, Otvos L Jr, Eckman C, Golde TE, Younkin SG (1994) An increased percentage of long amyloid beta protein secreted by familial amyloid beta protein precursor (beta APP717) mutants. Science 264:1336–1340

Taddei K, Clarnette R, Gandy SE, Martins RN (1997) Increased plasma apolipoprotein E (apoE) levels in Alzheimer's disease. Neurosci Lett 223:29–32

Taddei K, Fisher C, Laws SM, Martins G, Paton A, Clarnette RM, Chung C, Brooks WS, Hallmayer J, Miklossy J, Relkin N, St George-Hyslop PH, Gandy SE, Martins RN (2002) Association between presenilin-1 Glu318Gly mutation and familial Alzheimer's disease in the Australian population. Mol Psychiatr 7:776–781

Tagliavini F, Rossi G, Padovani A, Magoni M, Andora G, Sgarzi M, Bizzi A, Carella SMF, Morbin M, Giaccone G, Bugiani O (1999) A new βPP mutation related to hereditary cerebral haemorrhage. Alzheimer's Rep 2(Supp 1):S28

Talbot C, Lendon C, Craddock N, Shears S, Morris JC, Goate A (1994) Protection against Alzheimer's disease with apoE epsilon 2. Lancet 343:1432–1433

Tamaoka A, Odaka A, Ishibashi Y, Usami M, Sahara N, Suzuki N, Nukina N, Mizusawa H, Shoji S, Kanazawa I, et al (1994) APP717 missense mutation affects the ratio of amyloid beta protein species (A beta 1–42/43 and a beta 1–40) in familial Alzheimer's disease brain. J Biol Chem 269:32721–32724

Tang-Wai D, Lewis P, Boeve B, Hutton M, Golde T, Baker M, Hardy J, Michels V, Ivnik R, Jack C, Petersen R (2002) Familial frontotemporal dementia associated with a novel presenilin-1 mutation. Dement Geriatr Cogn Disord 14:13–21

Tang AC, Hasselmo ME (1996) Effect of long term baclofen treatment on recognition memory and novelty detection. Behav Brain Res 74:145–152.

Tanzi RE (1991) Gene mutations in inherited amyloidopathies of the nervous system. Am J Hum Genet 49:507–510

Tanzi RE, McClatchey AI, Lamperti ED, Villa-Komaroff L, Gusella JF, Neve RL (1988) Protease inhibitor domain encoded by an amyloid protein precursor mRNA associated with Alzheimer's disease. Nature 331:528–530

Tanzi RE, St George-Hyslop PH, Haines JL, Polinsky RJ, Nee L, Foncin JF, Neve RL, McClatchey AI, Conneally PM, Gusella JF (1987) The genetic defect in familial Alzheimer's disease is not tightly linked to the amyloid beta-protein gene. Nature 329:156–157

Terreni L, Fogliarino S, Franceschi M, Forloni G (2002) Novel pathogenic mutation in an Italian patient with familial Alzheimer's disease detected in APP gene. Neurobiol Aging 23:S319

Theuns J, Del-Favero J, Dermaut B, van Duijn CM, Backhovens H, Van den Broeck MV, Serneels S, Corsmit E, Van Broeckhoven CV, Cruts M (2000) Genetic variability in the regulatory region of presenilin 1 associated with risk for Alzheimer's disease and variable expression. Hum Mol Genet 9:325–331

Theuns J, Marjaux E, Vandenbulcke M, Van Laere K, Kumar-Singh S, Bormans G, Brouwers N, Van den Broeck M, Vennekens K, Corsmit E, Cruts M, De Strooper B, Van Broeckhoven C, Vandenberghe R (2006) Alzheimer dementia caused by a novel mutation located in the APP C-terminal intracytosolic fragment. Hum Mutat 27:888–896

Theuns J, Remacle J, Killick R, Corsmit E, Vennekens K, Huylebroeck D, Cruts M, Van Broeckhoven C (2003) Alzheimer-associated C allele of the promoter polymorphism -22C>T causes a critical neuron-specific decrease of presenilin 1 expression. Hum Mol Genet 12:869–877

Thinakaran G, Borchelt DR, Lee MK, Slunt HH, Spitzer L, Kim G, Ratovitsky T, Davenport F, Nordstedt C, Seeger M, Hardy J, Levey AI, Gandy SE, Jenkins NA, Copeland NG, Price DL, Sisodia SS (1996) Endoproteolysis of presenilin 1 and accumulation of processed derivatives in vivo. Neuron 17:181–190

Thomas P, Fenech M (2007) A review of genome mutation and Alzheimer's disease. Mutagenesis 22:15–33

Toji H, Maruyama H, Sasaki K, Nakamura S, Kawakami H (1999) Apolipoprotein E promoter polymorphism and sporadic Alzheimer's disease in a Japanese population. Neurosci. Lett 259:56–58

Tolia A, Chavez-Gutierrez L, De Strooper B (2006) Contribution of presenilin transmembrane domains 6 and 7 to a water-containing cavity in the gamma-secretase complex. J Biol Chem 281:27633–27642

Town T, Paris D, Fallin D, Duara R, Barker W, Gold M, Crawford F, Mullan M (1998) The -491A/ T apolipoprotein E promoter polymorphism association with Alzheimer's disease: independent risk and linkage disequilibrium with the known APOE polymorphism. Neurosci Lett 252:95–98

Tsubuki S, Takaki Y, Saido TC (2003) Dutch, Flemish, Italian, and Arctic mutations of APP and resistance of Abeta to physiologically relevant proteolytic degradation. Lancet 361:1957–1958

Tysoe C, Whittaker J, Xuereb J, Cairns NJ, Cruts M, Van Broeckhoven C, Wilcock G, Rubinsztein DC (1998) A presenilin-1 truncating mutation is present in two cases with autopsy-confirmed early-onset Alzheimer disease. Am J Hum Genet 62:70–76

Van Broeckhoven C, Genthe AM, Vandenberghe A, Horsthemke B, Backhovens H, Raeymaekers P, Van Hul W, Wehnert A, Gheuens J, Cras P, et al (1987) Failure of familial Alzheimer's disease to segregate with the A4-amyloid gene in several European families. Nature 329:153–155

Van Broeckhoven C, Haan J, Bakker E, Hardy JA, Van Hul W, Wehnert A, Vegter-Van der Vlis M, Roos RA (1990) Amyloid beta protein precursor gene and hereditary cerebral hemorrhage with amyloidosis (Dutch). Science 248:1120–1122

van Duijn CM, Cruts M, Theuns J, Van Gassen G, Backhovens H, van den Broeck M, Wehnert A, Serneels S, Hofman A, Van Broeckhoven C (1999) Genetic association of the presenilin-1 regulatory region with early-onset Alzheimer's disease in a population-based sample. Eur J Hum Genet 7:801–806

Verdile G, Fuller S, Atwood CS, Laws SM, Gandy SE, Martins RN (2004a) The role of beta amyloid in Alzheimer's disease: still a cause of everything or the only one who got caught? Pharmacol Res 50:397–409

Verdile G, Gandy SE, Martins RN (2007) The role of presenilin and its interacting proteins in the biogenesis of Alzheimer's beta amyloid. Neurochem Res 32:609–623

Verdile G, Gnjec A, Miklossy J, Fonte J, Veurink G, Bates K, Kakulas B, Mehta PD, Milward EA, Tan N, Lareu R, Lim D, Dharmarajan A, Martins RN (2004b) Protein markers for Alzheimer disease in the frontal cortex and cerebellum. Neurology 63:1385–1392

Verkkoniemi A, Somer M, Rinne JO, Myllykangas L, Crook R, Hardy J, Viitanen M, Kalimo H, Haltia M (2000) Variant Alzheimer's disease with spastic paraparesis: clinical characterization. Neurology 54:1103–1109

Wakutani Y, Watanabe K, Adachi Y, Wada-Isoe K, Urakami K, Ninomiya H, Saido TC, Hashimoto T, Iwatsubo T, Nakashima K (2004) Novel amyloid precursor protein gene missense mutation (D678N) in probable familial Alzheimer's disease. J Neurol Neurosurg Psychiatr 75:1039–1042

Walsh DM, Hartley DM, Condron MM, Selkoe DJ, Teplow DB (2001) In vitro studies of amyloid beta-protein fibril assembly and toxicity provide clues to the aetiology of Flemish variant (Ala692->Gly) Alzheimer's disease. Biochem J 355:869–877

Walsh DM, Hartley DM, Kusumoto Y, Fezoui Y, Condron MM, Lomakin A, Benedek GB, Selkoe DJ, Teplow DB (1999) Amyloid beta-protein fibrillogenesis. Structure and biological activity of protofibrillar intermediates. J Biol Chem 274:25945–25952

Walsh DM, Lomakin A, Benedek GB, Condron MM, Teplow DB (1997) Amyloid beta-protein fibrillogenesis. Detection of a protofibrillar intermediate. J Biol Chem 272:22364–22372

Wang PN, Liao SQ, Liu RS, Liu CY, Chao HT, Lu SR, Yu HY, Wang SJ, Liu HC (2000) Effects of estrogen on cognition, mood, and cerebral blood flow in AD: a controlled study. Neurology 54:2061–2066

Weisgraber KH, Rall SC Jr, Innerarity TL, Mahley RW, Kuusi T, Ehnholm C (1984) A novel electrophoretic variant of human apolipoprotein E. Identification and characterization of apolipoprotein E1. J Clin Invest 73:1024–1033

West HL, Rebeck GW, Hyman BT (1994) Frequency of the apolipoprotein E epsilon 2 allele is diminished in sporadic Alzheimer disease. Neurosci Lett 175:46–48

Wisniewski T, Frangione B (1992) Apolipoprotein E: a pathological chaperone protein in patients with cerebral and systemic amyloid. Neurosci Lett 135:235–238

Wolfe MS (2007) When loss is gain: reduced presenilin proteolytic function leads to increased Abeta42/Abeta40. Talking Point on the role of presenilin mutations in Alzheimer disease. EMBO Rep 8:136–140

Wolfe MS, Xia W, Ostaszewski BL, Diehl TS, Kimberly WT, Selkoe DJ (1999) Two transmembrane aspartates in presenilin-1 required for presenilin endoproteolysis and gamma-secretase activity. Nature 398:513–517

Wong PC, Zheng H, Chen H, Becher MW, Sirinathsinghji DJ, Trumbauer ME, Chen HY, Price DL, Van der Ploeg LH, Sisodia SS (1997) Presenilin 1 is required for Notch1 and DII1 expression in the paraxial mesoderm. Nature 387:288–292

Xia W, Zhang J, Kholodenko D, Citron M, Podlisny MB, Teplow DB, Haass C, Seubert P, Koo EH, Selkoe DJ (1997) Enhanced production and oligomerization of the 42-residue amyloid beta-protein by Chinese hamster ovary cells stably expressing mutant presenilins. J Biol Chem 272:7977–7982

Xie Z, Romano DM, Kovacs DM, Tanzi RE (2004) Effects of RNA interference-mediated silencing of gamma-secretase complex components on cell sensitivity to caspase-3 activation. J Biol Chem 279:34130–34137

Xu H, Gouras GK, Greenfield JP, Vincent B, Naslund J, Mazzarelli L, Fried G, Jovanovic JN, Seeger M, Relkin NR, Liao F, Checler F, Buxbaum JD, Chait BT, Thinakaran G, Sisodia SS, Wang R, Greengard P, Gandy S (1998) Estrogen reduces neuronal generation of Alzheimer beta-amyloid peptides. Nat Med 4:447–451

Xu PT, Gilbert JR, Qiu HL, Ervin J, Rothrock-Christian TR, Hulette C, Schmechel DE (1999) Specific regional transcription of apolipoprotein E in human brain neurons. Am J Pathol 154:601–611

Yaffe K (2001) Estrogens, selective estrogen receptor modulators, and dementia: what is the evidence? Ann N Y Acad Sci 949:215–222

Yaffe K, Haan M, Byers A, Tangen C, Kuller L (2000) Estrogen use, APOE, and cognitive decline: evidence of gene-environment interaction. Neurology 54:1949–1954

Yamada N, Shimano H, Yazaki Y (1995) Role of apolipoprotein E in lipoprotein metabolism and in the process of atherosclerosis. J Atheroscler Thromb 2(Suppl 1):S29–S33

Yamagata K, Urakami K, Ikeda K, Ji Y, Adachi Y, Arai H, Sasaki H, Sato K, Nakashima K (2001) High expression of apolipoprotein E mRNA in the brains with sporadic Alzheimer's disease. Dement Geriatr Cogn Disord 12:57–62

Yamamura T, Yamamoto A, Sumiyoshi T, Hiramori K, Nishioeda Y, Nambu S (1984) New mutants of apolipoprotein E associated with atherosclerotic diseases but not to type III hyperlipoproteinemia. J Clin Invest 74:1229–1237

Yanagisawa K, Ihara Y, Miyatake T (1992) Secretory pathway of beta/A4 amyloid protein precursor in familial Alzheimer's disease with Val717 to Ile mutation. Neurosci Lett 144:43–45

Yang DS, Small DH, Seydel U, Smith JD, Hallmayer J, Gandy SE, Martins RN (1999) Apolipoprotein E promotes the binding and uptake of beta-amyloid into Chinese hamster ovary cells in an isoform-specific manner. Neuroscience 90:1217–1226

Yang DS, Smith JD, Zhou Z, Gandy SE, Martins RN (1997) Characterization of the binding of amyloid-beta peptide to cell culture-derived native apolipoprotein E2, E3, and E4 isoforms and to isoforms from human plasma. J Neurochem 68:721–725

Yoshikai S, Sasaki H, Doh-ura K, Furuya H, Sakaki Y (1990) Genomic organization of the human amyloid beta-protein precursor gene. Gene 87:257–263

Yoshioka K, Miki T, Katsuya T, Ogihara T, Sakaki Y (1991) The 717Val–Ile substitution in amyloid precursor protein is associated with familial Alzheimer's disease regardless of ethnic groups. Biochem Biophys Res Commun 178:1141–1146

Zannis VI, Breslow JL, Utermann G, Mahley RW, Weisgraber KH, Havel RJ, Goldstein JL, Brown MS, Schonfeld G, Hazzard WR, Blum C (1982) Proposed nomenclature of apoE isoproteins, apoE genotypes, and phenotypes. J Lipid Res 23:911–914

Zekanowski C, Styczynska M, Peplonska B, Gabryelewicz T, Religa D, Ilkowski J, Kijanowska-Haladyna B, Kotapka-Minc S, Mikkelsen S, Pfeffer A, Barczak A, Luczywek E, Wasiak B, Chodakowska-Zebrowska M, Gustaw K, Laczkowski J, Sobow T, Kuznicki J, Barcikowska M (2003) Mutations in presenilin 1, presenilin 2 and amyloid precursor protein genes in patients with early-onset Alzheimer's disease in Poland. Exp Neurol 184:991–996

Zhang Y, Champagne N, Beitel LK, Goodyer CG, Trifiro M, LeBlanc A (2004) Estrogen and androgen protection of human neurons against intracellular amyloid beta1–42 toxicity through heat shock protein 70. J Neurosci 24:5315–5321

Zheng H, Xu H, Uljon SN, Gross R, Hardy K, Gaynor J, Lafrancois J, Simpkins J, Refolo LM, Petanceska S, Wang R, Duff K (2002) Modulation of A(beta) peptides by estrogen in mouse models. J Neurochem 80:191–196

Zubenko GS, Stiffler S, Stabler S, Kopp U, Hughes HB, Cohen BM, Moossy J (1994) Association of the apolipoprotein E epsilon 4 allele with clinical subtypes of autopsy-confirmed Alzheimer's disease. Am J Med Genet 54:199–205

Molecular Biology of Parkinson's Disease

Abbas Parsian and Biswanath Patra

Contents

1 Introduction 278
2 Molecular Mechanisms in the Development of Parkinson's Disease 279
   2.1 The Role of Iron and Oxidative Stress 279
   2.2 Role of Alpha-Synuclein 280
   2.3 Mitochondrial Polymorphism and Deletions in Parkinson's Disease 282
   2.4 Heat Shock Protein Chaperone (HSC-70) 283
   2.5 Dopamine Neurotransmission and Metabolism 283
   2.6 Association of LRRK2 Gene and Inappropriate Phosphorylation
      in Parkinson's Disease 284
   2.7 Role of Glutathione-S-Transferase Gene in the
      Development of Parkinson's Disease 285
3 Conclusions 285
References 286

Abstract Parkinson's disease is one of the most common neurodegenerative movement disorders of non specific etiology. It is characterized by tremor at rest, bradykinesia, rigidity, and in more advanced cases, postural instability. The most important environmental factors such as neurotropic infective agents, neurotoxins like heavy metals and pesticides are considered as major culprit to initiate the disease. So far, mutations in eight genes (SNCA, PRKN, PINK1, DJ-1, MAPT, UCH-L1, ATP13A2 and LRRK2) are reported for the familial form of Parkinson's disease. Genetic, neuropathological and neurochemical studies on substantia nigra from Parkinson patients and animal models have focused on several pathogenic processes at the time of neuronal death. A few potential contributing factors have been established to play crucial role in the development of Parkinson's disease. Most of these factors are involved in ongoing selective oxidative stress resulting from mitochondrial dysfunction, auto-oxidation or enzymatic (monoamine oxidase) oxidation of dopamine, excessive iron accumulation in the substantia nigra pars compacta and genetic susceptibility. In the following sections the role of each of these factors are explained in detail.

A. Parsian(✉)
International Scientific Consultation Firm, 20 Margeaux Dr., Little Rock, AR 72223, USA
parsian@sbcglobal.net

D.B. Wildenauer (ed.), *Molecular Biology of Neuropsychiatric Disorders*,
Nucleic Acids and Molecular Biology, © Springer-Verlag Berlin Heidelberg 2009

1 Introduction

Parkinson's disease (PD) is one of the most common devastating neurodegenerative movement disorders of nonspecific etiology in the world, characterized by tremor at rest, bradykinesia, rigidity, and, in more advanced cases, postural instability. Perhaps more than any other single feature, the presence of a typical resting tremor increases the likelihood of pathologically supported PD, although approximately 20% of patients fail to develop a typical resting tremor (Hughes et al. 1993). The presence of Lewy bodies (LB)/Lewy neuritis in association with dopaminergic nerve cell loss in the substantia nigra (SN) and various other regions of the nervous system (Lennox and Lowe 1997) is the pathogenic symptom of PD. It has been established that LB and neurites stain with antibodies to α-synuclein, ubiquitin, and a variety of other biochemical markers, and are found in many areas of the PD brain, not only the SN, but also the dorsal motor nucleus of the vagus, locus ceruleus, raphe and reticular formation nuclei, thalamus, amygdala, olfactory nuclei, pediculopontine nucleus, and cerebral cortex, among others (Braak et al. 2003; Jellinger 2003).

Today, the most important environmental factors like neurotropic infective agents, neurotoxins like heavy metals, and pesticides are considered as major culprits to initiate the disease. There is a possibility that a different viral exposure (encephalitis lethargica, intrauterine influenza, herpes simplex, Japanese B encephalitis, and coxsackie; Ben-Shlomo and Sieradzan 1995) may result in partial damage to the substantia nigra, which only appears much later after further age-related degeneration (the two-stage hypothesis; Calne and Langston 1983). However, most sero-epidemiologic studies (population-based prevalence estimates of sero-positivity) have failed to find any differences in antibody titers for a wide variety of viruses between cases with idiopathic PD and controls (Elizan and Casals 1983). The neurotoxic factors like manganese, a meperidine analogue 1-methyl-4-phenyl-1,2,3,6-tetrahydropyridine (MPTP), carbon disulphide, carbon monoxide, mercury, pesticides such as paraquat, well-water, and several occupation-related exposures have been associated with idiopathic PD (e.g., manganese). The serendipitous discovery of a group of drug addicts with a parkinsonian syndrome led to the detection of the neurotoxic meperidine analogue MPTP (Langston et al. 1983).

Alpha-synuclein (SNCA), the first familial PD gene, has been reported for missense mutation, polymorphic variability within promoter region, and increased expression in autosomal dominant PD families (Zarranz et al. 2004; Chartier-Harlin et al. 2004; Pals et al. 2004). In a German family with PD, a missense mutation in the ubiquitin carboxy-terminal hydrolase L1 (UCH-L1, Zhang et al. 2000) gene has been identified. It is reported (Leroy et al. 1998) that this mutation, Ile93Met, causes a partial loss of the catalytic activity of this thiol protease, which could lead to aberrations in the proteolytic pathway and aggregation of proteins. So far, mutations in eight genes (SNCA, PRKN, PINK1, DJ-1, MAPT, UCH-L1, ATP13A2, and LRRK2) are reported for familial PD.

2 Molecular Mechanisms in the Development of Parkinson's Disease

Genetic (gene expression, mutation, polymorphism), neuropathological, and neurochemical studies on SN from PD patients and animal models research have focused on several pathogenic processes or possibilities at the time of neuronal death, although the etiology of the disease remains mysterious. A few potential contributing factors have been established to play crucial role, that are involved in ongoing selective oxidative stress (OS), include mitochondrial dysfunction, auto-oxidation, or enzymatic (monoamine oxidase, MAO) oxidation of dopamine (DA), excessive iron accumulation in the SN pars compacta (pc) and genetic susceptibility (Youdim et al. 1993; Riederer et al. 1989; Gotz et al. 1994; Youdim and Riederer 1997; Jenner & Olanow 1996; Mandel et al. 2005).

2.1 The Role of Iron and Oxidative Stress

In pathogenesis of neurodegenerative brain disease, iron accumulation and alpha-synuclein aggregate are present in axonal spheroids, glial, and neuronal inclusions. The identification of the major components of Lewy bodies (LB, Lowe and Dickson 1997) suggests that a pathway leading from normal soluble to abnormal misfolded filamentous proteins is central for PD pathogenesis regardless of the primary disorder. There are conformational differences in alpha-synuclein between neuronal and glial aggregates, showing nonuniform mapping for its epitopes. Despite several cellular and transgenic models, it is not clear whether inclusion body formation is an adaptive/neuroprotective or a pathogenic reaction/process generated in response to different, mostly undetermined, functional triggers linked to neurodegeneration. Specifically, redox-active iron has been observed within the melanin-containing neurons that selectively die, and in the rim of LB, the morphological characteristic of PD. LB are composed of lipids, aggregated alpha-synuclein (concentrating in its peripheral halo), and ubiquitinated, hyperphosphorylated neurofilament proteins (Jellinger 2003). The gene expression changes in the substantia nigra pars compacta (SNpc) of postmortem parkinsonian brains when compared to age-matched controls showed the diminished expression of an essential component of protein catabolism, the SKP1A gene (Grunblatt et al. 2004). Its decline was accompanied by decreased expression in various subunits of the 26S proteasome, in energy pathways, and in signal transduction (Mandel et al. 2005).

Youdim et al. (2004) reported that, in PD and its neurotoxin-induced rat models, using 6-hydroxydopamine (6-OHDA) and N-methyl-4-phenyl-1,2,3,6-tetrahydropyridine (MPTP), significant accumulation of iron ($Fe^{+2}:Fe^{+3}$) occurs in the SNpc with a ratio of 1:3 as compared to 1:1 normally observed in control brains. The iron is thought to be in a labile pool, unbound to ferritin, and is thought to have a pivotal role to induce oxidative stress-dependent neurodegeneration of dopamine neurons via Fenton chemistry [H_2O_2 + Fe (II)]. The consequence of this process is the interaction with H_2O_2 to generate the most reactive radical oxygen species, the hydroxyl radical. This scenario is supported by studies in both human and neurotoxin-induced parkinsonism in rat showing that disposition of H_2O_2 is compromised via depletion of glutathione (GSH), the rate-limiting cofactor of glutathione peroxide, the major enzyme source to dispose H_2O_2 as water in the brain. Further, radical scavengers have been shown to prevent the neurotoxic action of the above neurotoxins and depletion of GSH. The Youdim et al. (2004) group was the first to demonstrate that the prototype iron chelator, desferal, is a potent neuroprotective agent in the 6-OHDA model.

Hashimoto et al. (1999) reported that the precursor of non-amyloid beta protein component of Alzheimer's disease amyloid (NACP/alpha-synuclein), found in LB of PD, is a pre-synaptic protein genetically linked to some familial types of PD. Since oxidative stress might play a role in PD pathogenesis, they investigated the role of iron and peroxide in NACP/alpha-synuclein aggregation. Immunoblot analysis showed that human NACP/alpha-synuclein (but not beta-synuclein) aggregated in the presence of ferric ion and was inhibited by the iron chelator, deferoxamine. Ferrous ion was not effective by itself, but it potentially aggregated NACP/alpha-synuclein in the presence of hydrogen peroxide.

The regional distributions of iron, copper, zinc, magnesium, and calcium in parkinsonian brains were compared with those of matched controls by Riederer et al. (1989). The most notable finding was a shift in the iron (II)/iron (III) ratio in favor of iron (III) in SN and a significant increase in the iron (III)-binding protein, ferritin. Significantly, lower glutathione content was present in pooled samples of putamen, globus pallidus, SN, nucleus basalis of Meynert, amygdaloid nucleus, and frontal cortex of PD brains with severe damage to SN. However, no significant changes were observed in clinically pathologically mild forms of PD. Upon high oxygen levels or iron overload, the EGLN hydroxylases target hypoxia-inducible factor-1 alpha (HIF) to proteasomal degradation. Interestingly, the free iron-induced proteasome-mediated degradation of iron regulatory protein (IRP2) also involves activation of 2-oxoglutarate-dependent dioxygenases and is inhibited by iron chelators (Hanson et al. 2003; Wang et al. 2004). Thus, it is possible that IRP2 is a substrate of EGLN1, which causes post-translational modification, signaling it for protein degradation. Excessive production of EGLN1 in the SNpc may lead to a fall in IRP2 and subsequent decrease in transferrin receptor (TfR) mRNA and an increase in ferritin levels, both subjected to positive and negative transcriptional regulation by IRP2, respectively (Ponka 2004; Meyron-Holtz et al. 2004). Deplazes et al. (2004) reported that in one PD patient a −74C > T variation in IRP2 gene was found, which was not present in the control group.

2.2 Role of Alpha-Synuclein

In 1997, Polymeropoulos et al. reported that a G209A mutation within the alpha-synuclein gene (SNCA) on chromosome 4, segregates with PD in an Italian kindred (Polymeropoulos et al. 1997). Sequence analysis of exon 4 revealed a single base pair change at position 209 from G to A (G209A) that causes an Ala to Thr substi-

Molecular Biology of Parkinson's Disease 281

tution at position 53 of the protein (Ala53Thr) and creates a Tsp45I restriction site. This mutation, also found in five Greek families with early onset PD (Polymeropoulos et al. 1997; Papadimitriou et al. 1999), has not been replicated in other samples (Parsian et al. 1998, 2007; Scott et al. 1997; Gasser et al. 1997; Chan et al. 1998). Two other important mutations in this gene have been found, a G–C transversion at position 88 of the coding sequence (Ala30Pro) in a German family (Kruger et al. 1998) and an E46K mutation in a Spanish family with autosomal dominant parkinsonism, dementia, and visual hallucination (Zarranz et al. 2004). Triplication of the SNCA gene has also been reported to be associated with autosomal dominant PD with dementia (Singleton et al. 2003).

It has been reported (Ebadi et al. 2001; Ostrerova-Golts et al. 2000; Turnbull et al. 2001) that α-synuclein forms toxic aggregates in the presence of iron, and this is considered as a contributing factor to the formation of LB via oxidative stress (OS). Disturbances of brain iron metabolism has taken critical role in neurodegenerative diseases since a significant number of mutated iron metabolism genes have now been shown to be directly involved in neurodegeneration (Felletschin et al. 2003; Youdim and Riederer 2004; Zecca et al. 2004). The iron redox status may constitute a crucial factor contributing to the extent of protein misfolding and aggregation in the aging and disease-affected brain. It has also been observed (Morfini et al. 2007) that neurotoxins such as MPTP and its metabolite 1-methyl-4-phenylpyridinium (MPP+) induce PD symptoms and recapitulate major pathological hallmarks of PD in human and animal models. Both sporadic and MPP + -induced forms of PD proceed through a "dying-back" pattern of neuronal degeneration in affected neurons, characterized by early loss of synaptic terminals and axonopathy. However, axonal and synaptic-specific effects of MPP + are poorly understood. Using isolated squid axoplasm, it has been shown that MPP + produces significant alterations in fast axonal transport (FAT) through activation of a caspase and a previously undescribed protein kinase C (PKC-delta) isoform (Morfini et al. 2007).

2.2.1 The A53T Alpha-Synuclein Mutation Increases Iron-Dependent Aggregation and Toxicity

In human BE-M17 neuroblastoma cells, overexpression of wild-type and alphasynuclein with A53T or A30P mutation showed that iron and free radical generators, such as dopamine or hydrogen peroxide, stimulate the production of intracellular aggregates that contain alpha-synuclein and ubiquitin (Ostrerova-Golts et al. 2000). The amount of aggregation occurring in the cells is dependent on the amount and type of alpha-synuclein expressed. The amount of alpha-synuclein aggregation was following a rank order of A53T > A30P > wild-type > untransfected. In addition to stimulating aggregate formation, alpha-synuclein also appears to induce toxicity. BE-M17 neuroblastoma cells overexpressing alpha-synuclein show up to a fourfold increase in vulnerability to toxicity induced by iron. The vulnerability follows the same rank order as for aggregation. These data raise the possibility that alpha-synuclein acts in concert with iron and dopamine to induce formation of Lewy body pathology and cell death in PD (Ostrerova-Golts et al. 2000).

2.2.2 Autoantibodies Against Alpha-Synuclein in Inherited Parkinson's Disease

Papachroni et al. (2007) examined the presence of autoantibodies (AAb) against synuclein family members in the peripheral blood serum of PD patients and control individuals. Presence of AAb against beta-synuclein or gamma-synuclein showed no association with PD. Multi-epitopic AAb against alpha-synuclein were detected in 65% of all patients tested and their presence strongly correlated with an inherited mode of the disease, but not with other disease-related factors. The frequency of the presence of AAb in the group of patients with sporadic form of PD was not significantly different from the control group. However, a very high proportion (90%) of patients with familial form of the disease was positive for AAb against alpha-synuclein. The authors hypothesize that these AAb could be involved in pathogenesis of the inherited form of PD (Papachroni et al. 2007).

2.3 Mitochondrial Polymorphism and Deletions in Parkinson's Disease

Bender et al. (2006) reported that, in substantia nigra neurons from both aged controls and individuals with Parkinson's disease, there was a high level of deleted mitochondrial DNA (mtDNA) (controls, 43.3% ± 9.3%; individuals with Parkinson disease, 52.3% ± 9.3%). These mtDNA mutations are somatic, with different clonally expanded deletions in individual cells, and high levels of these mutations are associated with respiratory chain deficiency. They suggested that somatic mtDNA deletions are important in the selective neuronal loss observed in aging brain and in PD. Kraytsberg et al. (2006) quantified the total burden of mtDNA molecules with deletions. They showed that a high proportion of individual pigmented neurons in the aged human substantia nigra contain very high levels of mtDNA deletions. Molecules with deletions are largely clonal within each neuron; that is, they originate from a single deleted mtDNA molecule that has expanded clonally. The fraction of mtDNA deletions is significantly higher in cytochrome c oxidase (COX)-deficient neurons than in COX-positive neurons, suggesting that mtDNA deletions may be directly responsible for impaired cellular respiration. The genetic polymorphism in the mitochondrial translation initiation factor 3 (MTIF3) gene has been identified to be associated with PD (Abahuni et al. 2007). Mitochondrial dysfunction occurs early in late onset sporadic PD, but the mitochondrial protein network mediating PD pathogenesis is largely unknown. Mutations in the mitochondrial serine-threonine kinase, PINK1 have recently been shown to cause the early onset autosomal recessive PD (PARK6). Abahuni et al. (2007) tested a candidate interactor protein of PINK1, the MTIF3 for involvement in PD pathogenesis. In two independent case-control studies, the c.798C > T polymorphism of the MTIF3 gene showed allelic association with PD ($p = 0.0073$). An altered function of variant MTIF3 may affect the availability of mitochondrial encoded proteins, lead to oxidative stress and create vulnerability for PD.

Molecular Biology of Parkinson's Disease 283

2.4 Heat Shock Protein Chaperone (HSC-70)

Grunblatt et al. (2004) reported that a significant reduction in gene expression of HSP8 (heat shock protein), coding for HSC-70 (heat shock protein chaperone-70), was observed specifically in the SNpc of 5 out of 6 parkinsonian patients. This was further confirmed by real-time quantitative PCR. There were no significant alterations in SNr (SN pars reticulate) or cerebellum between PD patients and controls, suggestive of tissue specificity. Recently, a functional polymorphism (−110 A/C) in the 5′ promoter region of HSP70–1 has been reported in 274 PD patients, which may increase susceptibility to PD (Wu et al. 2004).

It has been shown that overexpression of HSP70 reduces the amount of misfolded, aggregated alpha-synuclein species in vitro and in vivo in transgenic mouse models (Klucken et al. 2004) and prevents the loss of dopaminergic neurons. Auluck et al. (2002) reported that directed expression of the molecular chaperone Hsp70 prevented dopaminergic neuronal loss associated with alpha-synuclein in Drosophila and that interference with endogenous chaperone activity accelerated alpha-synuclein toxicity. Furthermore, LB in human postmortem tissue immunostained for molecular chaperones, also suggesting that chaperones may play a role in PD progression.

Scherzer et al. (2007) reported downregulation of cochaperone ST13 in blood sample of PD patient. This gene stabilizes heat shock protein 70, which is a modifier of alpha-synuclein misfolding and toxicity. ST13 messenger RNA copies were lower in patients with PD (mean ±SE; 0.59 ± 0.05) than in controls (0.96 ± 0.09) ($p = 0.002$) in two independent populations. Thus, gene expression signals measured in blood can facilitate the development of biomarkers for PD.

2.5 Dopamine Neurotransmission and Metabolism

Real-time PCR confirmatory analysis revealed a specific reduction of vesicular monoamine transporter VMAT2 (SCL18A2) mRNA in the SNpc, while the changes in the SNr were less pronounced. This finding is in agreement with previous reports on postmortem SN of controls and PD showing a marked reduction of VMAT2 mRNA in PD. This was associated with marked reductions in both dopamine (DA) transporter and VMAT2 signal per cell in the remaining pigmented neurons (Harrington et al. 1996; Brooks 2003). Three other striking changes, also related to DA transmission and metabolism, are the reduced expression of ALDH1A1, ADH5, and ARPP-21, mRNAs coding for aldehyde dehydrogenase (ALDH), alcohol dehydrogenase (ADH), and cAMP-regulated phosphoprotein, respectively. ARPP-21 is specifically enriched in DA-innervated brain regions of the basal ganglia (e.g., caudate-putamen) and in the SN (Ouimet et al. 1989; Tsou et al. 1993; Mandel et al. 2005).

DA is equally well metabolized in human brain striatum by MAO A and B (Collins et al. 1970; O'carroll et al. 1983) to a very reactive aldehyde derivatives (3,4-dihydroxyphenylacetaldehyde and 4-hydroxy-3-methoxyphenylacetaldehyde) that rarely accumulate in the striatum. ALDHs and ADHs are involved in the degradation of these aldehyde derivatives (Mardh and Vallee 1986), which are then metabolized to homovanillic acid and 3,4-dihydroxyphenylacetic acid, and in detoxification of aldehydes that are highly reactive and neurotoxic (Hjelle and Petersen 1983). Thus, alteration in DA transmission may alter ALDH and ADH activities; conversely, changes in ALDH-ADH-mediated metabolism may affect DA levels in nerve cell bodies and terminal fields in basal ganglia and the limbic system. This is of major relevance in light of recent evidence for mutations in genes encoding ADHs as genetic risk factors for PD (Buervenich et al. 2005). These proteins (ALDH, ADH, and ARPP-21), in conjunction with VMAT2, may now be considered as new markers for PD. For instance, SKP1 and ALDH1 protein levels are extremely low within the TH-positive dopaminergic neurons of the ventral-caudal SNpc (the major affected area in PD) from PD subjects, confirming the gene expression results (Mandel et al. 2005).

2.6 Association of LRRK2 Gene and Inappropriate Phosphorylation in Parkinson's Disease

LRRK2 (PARK8; OMIM*609007) is located on chromosome 12q12. The locus was first mapped in the Japanese "Sagamihara family" by Funayama et al. (2002). LRRK2 is encoded by 51 exons and seems to be expressed in most brain regions; the protein has a predicted molecular weight of 286 kDa. It is highly conserved among vertebrates and shares homology to the ROCO protein family (Bosgraaf and Haastert 2003; Zimprich et al. 2004). It is reported to have five conserved domains: a leucine-rich repeat (LRR), a Roc (Ras in complex proteins; Rab GTPase), a COR (domain C-terminal of Roc), a catalytic core common to both tyrosine and serine-threonine kinases (mixed-lineage MAPKKK), and a WD40 domain.

Galter et al. (2006) analyzed the LRRK2 gene activity at the cellular level using in situ hybridization. They found a high and strikingly specific expression of LRRK2 mRNA in rodent striatum and parts of cortex and no signals in dopamine neurons. LRRK2 is thus the first "PD gene" identified that shows an expression pattern that directly relates to the brain regions, most damaged in PD. The gene is being expressed in the target areas of the mesencephalic dopamine system, striatum, and frontal cortex.

It has recently been found that LRRK2 mRNA is highly enriched in motor systems as well as in other systems (Simon-Sanchez et al. 2006). Therefore, it is expected that mutations in LRRK2 may affect several motor and nonmotor structures that may play an important role in the development of PD. Simon-Sanchez et al. (2006) detected a moderate expression of this PD-related gene throughout the adult B2B6 mouse brain. A stronger hybridization signal was observed in deep cerebral cortex layers, superficial cingulate cortex layers, the piriform cortex, hippocampal formation, caudate putamen, substantia nigra, the basolateral and basomedial anterior amygdala nuclei, reticular thalamic nucleus, and also in the cerebellar granular cell layer. The discovery of a strong LRRK2 expression in structures such as the hippocampus, amygdala, or

Molecular Biology of Parkinson's Disease 285

olfactory tubercle may indicate that LRRK2 dysfunction leads to the appearance of the nonmotor symptoms that are characteristic of PD.

2.7 Role of Glutathione-S-Transferase Gene in the Development of Parkinson's Disease

It has been observed that an increased level of lipid peroxidation and the depletion of antioxidant molecules like glutathione (GSH) play an important role in the pathogenesis of PD and Alzheimer's disease (AD). A significant increase in the level of DNA oxidative damage in peripheral blood cells of PD patients with respect to controls, and the higher activity of glutathione transferases (GSTs) measured in circulating plasma in controls than in PD patients, suggest a lower enzymatic protection in PD individuals (Coppedè et al. 2005). An association between the GSTP1*B allele and PD has been detected in Caucasian Portuguese (Vilar et al. 2007). This association was particularly strong in the elder patient group (>69 years), who showed double PD risk for GSTP1*B heterozygous, whilst GSTP1*B/*B homozygous were exclusively detected amongst PD patients. An interaction between GSTM1 and GSTP1 was observed in this late onset PD group. The present results suggest that native GSTP1 encoding the fully active transferase variant should play a relevant role in dopaminergic neuroprotection. Ahmadi et al. (2000) reported that a significantly elevated median age for the onset of PD was found among GSTM1 gene carriers (median age = 68 years) compared to PD patients being GSTM1 null genotypes (median age = 57 years). They concluded that their observations indicate that (H) 113 isoform of microsomal epoxide hydrolase (mEPHX), which has been suggested as a low activity isoform, is overrepresented in PD patients and that inherited carriers of the GSTM1 gene postpone the onset of PD (Ahmadi et al. 2000). These detoxification pathways may represent important protective mechanisms against reactive intermediates modifying the susceptibility and onset of PD. An association of GSTM1*0/0 with Parkinson's disease in a Chilean population supports the hypothesis that Glutathione Transferase M1 plays a role in protecting astrocytes against toxic dopamine oxidative metabolism, and most likely by preventing toxic one-electron reduction of aminochrome (Perez-Pastene et al. 2007).

3 Conclusions

As can be seen from the above sections, there are several genes in several pathways that are contributing to the etiology or risk of PD. The variations in some of these genes are associated or are more frequent in PD cases compared to normal controls. In other cases, the gene expression is different between PD and controls. Of course, these genetic variations are observed only in a fraction of FPD or SPD. These data truly indicate the complexity of the genetic and molecular biology of PD and that the gene effect in any pathway is very small. All these findings are due to the characteristics of complex genetic diseases that are incomplete penetrance, phenocopy, genetic heterogeneity, polygenic inheritance, high frequency of disease-causing alleles, and other transmission mechanisms (mitochondrial inheritance, imprinting, trinucleotide repeats).

Acknowledgment This study was supported in part by the Arkansas Biosciences Institute, the major research component of the Tobacco Settlement Proceeds Act of 2000. We appreciate support from the National Institutes of Health (NIAAA).

References

Abahuni N, Gispert S, Bauer P, Riess O, Kruger R, Becker T, Auburger G (2007) Mitochondrial translation initiation factor 3 gene polymorphism associated with Parkinson's disease. Neurosci Lett 414:126–129

Ahmadi A, Fredrikson M, Jerregård H, Akerbäck A, Fall PA, Rannug A, Axelson O, Söderkvist P (2000) GSTM1 and mEPHX polymorphisms in Parkinson's disease and age of onset. Biochem Biophys Res Commun 269:676–680

Auluck PK, Chan HY, Trojanowski JQ, Lee VM, Bonini NM (2002) Chaperone suppression of alpha-synuclein toxicity in a Drosophila model for Parkinson's disease. Science 295:865–868

Bender A, Krishnan KJ, Morris CM, Taylor GA, Reeve AK, Perry RH, Jaros E, Hersheson JS, Betts J, Klopstock T, Taylor RW, Turnbull DM (2006) High levels of mitochondrial DNA deletions in substantia nigra neurons in aging and Parkinson disease. Nat Genet 38:515–517

Ben-Shlomo Y, Sieradzan K (1995) Idiopathic Parkinson's disease: epidemiology, diagnosis and management. Br J Gen Pract 45:261–268

Bosgraaf L, Van Haastert PJ (2003) Roc, a Ras/GTPase domain in complex proteins. Biochim Biophys Acta 1643:5–10

Braak H, Del Tredici K, Bratzke H, Hamm-Clement J, Sandmann-Keil D, Rüb U (2002) Staging of the intracerebral inclusion body pathology associated with idiopathic Parkinson's disease (preclinical and clinical stages). J Neurol 249(Suppl 3):1–5

Brooks DJ (2003) Imaging end points for monitoring neuroprotection in Parkinson's disease. Ann Neurol 53(suppl. 3):S110–S118; discussion, pp. S118–S119

Buervenich S, Carmine A, Galter D, Shahabi HN, Johnels B, Holmberg B, Ahlberg J, Nissbrandt H, Eerola J, Hellstrom O, Tienari PJ, Matsuura T, Ashizawa T, Wullner U, Klockgether T, Zimprich A, Gasser T, Hanson M, Waseem S, Singleton A, McMahon FJ, Anvret M, Sydow O, Olson L (2005) A rare truncating mutation in ADH1C (G78Stop) shows significant association with Parkinson disease in a large international sample. Arch Neurol 62:74–78

Calne DB, Langston JW (1983) Aetiology of Parkinson's disease. Lancet 8365:1457–1459

Chan P, Tanner CM, Jiang Z, Langston JW (1998) Failure to find the a-synuclein gene missense mutation (G209A) in 100 patients with younger onset Parkinson's disease. Neurology 50:513–514

Chartier-Harlin MC, Kachergus J, Roumier C, et al (2004) α- Synuclein locus duplication as a cause of familial Parkinson's disease. Lancet 364:1167–1169

Collins GG, Sandler M, Williams ED, Youdim MB (1970) Multiple forms of human brain mitochondrial monoamine oxidase. Nature 225:817–820

Molecular Biology of Parkinson's Disease 287

Coppedè F, Armani C, Bidia DD, Petrozzi L, Bonuccelli U, Migliore L (2005) Molecular implications of the human glutathione transferase A-4 gene (hGSTA4) polymorphisms in neurodegenerative diseases. Mutat Res 579:107–114

Deplazes J, Schobel K, Hochstrasser H, Bauer P, Walter U, Behnke S, Spiegel J, Becker G, Riess O, Berg D (2004) Screening for mutations of the IRP2 gene in Parkinson's disease patients with hyperechogenicity of the substantia nigra. J Neural Transm 111:515–521

Ebadi M, Govitrapong P, Sharma S, Muralikrishnan D, Shavali S, Pellett L, Schafer R, Albano C, Eken J (2001) Ubiquinone (coenzyme q10) and mitochondria in oxidative stress of Parkinson's disease. Biol Signals Recept 10:224–253

Elizan TS, Casals J (1983) The viral hypothesis in parkinsonism. J Neural Transm Suppl 19:75–88

Felletschin B, Bauer P, Walter U, Behnke S, Spiegel J, Csoti I, Sommer U, Zeiler B, Becker G, Riess O, Berg D (2003) Screening for mutations of the ferritin light and heavy genes in Parkinson's disease patients with hyperechogenicity of the substantia nigra. Neurosci Lett 352:53–56

Funayama M, Hasegawa K, Kowa H, Saito M, Tsuji S, Obata F (2002) A new locus for Parkinson's disease (PARK8) maps to chromosome 12p11.2-q13.1. Ann Neurol 51:296–301

Galter D, Westerlund M, Carmine A, Lindqvist E, Sydow O, Olson L (2006) LRRK2 expression linked to dopamine-innervated areas. Ann Neurol 59:714–719

Gasser T, Muller-Myhsok B, Wszolek ZK, Dorr A, Vaughan JR (1997) Genetic complexity and Parkinson's disease. Science 277:388–390

Gotz ME, Kunig G, Riederer P, Youdim MB (1994) Oxidative stress: free radical production in neural degeneration. Pharmacol Ther 63:37–122

Grunblatt E, Mandel S, Jacob-Hirsch J, Zeligson S, Amariglo N, Rechavi G, Li J, Ravid R, Roggendorf W, Riederer P, Youdim MB (2004) Gene expression profiling of parkinsonian substantia nigra pars compacta; alterations in ubiquitin-proteasome, heat shock protein, iron and oxidative stress regulated proteins, cell adhesion/cellular matrix, and vesicle trafficking genes. J Neural Transm 111:1543–1573

Hanson ES, Rawlins ML, Leibold EA (2003) Oxygen and iron regulation of iron regulatory protein 2. J Biol Chem 278:40337–40342

Harrington KA, Augood SJ, Kingsbury AE, Foster OJ, Emson PC (1996) Dopamine transporter (Dat) and synaptic vesicle amine transporter (VMAT2) gene expression in the substantia nigra of control and Parkinson's disease. Brain Res Mol Brain Res 36:157–162

Hashimoto M, Hsu LJ, Xia Y, Takeda A, Sisk A, Sundsmo M, Masliah E (1999) Oxidative stress induces amyloid-like aggregate formation of NACP/alpha-synuclein in vitro. Neuroreport 10:717–721

Hjelle JJ, Petersen DR (1983) Hepatic aldehyde dehydrogenases and lipid peroxidation. Pharmacol Biochem Behav 18(suppl 1):155–160

Hughes AJ, Daniel SE, Lees AJ (1993) The clinical features of Parkinson's disease in 100 histologically proven cases. Adv Neurol 60:595–599

Jellinger KA (2003) Neuropathological spectrum of synucleinopathies. Mov Disord 18(suppl 6): S2–S12

Jenner P, Olanow CW (1996) Oxidative stress and the pathogenesis of Parkinson's disease. Neurology 47:S161–S170

Klucken J, Shin Y, Masliah E, Hyman BT, McLean PJ (2004) Hsp70 reduces alpha-synuclein aggregation and toxicity. J Biol Chem 279:25497–25502

Kraytsberg Y, Kudryavtseva E, McKee AC, Geula C, Kowall NW, Khrapko K (2006) Mitochondrial DNA deletions are abundant and cause functional impairment in aged human substantia nigra neurons. Nat Genet 38:518–520

Kruger R, Kuhn W, Muller T, Woitalla D, Graeber M, Kosel S, Przuntek H, Epplen JT, Schols L, Riess O (1998) Ala30Pro mutation in the gene encoding alpha-synuclein in Parkinson's disease. Nat Genet 18:106–108

Langston JW, Ballard P, Tetrud JW, Irwin I (1983) Chronic Parkinsonism in humans due to a product of meperidine-analog synthesis. Science 219:979–980

Lennox GG, Lowe JS (1997) Dementia with Lewy bodies. Baillieres Clin Neurol 6:147–166

Leroy E, Boyer R, Auburger G, Leube B, Ulm G, Mezey E, Harta G, Brownstein MJ, Jonnalagada S, Chernova T, Dehejia A, Lavedan C, Gasser T, Steinbach PJ, Wilkinson KD, Polymeropoulos MH (1998) The ubiquitin pathway in Parkinson's disease. Nature 395:451–452

Lowe J, Dickson D (1997) Pathological diagnostic criteria for dementia associated with cortical Lewy bodies: review and proposal for a descriptive approach. J Neural Transm Suppl 51:111–120

Mandel S, Grunblatt E, Riederer P, Amariglio N, Jacob-Hirsch J, Rechavi G, Youdim MB (2005) Gene Expression Profiling of Sporadic Parkinson's Disease Substantia Nigra Pars Compacta Reveals Impairment of Ubiquitin-Proteasome Subunits, SKP1A, Aldehyde Dehydrogenase, and Chaperone HSC-70. Ann N Y Acad Sci 1053:356–375

Mardh G, Vallee BL (1986) Human class I alcohol dehydrogenases catalyze the interconversion of alcohols and aldehydes in the metabolism of dopamine. Biochem 18:7279–7282

Meyron-Holtz EG, Ghosh MC, Iwai K, LaVaute T, Brazzolotto X, Berger UV, Land W, Ollivierre-Wilson H, Grinberg A, Love P, Rouault TA (2004) Genetic ablations of iron regulatory proteins 1 and 2 reveal why iron regulatory protein 2 dominates iron homeostasis. EMBO J 23:386–395

Morfini G, Pigino G, Opalach K, Serulle Y, Moreira JE, Sugimori M, Llinas RR, Brady ST (2007) 1-Methyl-4-phenylpyridinium affects fast axonal transport by activation of caspase and protein kinase C. Proc Natl Acad Sci U S A 104:2442–2447

O'Carroll AM, Fowler CJ, Phillips JP, Tobbia I, Tipton KF (1983) The deamination of dopamine by human brain monoamine oxidase: specificity for the two enzyme forms in seven brain regions. Naunyn-Schmiedeberg's Arch Pharmacol 322:198–202

Ostrerova-Golts N, Petrucelli L, Hardy J, Lee JM, Farer M, Wolozin B (2000) The A53T alpha-synuclein mutation increases irondependent aggregation and toxicity. J Neurosci 20:6048–6054

Ouimet CC, Hemmings HC, Jr Greengard P (1989) ARPP-21, a cyclic AMP–regulated phospho-protein enriched in dopamine-innervated brain regions. II. Immunocytochemical localization in rat brain. J Neurosci 9:865–875

Pals P, Lincoln S, Manning J, Heckman M, Skipper L, Hulihan M, Van den Broeck M, De Pooter T, Cras P, Crook J, Van Broeckhoven C, Farrer MJ (2004) alpha-Synuclein promoter confers susceptibility to Parkinson's disease. Ann Neurol. 56:591–595

Papadimitriou A, Veletza V, Hadjigeorgiou GM, Partikiou A, Hirano M, Anastasopoulos I (1999) Mutated a-synuclein gene in two Greek kindreds with familial PD: incomplete penetrance? Neurology 52:651–654

Papachroni KK, Ninkina N, Papapanagiotou A, Hadjigeorgiou GM, Xiromerisiou G, Papadimitriou A, Kalofoutis A, Buchman VL (2007) Autoantibodies to alpha-synuclein in inherited Parkinson's disease. Journal of Neurochemistry 101:749–756

Parsian A, Racette B, Zhang ZH, Rundle M, Goate AM, Perlmutter JS (1998) Mutation, sequence analysis and association studies of a-synuclein in Parkinson's disease. Neurology 51:1757–1759

Parsian AJ, Racette BA, Zhao JH, Sinha R, Patra B, Perlmutter JS, Parsian A (2007) Association of alpha-synuclein gene haplotypes with Parkinson's disease. Parkinsonism Relat Disord 13:343–347

Perez-Pastene C, Graumann R, Díaz-Grez F, Miranda M, Venegas P, Godoy OT, Layson L, Villagra R, Matamala JM, Herrera L, Segura-Aguilar J (2007) Association of GST M1 null polymorphism with Parkinson's disease in a Chilean population with a strong Amerindian genetic component. Neurosci Lett 418:181–185

Polymeropoulos MH, Lavedan C, Leroy E, Ide SE, Dehejia A, Dutra A, et al (1997) Mutation in the a-synuclein gene identified in families with Parkinson's disease. Science 276:2045–2047

Ponka P (2004) Hereditary causes of disturbed iron homeostasis in the central nervous system. Ann NY Acad Sci 1012:267–281

Riederer P, Sofic E, Rausch WD, Schmidt B, Reynolds GP, Jellinger K, Youdim MB (1989) Transition metals, ferritin, glutathione, and ascorbic acid in parkinsonian brains. J Neurochem 52:515–520

Scherzer CR, Eklund AC, Morse LJ, Liao Z, Locascio JJ, Fefer D, Schwarzschild MA, Schlossmacher MG, Hauser MA, Vance JM, Sudarsky LR, Standaert DG, Growdon JH, Jensen RV, Gullans SR (2007) Molecular markers of early Parkinson's disease based on gene expression in blood. Proc Natl Acad Sci U S A 104:955–960

Scott WK, Stajich JM, Yamaoka LH, Spur MC, Vance JM, Roses AD, et al (1997) Genetic complexity and Parkinson's disease. Science 277:387

Simon-Sanchez J, Herranz-Perez V, Olucha-Bordonau F, Perez-Tur J (2006) LRRK2 is expressed in areas affected by Parkinson's disease in the adult mouse brain. Eur J Neurosci 23:659–666

Singleton AB, Farrer M, Johnston J, Singleton A, Hague S, Kachergus J, et al (2003) a-synuclein locus triplication causes Parkinson's disease. Science 302:841

Tsou K, Girault JA, Greengard P (1993) Dopamine D1 agonist SKF 38393 increases the state of phosphorylation of ARPP-21 in substantia nigra. J Neurochem 60:1043–1046

Turnbull S, Tabner BJ, El-Agnaf OM, Moore S, Davies Y, Allsop D (2001) α-Synuclein implicated in Parkinson's disease catalyses the formation of hydrogen peroxide in vitro. Free Radical Biol Med 30:1163–1170

Vilar R, Coelho H, Rodrigues E, Gama MJ, Rivera I, Taioli E, Lechner MC (2007) Association of A313 G polymorphism (GSTP1'B) in the glutathione-S-transferase P1 gene with sporadic Parkinson's disease. Eur J Neurol 14:156–161

Wang J, Chen G, Muckenthaler M, Galy B, Hentze MW, Pantopoulos K (2004) Iron-mediated degradation of IRP2, an unexpected pathway involving a 2-oxoglutarate-dependent oxygenase activity. Mol Cell Biol 24:954–965

Wu YR, Wang CK, Chen CM, Hsu Y, Lin SJ, Lin YY, Fung HC, Chang KH,Lee-Chen GJ (2004) Analysis of heat-shock protein 70 gene polymorphisms and the risk of Parkinson's disease. Hum Genet 114:236–241

Youdim MB, Stephenson G, Ben Shachar D (2004) Ironing iron out in Parkinson's disease and other neurodegenerative diseases with iron chelators: a lesson from 6-hydroxydopamine and iron chelators, desferal and VK-28. Ann N Y Acad Sci 1012:306–325

Youdim MBH, Ben-Shachar D Riederer P (1993) The possible role of iron in the etiopathology of Parkinson's disease. Mov Disord 8:1–12

Youdim MBH, Riederer P (1997) Understanding Parkinson's disease: the smoking gun is still missing, but growing evidence suggests highly reactive substances called free radicals are central players in this common neurological disorder. Sci Am 276:52–59

Youdim MBH, Riederer P (2004) Iron in the brain, normal and pathological. In: Encyclopedia of neuroscience. Elsevier, Amsterdam, New York

Zarranz JJ, Alegre J, Ge mez-Esteban J, Lezcano E, Ros R, Ampuero I, et al (2004) The new mutation, E46K, of alpha-synuclein causes Parkinson and Lewy body dementia. Ann Neurol 55:164–173

Zecca L, Youdim MB, Riederer P, Connor JR, Crichton RR (2004) Iron, brain ageing, and neurodegenerative disorders. Nat Rev Neurosci 5:863–873

Zhang J, Hattori N, Leroy E, Morris HR, Kubo S, Kobayashi T, Wood NW, Polymeropoulos MH, Mizuno Y (2000) Association between a polymorphism of ubiquitin carboxy-terminal hydrolase L1 (UCH-L1) gene and sporadic Parkinson's disease. Parkinsonism Relat Disord 6:195–197

Zimprich A, Biskup S, Leitner P, et al (2004) Mutations in LRRK2 cause autosomal-dominant parkinsonism with pleomorphic pathology. Neuron 44:601–607

Index

A

Abnormalities, chromosomal, 61, 67–68
Addiction, 187–199
ADHD. *See* Attention-deficit hyperactivity
    disorder
Adoption studies, 55
Age of onset, 52
AKT1 (protein kinase B), 63–64
Alcohol dehydrogenase, 283
Alpha-Synuclein, 278–283
    mutations, 280, 281
Animal models, 51, 72
Antidepressant, 6, 23, 27–36
Antioxidant, 284
    association, 285
    GST, 285
Antisocial personality disorder, 190–191, 194
Apolipoprotein E, 246–254
Asperger syndrome, 82–86
Association, 81–84, 86–91
    direct, 59
    genetic, 51, 56, 58–71
    indirect, 59
Attention-deficit hyperactivity disorder
    (ADHD), 99–139
Autism spectrum disorders, 81–92

B

Beta amyloid, 229, 331
Biomarker, 71
BLOC-1 protein complex, 66
Brain imaging, 16, 31, 37

C

cAMP response element-binding (CREB), 193
Candidate genes, 4, 7, 9–10, 12–14, 16, 23,
    56, 60, 61–70, 107, 132–133

BDNF, 13–14, 17, 23, 26, 28–31
DISC1, 9–10, 35
G72 (DAOA), 8–9, 26
Catechol-O-methyltransferase (COMT), 61,
    63, 67–69, 71–72, 169, 173
Childhood disintegrative disorder, 82
Chips, 10, 12–13
    Affymetrix, 4–5, 10, 12, 23–24
    GeneChip, 10, 23–24, 35
    Illumina, 4–5, 10, 12, 14, 38
Chromosome, 8, 14, 34–35
Circadian rhythm, 22, 24
    CLOCK, 22
    DBP, 22
Collaborative Studies on the Genetics of
    Alcoholism, 189–190
Complications, obstetric, 54, 56
Conduct disorder, 190–191
Copy number variants (CNVs), 61, 68, 83, 88,
    89, 91
Course of schizophrenia, 52, 56
CREB. *See* cAMP response
    element-binding

D

Dependence
    alcohol, 189–191, 198, 199
    heroin, 194–195
    nicotine, 190, 191, 197–199
Diagnosis, 51–52, 56, 62, 67, 71–72
Diagnostic and Statistical Manual of Mental
    Disorders (DSM), 2–3, 6
Disease burden, 189–191, 194, 197
Disrupted in schizophrenia 1 (DISC1), 61,
    67, 69
DNA methylation, 71–72
DNA Methyltransferase I, 72
DNA pooling, 69

292 Index

Dopamine, 99–103, 107–116, 119, 122, 124–127, 129, 130, 132, 136, 137, 191–193, 196, 197–199, 277–279, 281, 283–285
Dopamine hypothesis, 57, 63
Dopamine receptor, 61, 63
D1 receptor, 57
D2 receptor, 57, 61, 66
DSM. *See* Diagnostic and Statistical Manual of Mental Disorders
DSM-IV criteria, 52
Dysbindin (dystrobrevin binding protein1, DTNBP1), 59, 61, 65–66
Dystrophin protein complex, 66

E
Endophenotypes, 15–16, 72, 87, 91
Epigenetic(s), 3, 16–18, 37, 51, 71–72
  acetylation, 16–17, 37
  histone, 16–19, 37
  methylation, 16–17, 37

F
Fagerstrom test, 197–198

G
GABA hypothesis, 58
Gene-environment interaction (gene x environment interaction), 17, 26–27, 37
Gene expression, 51, 64–65, 69–72
Gene expression profiling, mRNA profiling, 23
  gene expression, 15–16, 23–24, 28, 33, 35–36
  microarray, 15, 22–23, 35–36
Generalized anxiety disorder, 166, 168, 170–171
Genes, 99–139
Genetic architecture, 3
  common disease-common variant (CDCV), 3
  rare allele, 4
  rare variant, 3–4
Genetics, 1–38, 99–139
  genotype, 10, 12, 15, 26, 37
  haplotype, 4, 9, 12–13
  heritability, 3–4, 6, 16, 25
  phenotype, 3, 7, 13, 15–16, 22, 34–35, 37
  prevalence, 1, 6, 25
  risk allele, 37
  susceptibility allele, 12–13
Genetics of suicide, 214–219

epidemiology, 213
Genome screens, 81, 83, 84, 87, 90, 91
Genome-wide association (GWA), 62, 69, 83, 174–176
  neuroticism, 174–175
Genome-wide association studies (GWAS), 4–5, 10–13, 26, 37–38
  DGKH, 11
  GAIN, 5, 10, 26
  impute, 12
  protein kinase, 11, 18–19, 29
  WTCCC, 10–12, 38
Genome-wide gene expression, 70
Glutamate hypothesis, 57–58
Glycogen-synthasekinase- 3 (GSK3), 63–64

H
Heat shock protein, 282–283
Heritability, 55, 72, 187, 188–191, 193, 199
Heterogeneity, 7–8, 12–13, 18, 28, 37–38
  allelic heterogeneity, 7–8, 12–13, 37–38
  genetic heterogeneity, 18
  locus heterogeneity, 7, 37
  phenotype heterogeneity, 7, 37
Hippocampus, 10, 17–18, 22, 28, 30–32, 35
Hypothalamic-pituitary axis (HPA), 1, 31–32, 34
  ACTH, 32
  CRH, 31–32
  glucocorticoid receptor (GR), 32–33
  mineralocorticoid receptor (MR), 32

I
Identity by descent, 84, 87
Incidence, 51, 53
Inositol depletion hypothesis, 18–19
  IMPA1, Impa1, 19–20
  Inositol, 18–20
Iron regulatory protein, 280

L
Lewy body, 277, 279, 281
Lifetime risk, 54–55
Linkage, 4, 6–13, 18, 25–26, 34–35, 37, 83–90
  disequilibrium, 59, 84, 88
  genetic, 56, 58–59
Linkage analysis, 167–168, 170
  neuroticism, 174–175
  obsessive-compulsive disorder, 172–173
  panic disorder, 167–168
  phobia, 170
Lod score, 84, 85

Index 293

M

Matrix-based comparative genomic hybridization, 88
Maximum multipoint logarithm of odds score, 84
Meta-analysis, 7, 8, 12–14, 16, 37, 62–64
Microdeletions, 61, 68–69
MicroRNA (miRNA), 15, 37, 51, 70–71
Mitochondria, 282
Molecular genetics, 213
    dopamine, 217
    interaction gene/environment, 218–219
    monoamine oxydase, 216
    serotonin receptors, 216
    serotonin transporter, 215–216
    tryptophanhydroxylase, 213–215
Monoamine
    norepinephrine (NE), 27–28
    serotonin (HT), 27
        5-HT$_{1A}$ receptor, 28
        5-HT$_{2A}$ receptor (HTR2A), 28
Monoamine oxidase A, 171
Mood disorder, 1–38
    bipolar disorder, 1–3, 5–8, 10–24, 27, 32–33, 35–38
    depression, 1–3, 5–7, 13, 22, 24–36
Mood stabilizer, 14, 17–18, 21–22
    carbamazepine, 17, 19–20
    lithium, 17–19, 21
    valproate, 14
    valproic acid, 17, 19, 21
mRNA expression, 51, 66, 70–71

N

Neuregulin (NRG1), 56, 59, 61, 64–65, 68
Neurexin, 198
Neurochemical basis of suicide, 208
    dopamine, 210
    5-hydroxyindolacetic acid, 209
    noradrenalin, 210, 211, 213, 220
    serotonin, 209
    serotonin receptors, 216
    serotonin transporter, 209
Neurodegeneration, 279, 280
Neurodevelomental hypothesis, 56
Neurogenesis, 1, 13, 17, 29–30, 35–36
Neuroligin, 90, 91
Neuron, 10, 13–15, 17, 19, 21–22, 29–31
Neuronal, 9, 10, 13–15, 17–18, 21, 28, 29, 31, 36
Neurotoxin, 278, 279, 281
Neurotransmitters, 99–102, 107, 108, 116, 119, 124–129, 134

Neurotrophic factors, 213, 218
Nicotine, 188–191, 194, 196–199
Nicotinic receptor, 197–198
NIMH, 10–11, 24, 38
NIMH Genetics Initiative, 10–11

O

Obsessive-compulsive disorder, 166, 171–173

P

Panic disorder, 166–170, 174
PARK8, mutations, 284
Parkinson, 277–285
PDD-NOS, 82, 86
Pervasive developmental disorder-not otherwise specified, 82
Pharmacogenetics, 1, 17–21, 27–29
Post mortem brain, 56, 65
Presenilin, 230, 231, 239–246, 254
Prevalence, 51, 53
Prodynorphin, 195
Proopiomelanocortin, 195
Protein expression, 51, 58, 71
Proteomics, 71
Psychiatric disorder, 3, 16, 37
    schizophrenia, 2, 4, 8, 9

Q

22q deletion syndrome (22q11DS), 67–68

R

Reelin, 56, 67, 69, 71–72
Regulator of G-protein signaling4 (RGS4), 61, 66
Replication, 62–63, 67–69
Representational oligonucleotide microarray analysis, 88
Rett syndrome, 82, 83
Risk factors, 54

S

Schizophrenia, course of, 52, 56
Serotonin transporter, 90, 173, 175
Serotonin transporter (SERT, 5-HTT, SLC6A4), 26–27, 37
    HTTLPR, 26–27
Signal transduction system

Single nucleotide polymorphism, 83, 87, 90
Social anxiety disorder, 166, 169, 170, 174
Specific phobia, 166, 169–170, 174
Stress and HPA-axis, 211–212
   Dex/CRH Test, 211
Stress-based animal models of depression-like behavior, 33
   FST, 22, 33–34
   LH, 33–35
   TST, 33–34
Studies, epidemiological, 187, 189, 191
Subphenotype, 197, 199
Suicide
   clinical phenotype, 206–208
   pathophysiology, 209, 211–213, 217, 220
Susceptibility
   genes, 51, 58–59, 61–63, 81, 83, 84, 86, 88, 91
   loci, 81, 86, 87, 89
Synaptic plasticity, 193

T
Transmission
   genetic, 53–55
   mode of, 55–56
Transmission Disequilibrium Test (TDT), 61
Twin studies, 55

V
Variation, 2–4, 9–10, 13–14, 19–20, 22–23, 26–27, 37
   copy number variation (CNV), 14–15, 37
   single nucleotide polymorphism (SNP), 14
      nonsynonymous SNP, 4, 8
Ventricle size, 56
Vietnam Era Twin Registry, 189
Volume, cortical, 56

W
Wnt, 11, 17, 20–21
   GSK3, 20–21